Materials Characterisation VI

WIT*PRESS*

WIT Press publishes leading books in Science and Technology.
Visit our website for new and current list of titles.
www.witpress.com

WIT*eLibrary*

Home of the Transactions of the Wessex Institute.
Papers presented at Materials Characterisation VI are archived in the WIT eLibrary in volume 77 of WIT Transactions on Engineering Sciences (ISSN 1743-3533).
The WIT eLibrary provides the international scientific community with immediate and permanent access to individual papers presented at WIT conferences.
http://library.witpress.com

Sixth International Conference on
Computational Methods and Experiments in
Materials Characterisation

MATERIALS CHARACTERISATION 2013

Conference Chairmen

C.A. Brebbia
Wessex Institute of Technology, UK

A.A. Mammoli
University of New Mexico, USA

A. Klemm
Glasgow Caledonian University, UK

International Scientific Advisory Committee

Organised by

Wessex Institute of Technology, UK
University of New Mexico, USA
Glasgow Caledonian University, UK

Sponsored by

WIT Transactions on Engineering Sciences
International Journal of Computational Methods and Experimental Measurements

WIT Transactions

Transactions Editor

Materials Characterisation VI

Computational Methods and Experiments

Editors

C.A. Brebbia
Wessex Institute of Technology, UK

A. Klemm
Glasgow Caledonian University, UK

Editors:

C.A. Brebbia
Wessex Institute of Technology, UK

A. Klemm
Glasgow Caledonian University, UK

Published by

WIT Press
Ashurst Lodge, Ashurst, Southampton, SO40 7AA, UK
Tel: 44 (0) 238 029 3223; Fax: 44 (0) 238 029 2853
E-Mail: witpress@witpress.com
http://www.witpress.com

For USA, Canada and Mexico

Computational Mechanics Inc
25 Bridge Street, Billerica, MA 01821, USA
Tel: 978 667 5841; Fax: 978 667 7582
E-Mail: infousa@witpress.com
http://www.witpress.com

British Library Cataloguing-in-Publication Data

A Catalogue record for this book is available
from the British Library

ISBN: 978-1-84564-720-9
eISBN: 978-1-84564-721-6
ISSN: 1746-4471 (print)
ISSN: 1743-3533 (on-line)

The texts of the papers in this volume were set individually by the authors or under their supervision. Only minor corrections to the text may have been carried out by the publisher.

No responsibility is assumed by the Publisher, the Editors and Authors for any injury and/ or damage to persons or property as a matter of products liability, negligence or otherwise, or from any use or operation of any methods, products, instructions or ideas contained in the material herein.

Printed in Great Britain by Lightning Source, UK

Preface

This volume contains most of the papers presented at the sixth international conference on Computational Methods and Experiments in Materials Characterisation held in La Certosa of Pontignano of the University of Siena in 2013. It follows the success of five previous meetings in this series which started in Santa Fe, New Mexico in 2003; followed by conferences in Portland, Maine (2005); Bologna (2007); the New Forest Campus of the Wessex Institute of Technology (2009), and Kos, Greece (2011).

The aim of the Conference is to facilitate interactions within the research community and discuss the latest developments in this rapidly advancing field.

The meeting responds to the demand for high quality products for both industry and consumers, which has led to rapid developments in materials science and engineering. This requires the characterisation of the physical and chemical properties of the materials. Consideration of different experimental techniques as well as computer simulation methods is essential to achieve a proper analysis. A very wide range of materials, starting with metals through polymers and semiconductors to composites, necessitates a whole spectrum of material characterisation experimental techniques and numerical methods.

All papers published in this volume as well as those from previous conferences are permanently archived at http://library.witpress.com/ where they are available to the international community.

The books are published in paper and digital format and widely distributed throughout the world. As other papers presented at the Wessex Institute conferences they are referenced and regularly appear in notable reviews, publications and databases.

The Editors are grateful to all authors for the presentations and particularly to the members of the Scientific Advisory Committee who have reviewed the contributions, hence ensuring the quality of this volume.

Carlos A. Brebbia
Agnieszka Klemm
Siena, 2013

Contents

Section 2: Mechanical characterisation and testing

Section 3: Computational models and experiments

Section 4: Corrosion problems

Section 1
Micro and macro materials characterisation

Characterisation of the mechanical behaviour of a polyurethane elastomer based on indentation and tensile creep experiments

B. Buffel[1], K. Vanstreels[2], F. Desplentere[1], B. Dekeyser[3] & I. Verpoest[4]
[1]*KHBO Expertisecentrum Kunststoffen, KU Leuven-KHBO, Belgium*
[2]*imec, Belgium*
[3]*International Development Centre, Recticel N.V., Belgium*
[4]*Department of Metallurgy and Materials Engineering, KULeuven, Belgium*

Abstract

This research focuses on the determination of the mechanical properties of a viscoelastic polyurethane material with 2 different measuring techniques on 2 different length scales. Instrumented Indentation Testing (IIT) was used to test the material on a micro scale while tensile creep experiments characterised the macro scale material behaviour. All experimental data were processed by means of a fitting procedure based on the standard linear solid material model. The experiments were performed with different loading rates and hold values. The developed fitting procedure proved to be applicable to analyse the experimental data on both length scales. FEM was used to coordinate the applied strains of both measuring techniques. A comparison between the results originating from the experiments with both techniques indicated a stiffer material response on the micro scale (up to 4x). The more complex strain field inside the material during indentation compared to the uniform tensile loading on macro scale is responsible for this large discrepancy. For this reason comparing the results of IIT with tensile creep results should be done with great care.
Keywords: instrumented indentation testing, creep experiments, standard linear solid material model, polyurethane elastomer.

www.witpress.com, ISSN 1743-3533 (on-line)
doi:10.2495/MC130011

1 Introduction

The worldwide interest in elastomeric materials has grown significantly during the last decades. Applications can be found in various areas such as automotive and construction. Together with the increasing importance of elastomers, a higher demand for quantitative material data arose. Despite the fact that the first non-linear material models date back to the 19th century, the measurement of quantitative material parameters still remains a key issue regarding the modelling of viscoelastic materials.

Elastomeric materials can be produced in a large window of material properties. These properties originate from the chemical and physical interactions between the polymer chains which depend on the structure and composition of the chains. In this way not only the raw materials but also the recipe and preparation method influences the final properties of elastomeric materials. Because of this high number of influencing parameters High Throughput Experiments (HTE) are helpful to accelerate the investigations on different elastomeric formulation and preparation methods. HTE allows a high degree of automation of the production and testing of multiple elastomers which is essential for fast and sustainable research (Kranenburg *et al.* [1], Majoros *et al.* [2]).

Instrumented Indentation Testing (IIT) uses a probe with a spherical or pyramidal tip to indent a sample on a micro or nano scale. During the indentation a force-displacement curve is recorded which is processed afterwards to extract the material parameters (Fischer-Cripps [3]). IIT can be used to characterize the mechanical properties of a wide range of materials. This range includes polymers and elastomers with a distinct time dependent mechanical response. (Galli *et al.* [4], Vanlandingham *et al.* [5]). For this type of materials the standard elastic-plastic analysis (Oliver and Pharr [6]) is insufficient and can be replaced by more powerful methods to characterize linear viscoelastic material behaviour (Fischer-Cripps [7], Oyen [8, 9]). These methods were used within the scope of this research and adapted to tensile creep experiments on macro scale.

Conventional methods like bending or tensile testing are considered as the bottleneck in HTE. To overcome this problem IIT is often applied in a HTE workflow as it is capable of capturing experimental data in a fast and accurate way. In combination with the ability to test viscoelastic materials, IIT was selected to test the polyurethane elastomer within this research. This research focuses on the comparison of IIT to conventional tensile creep experiments in order to validate the measuring technique. The standard linear solid model was used to capture the non-linear material behaviour (Fischer-Cripps [7], Powel [10]).

2 Experimental methods

2.1 Micro scale testing: instrumented indentation testing

Indentation tests were performed at the imec Research Institute with a NanoIndenter XP (MTS Systems) on a polyurethane elastomer film produced by

Recticel (IDC Wetteren Belgium). The film had an average thickness of 1,7mm. The indenter tip was a sphere with a radius R=250μm. The indentation experiments consisted out of a force controlled loading phase with a loading rate of 0,5mN/s up to a varying peak force (F_{max}= 0,5 – 1 – 2 – 5 – 10 – 20mN). In another set of experiments the peak force was fixed (F_{max} = 2mN) with varying loading rates (LR = 0,05 – 0,3 – 0,5 – 0,75 – 1mN/s).

The governing equation for spherical indentation in an elastic medium is (Johnson [11]):

$$h^{3/2} = \frac{3}{4\sqrt{R}} F \frac{1-\nu}{E} \tag{1}$$

where h and F are indentation depth and load respectively. During a force controlled loading phase of an indentation test (F(t)=k.t) and in the case of an incompressible solid (υ=0,5) this equation can be written as:

$$h^{3/2}(t) = \frac{3k}{8\sqrt{R}} \frac{t}{2G} \tag{2}$$

in which G is the shear modulus of the material. Due to the non-linear material behaviour the experimental time–displacement data were processed based on the procedure developed by M. Oyen (Oyen [8, 9]) in which the standard linear solid model describes the non-linear material behaviour by means of a parallel spring-dashpot system in series with a second spring. The procedure resulted in the fitting of the loading curve with the following equation:

$$h^{3/2}(t) = B_0 t - B_1 \left(1 - e^{-t/\tau_1}\right) \tag{3}$$

$$\text{with } B_0 = \frac{3k}{8\sqrt{R}} C_0 \quad \text{and} \quad B_1 = \frac{3k}{8\sqrt{R}} C_1 \tau_1 \qquad \text{(4) and (5)}$$

The C_0 end C_1 values obtained from the fitting procedure are used to calculate the shear and Young's modulus of the polyurethane elastomer:

$$G_0 = \frac{1}{2(C_0 - C_1)} \quad \text{and} \quad G_\infty = \frac{1}{2C_0} \qquad \text{(6) and (7)}$$

$$E_0 = 2G_0(1+\nu) \quad \text{and} \quad E_\infty = 2G_\infty(1+\nu) \qquad \text{(7) and (8)}$$

These values are limiting values for the immediate response (G_0 and E_0) and the response with the lapse of time (G_∞ and E_∞). The ratio of both is a measure for the viscoelasticity of the material. The more this ratio approaches one, the more elastic the material behaves.

2.2 FE-simulations

In order to coordinate the experimental conditions of the micro scale testing on the macro scale tensile testing a FE model of the indentation experiment was build. Siemens NX 8.0 software coupled to a NASTRAN solver was used to build and solve the FE model. Due to symmetry, the model was limited to a quarter of the specimen and a part of the indenter sphere (fig. 1). The main goal of the model was to estimate the occurring strains in the indentation test in order to apply analogue strains in the tensile creep experiments on macro scale.

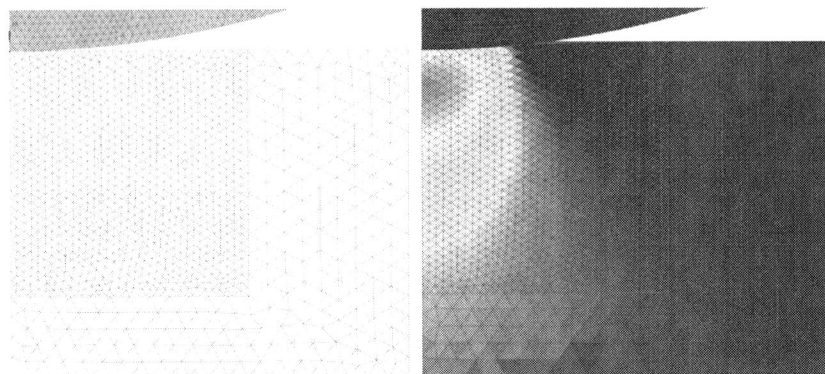

Figure 1: FE indentation model with refined mesh in the high strain zone (left). Strains in the vertical direction inside the specimen reaching a maximum and fading out deeper in the material (right).

Multiple simulations with varying indentation depths were performed with different coefficients of friction between the indenter tip and the sample surface. The highest strains were observed in the vertical direction of indentation. These strains imply the highest loading conditions on the material and were transferred to the macro tensile test. The results of these simulations are presented in figure 2. No significant influence of the coefficient of friction was observed.

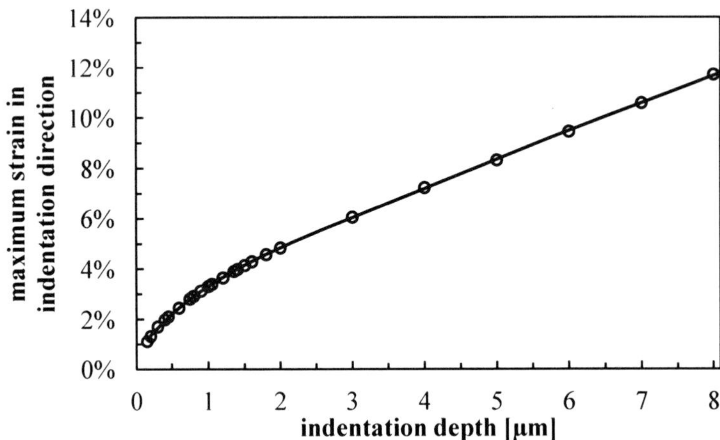

Figure 2: Maximum occurring strain in the indentation direction as a function of the indentation depth. Radius of the indenter sphere = 250µm.

2.3 Macro scale testing: tensile creep experiments

The tensile experiments were performed on an Instron 3345 single column testing system equipped with a 5kN load cell and an initial clamping distance of

55mm. Dog bone shaped tensile test specimens were punched out of the same film as the indentation test sample. Experiments were performed at a fixed loading rate of 10N/min up to a varying hold value of 0,33 – 0,88 – 1,1 – 2,64 – 3,63mm. These displacements can be linked to the occurring strains in the indentation direction via the FE-simulations, fig. 2. In a second series of tensile tests the peak force was fixed at = 50N) and the loading rate was varied (LR= 5 – 10 – 50 – 100 – 500 N/min). In both series the loading phase was followed by a 300s long holding phase in which the force was kept constant at the peak force F_{max}.

The experimental macro scale tensile creep data were fitted to the same viscoelastic material model as the indentation test data. Therefore, the fitting procedure presented in equations (3)–(5) was adapted for the loading phase of the tensile test:

$$\Delta l(t) = \frac{l_0 k}{A(1+\upsilon)}\left[C_0 - C_1\tau_1\left(1 - e^{-t/\tau_1}\right)\right] \tag{9}$$

In which l_0, Δl and A are respectively the initial clamping distance, displacement and cross sectional area of the specimen. During the constant force holding phase the material model is adapted to:

$$\Delta l(t) = \frac{l_0 k}{A(1+\upsilon)}\left[C_0 t_{rise} - C_1\tau_1 e^{-t/\tau_1}\left(e^{t_{rise}/\tau_1} - 1\right)\right] \tag{10}$$

with t_{rise} = time to reach the hold value. Equations (6)–(8) can again be used to calculate the material stiffness values based on the outcome of the fitting procedure.

2.4 Temperature dependency

Due to different laboratory conditions the dependency of the mechanical properties on temperature was taken into account in order to make a comparison between macro and micro testing results. The glass transition temperature region was determined with a Q200 MDSC (TA Instruments) apparatus. A constant temperature increase of 10°C/min from -50°C up to 80°C was used. The results indicated a broad glass transition region between 5°C and 35°C.

The dependency of material stiffness on temperature was measured using a tensile set up inside a Q800 DMA (TA Instruments). 3 samples were tested with a clamping distance and width of respectively 17,6 and 2,6mm. A frequency of 1Hz and an amplitude of 15μm were applied. The results of the DMA temperature sweep are presented in figure 3.

The stiffness of the polyurethane film possesses a very high dependency on temperature in the region around room temperature. Since the indentation laboratory is conditioned at 23°C this was taken as the reference situation and all experimental tensile creep results will be corrected with respect to this condition.

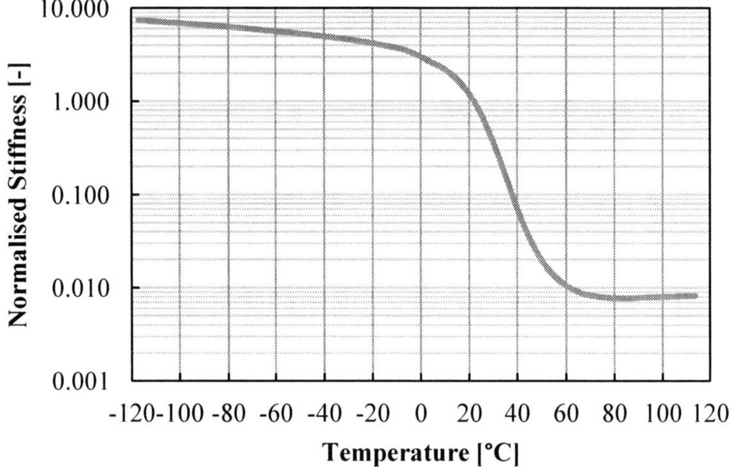

Figure 3: Normalised stiffness of the polyurethane film as a function of temperature. The stiffness result at 23°C was taken as a reference.

3 Results and discussion

3.1 Micro scale testing: instrumented indentation testing

In contrast to the macro scale testing, no holding phase results could be obtained during the IIT due to practical limitations of the measuring device. Therefore E_0 and E_∞ were extracted from the loading phase of the measurement. The results are presented as a function of hold value and loading rate in figures 4 and 5 respectively. E_0 is independent of the hold value up to a strain of 4%. At larger strains the value of E_0 decreases because the mathematical curve fitting optimisation is influenced by the data and larger strains which do not anymore correlate to the initial stiffness E_0. The decreasing trend in E_∞ as a function of the hold value can be explained by the increasing viscous-plastic material behaviour as the indenter moves into the sample.

Figure 5 shows an increase of E_0 and E_∞ with increasing loading rate. This observation, a stiffer material response with increasing loading rate, is often induced by the higher resistance to make polymer chains move across each other at higher rates. Comparing the values of E_0 versus E_∞ clearly indicates the need for a non-linear material model. This proves that the selected loading rates were not too fast with respect to the material time constants. Due to the dependency of the fitted parameters on experimental conditions, future work is required to optimise the measuring methods and obtain absolute material properties. The large standard deviations presented by the error bars in the graphs will be discussed in section 3.3.

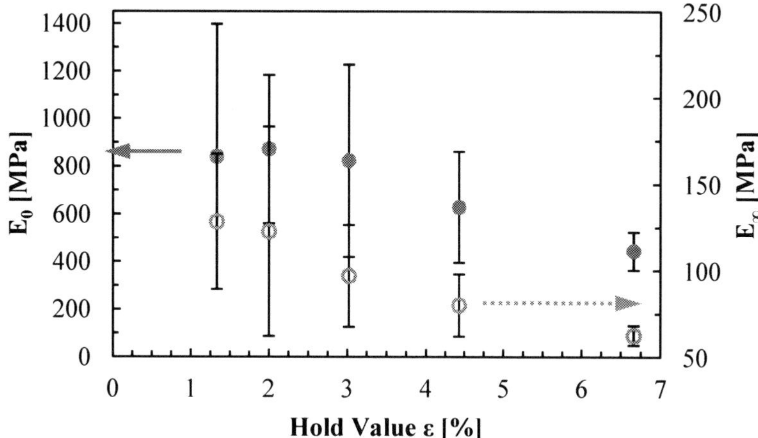

Figure 4: E_0 and E_∞ values as a function of the hold value applied during the indentation experiments. This value is expressed as the corresponding strain which was calculated via FEM. Filled circles represent E_0; Empty circles represent E_∞.

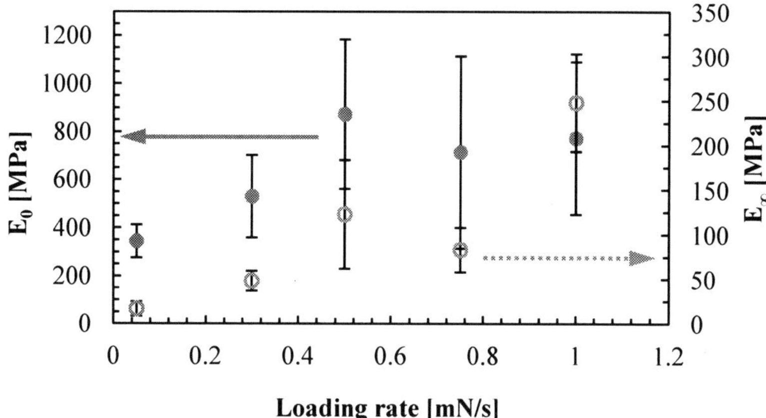

Figure 5: E_0 and E_∞ values as a function of the loading rate of the indentation experiments. Filled circles represent E_0 while empty circles represent E_∞.

3.2 Macro scale testing: tensile creep experiments

Figures 6 and 7 present E_0 and E_∞ as a function of the hold value. In both figures data obtained from the loading phase (eqn. (9)) are distinguished from the data obtained from the holding phase (eqn. (10)). All data were corrected for the laboratory conditions based on the temperature dependency presented in fig. 3. The error bars in the figures indicate the standard deviations out of 5 independent

measurements. In almost all cases the results from the holding phase are more accurate. Moreover, the total sum of the residual values of the fitted curve vs. the experimental curve proved to be much lower in the case the holding phase of the curve was used to determine the stiffness values.

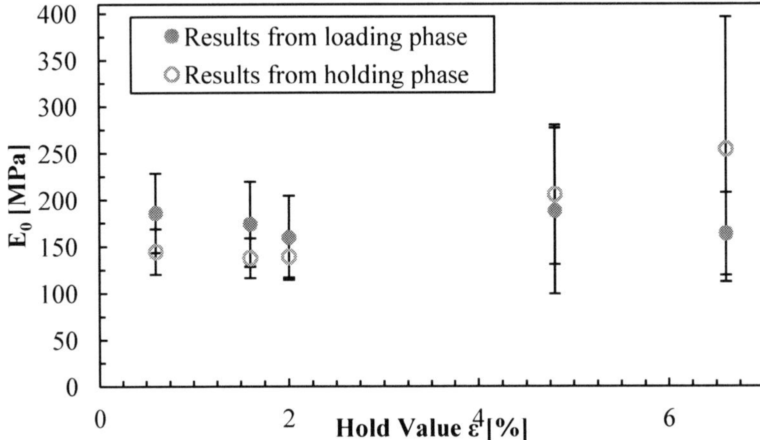

Figure 6: E_0 values as a function of the hold value based on experimental tensile test data originating from the loading or holding phase. All experiments were performed at 10N/min. The hold value is expressed as the corresponding strain (fig. 2) which was calculated via FEM.

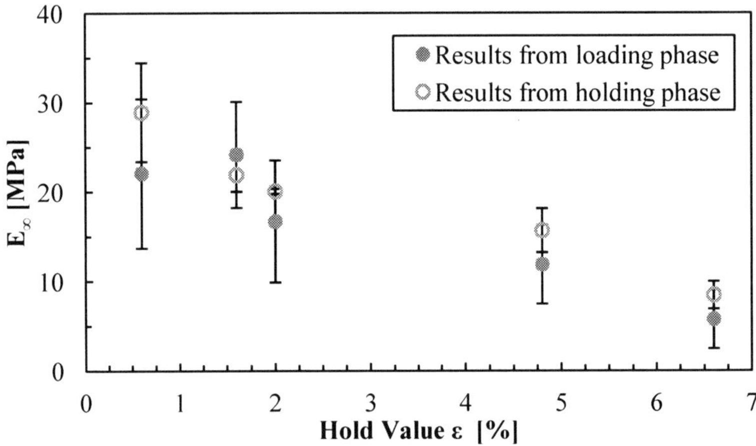

Figure 7: E_∞ values as a function of the hold value based on experimental tensile test data originating from the loading or holding phase. All experiments were performed at 10N/min. The hold value is expressed as the corresponding strain (fig. 2) which was calculated via FEM.

In correspondence to the IIT results, there is no significant dependency of E_0 on the hold value. Due to an increasing viscous-plastic behaviour of the elastomeric material, the values of E_∞ decrease with increasing hold value. The ratio of E_∞ on E_0 yields values in the range of 0,075, which indicates the need for a non-linear material model.

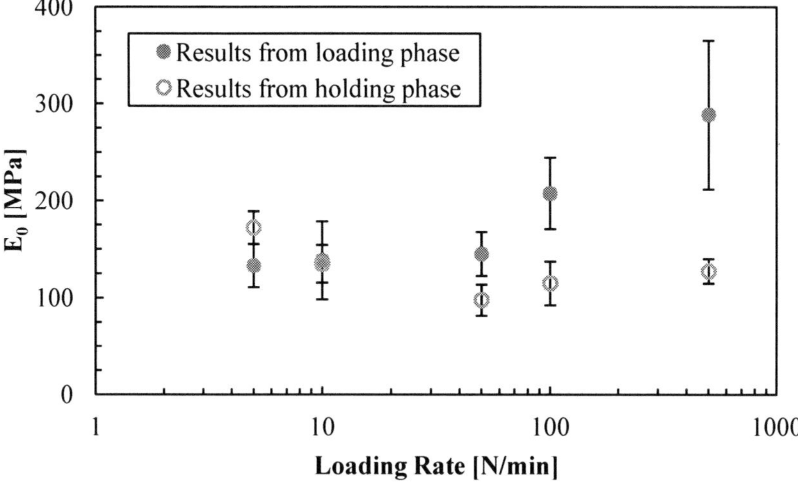

Figure 8: E_0 values as a function of the loading rate based on experimental tensile test data originating from the loading or holding phase. All tests were performed up to a hold value of 50N.

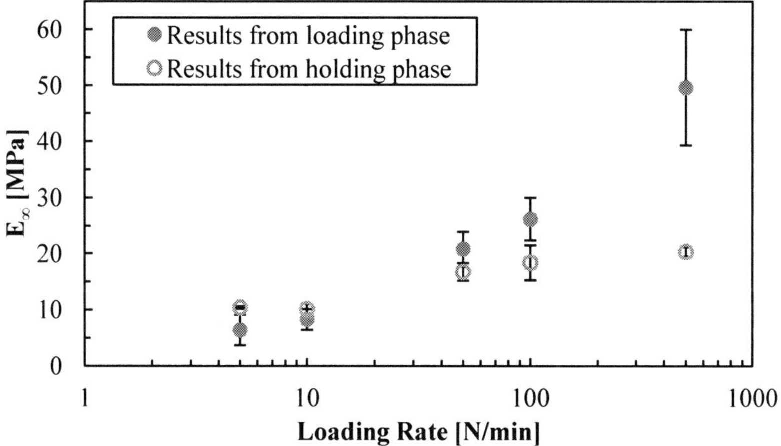

Figure 9: E_∞ values as a function of the loading rate based on experimental tensile test data originating from the loading or holding phase. All tests were performed up to a hold value of 50N.

The dependency of E_0 and E_∞ on a wide range of loading rates is shown in figures 8 and 9. The results based on the loading phase of the curve show an increase of E_0 and E_∞ with increasing loading rate. These results are again less accurate compared to the results from the holding phase as can be seen on the size of the error bars.

When E_0 and E_∞ are extracted from the loading phase of the curve an increase in stiffness with increasing loading rate is observed. The increased resistance of polymer flow at higher rates explains this behaviour. This tendency is however not observed when data are extracted from the holding phase of the curve. This suggests that the stiffness values obtained from the holding part the curve are not sensitive to the preceding loading rate. The values for E_∞/E_0 are situated in the range of 0,1, which again indicates a distinct non-linear material behaviour.

3.3 Comparison between both length scales

When comparing the trends of E_0 an E_∞ as a function of the hold value during the indentation experiments to the results of the macro tensile creep experiments a quite similar trend can be seen. When only the results from the loading phase are taken into account, the dependency of E_0 an E_∞ on the loading rate is also quite comparable.

Considering the absolute stiffness values, the values obtained on the micro scale with IIT are significantly larger than the values obtained with tensile creep experiments on the macro scale. Figure 10 is an illustrative graph in which the average E_0 values determined on both length scales are compared to each other as a function of the hold value. The micro scale results are up to 4 times as stiff as the macro scale results. This stiffer response originates from the much more complex strain-fields during indentation testing compared to tensile testing (Johnson [7]). While the former method combines shear, tension and compression in one single indentation, tensile testing applies a single uniform loading condition. Comparing results of both experimental techniques should thus be done with great care.

The results obtained from the measurements on the macro scale are also much less dependent on experimental conditions than the results from IIT. The macro scale tests have the advantage of testing a larger volume of the sample material in one test and thus possess a higher averaging potential. This is in contrast to the IIT which tests a very small volume at the surface of the sample. This difference in test set up leads to a higher dependency of the fitted parameters on the testing conditions.

The standard deviations presented by the error bars in figures 4 and 5 are remarkably large. This high uncertainty can be explained by the sensitivity of the stiffness values on the fitted parameters. As can be seen in figure 11, a small change in C_0-C_1 has a large impact on E_0 in the IIT-results region. Together with the inevitable mathematical error in the curve fitting algorithm this results in a large spread of the experimental data. Due to the different loading conditions and subsequent lower stiffness values, E_0 is less sensitive for the variation in C_0-C_1 in the tensile creep experiments.

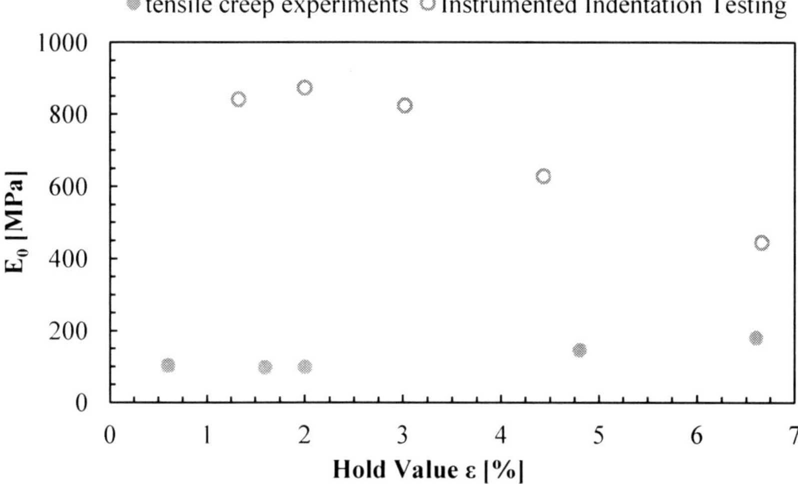

Figure 10: Comparison of E_0 based on the holding phase of the tensile creep experiment on the macro scale (filled circle) and E_0 determined with IIT on the micro scale (hollow circle).

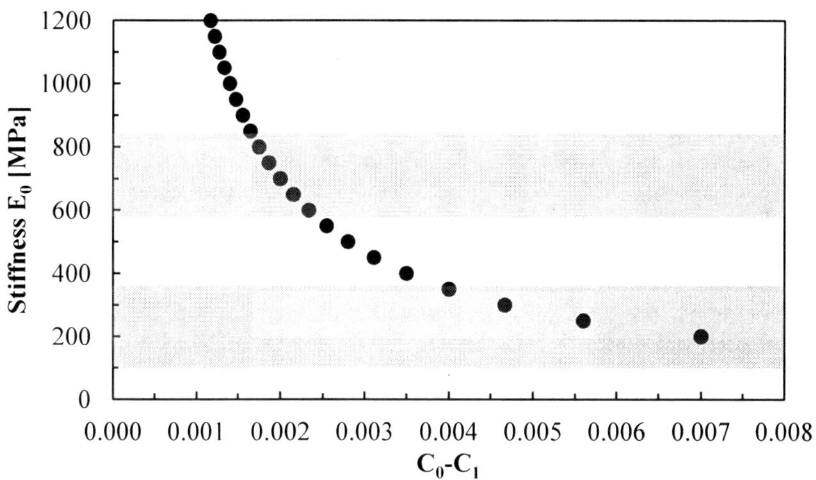

Figure 11: Sensitivity of E_0 on the difference between C_0 and C_1. The blue and red regions indicate the stiffness ranges in which the results of respectively the indentation and tensile experiments are situated.

www.witpress.com, ISSN 1743-3533 (on-line)

4 Conclusions

This research focuses on the characterisation of the stiffness properties of a polyurethane elastomer with 2 different experimental methods on 2 different length scales. As there is no single-valued elastic modulus for any elastomer, the limiting values of the elastic modulus E_0 and E_∞ were determined within this research. In order to find these values, a curve fitting algorithm was used to fit experimental data to the standard linear solid material model. This was done on micro scale by making use of IIT and with macro scale tensile creep experiments.

FEM proved to be useful to coordinate the loading conditions on both length scales. Nonetheless the loading conditions of an indentation test can-not be transferred unambiguously to a tensile experiment. In an indentation test; the nature of the loading is a complex mixture of hydrostatic compression, tension and shear (Johnson [7]) while a tensile test applies a uniform tensional strain. This difference in loading condition induces a difference of a factor up to 4 between the results on both length scales.

In order to exclude the loading condition dependency, future research should focus on the optimization of the IIT. Useful steps would be an increase in data sampling rate or hold value and lower loading rates. In correspondence to the good results on macro scale, a creep measurement at constant load during an IIT is advisable.

References

[1] Kranenburg, J.M., van Duin, M., and Schubert S., Screening of EPDM cure states using depth-sensing indentation, *Macromolecular Chemistry and Physics*, **208**, pp. 915-923, 2007.

[2] Majoros, L.I., Dekeyser, B., Haucourt, N., Castelein, P., Paul, J., Kranenburg, J.M., Rettler, E., Hoogenboom and R., Schubert, S., Preparation of polyurethane elastomers (PUEs) in a high-throughput workflow, *Journal of Polymer Science: Part A: Polymer Chemistry*, **49**, pp. 301-313, 2011.

[3] Fischer-Cripps, A.C., *Nanoindentation*, Springer: NY, pp. 21-38, 2002.

[4] Galli, M., Comley, K.S.C., Shean, T.A.V. and Oyen, M.L., Viscoelastic and poroelastic mechanical characterization of hydrated gels. *Journal of Materials Research*, **24(3)**, pp.973-979, 2008.

[5] Vanlandingham, M.R., Chang N.-K., Drzal, P.L., White, C.C. and Chang, S.-H., Viscoelastic characterization of polymers using instrumented indentation. I. quasi-static testing. *Journal of Polymer Science: Part B: Polymer Physics*, **43**, pp. 1794-1811, 2005.

[6] Oliver, W.C. and Pharr, G.M., An improved technique for determining hardness and elastic modulus using load and displacement sensing indentation experiments, *Journal of Materials Research,* **7(6)**, pp. 1564-1583, 1992.

[7] Fischer-Cripps, A.C., A simple phenomenological approach to nanoindentation creep, *Material Science and Engineering*, **385**, pp. 74-82, 2004.
[8] Oyen, M.L., Spherical indentation creep following ramp loading, *Journal of Materials Research*, **20(8)**, pp. 2094-2100, 2005.
[9] Oyen, M.L., Sensitivity of polymer nanoindentation creep measurements to experimental variables, *Acta Materialia*, **55**, pp. 3633-3639, 2007.
[10] Powel, P.C., *Engineering with polymers*, Chapman and Hall London, 1983.
[11] Johnson, K.L., *Contact Mechanics*, Cambridge University Press, 1985.

The micro-macroscale correlation of NiTi mechanical behavior: a finite element analysis

F. M. Weafer & M. S. Bruzzi
Department of Mechanical & Biomedical Engineering, National University of Ireland, Galway, Ireland

Abstract

One material that has found particular favor for use in biomedical endovascular stents is the near equi-atomic NiTi alloy, Nitinol. One remarkable trait exhibited by this superelastic material is the improvement of its fatigue performance with increasing mean strain [1, 2]. Clarification into this phenomenon still remains incomplete in literature. This study proposes a microstructural explanation for this unique macroscopic behavior; it is hypothesized stress-induced martensite (SIM) will stabilize with increasing strain which, in turn, leads to the observed increase in fatigue life. Finite Element Analysis (FEA) is employed to investigate the behavior of a 'v-strut' stent subcomponent under various strain levels. The volume fraction of SIM is analyzed to identify its potential influence on macroscopic response and, ultimately, fatigue behavior. In addition, a computational investigation is performed on the effect of crystallographic texture on macroscopic response. Granular transformational behavior is analyzed using FEA models with realistic and idealized grain structures, specifically evaluating the effect of individual grain orientations on the stress-induced martensite transformation (SIMT).
Keywords: Nitinol, microstructure, superelasticity, fatigue.

1 Introduction

1.1 Medical use of Nitinol

Nitinol possesses a unique combination of properties, which renders it suitable in a broad range of engineering applications. In particular, characteristics such as superelasticity, biocompatibility, flexibility, and compatibility with magnetic

WIT Transactions on Engineering Sciences, Vol 77, © 2013 WIT Press
www.witpress.com, ISSN 1743-3533 (on-line)
doi:10.2495/MC130021

resonance imaging (MRI) procedures make it particularly appealing in the biomedical industry for use as self-expanding endovascular stents [3]. Such devices have proven effective in the treatment of atherosclerosis in a variety of vessels and arteries. However, fracture rates of up to 65.4% in such stents used in the superior femoral artery have been reported in clinical studies [4]. Due to physiological movement from the cardiac systolic-diastolic cycle, in addition to the muscular movement associated with the anatomy in which they are placed, such failures can be attributed to cumulative fatigue damage. Accurate characterization of the fatigue behavior of such stents is therefore essential for their prolonged safe use in human arteries.

1.2 Superelasticity

Superelasticity is the term given to the first-order phase transformation, from a parent austenite to a daughter martensite phase, which underpins Nitinol's unique performance. Nitinol's phase transformation can be induced by a change in temperature or stress; via an application of load, or upon cooling below the martensite-start temperature (M_s). Austenitic Nitinol is a hard, stiff material whereas martensitic Nitinol is a softer, more ductile material with a lower yield stress [5]. These vast differences in material properties can be explained by their different microstructural crystallographic structures; austenite has an ordered cubic B2 structure while martensite has a more complex twinned monoclinic B19' structure.

In a crystallographic context, the stress-induced martensite transformation (SIMT) occurs by rearrangement of atomic planes via Bain strain and lattice invariant shear. The twin boundaries in martensite readily shift such that the twins are predominantly oriented in one preferential direction; this process is known as de-twinning. By this microstructural process, the material can withstand approximately 6% strain without permanent deformation. This ability to accommodate such significant strains is highly desirable in stent device design for stent deliverability, durability and conformance. The SIMT has no associated breaking of atomic bonds, therefore, no macroscopic change is associated with the transformation; this can be explained by the self-accommodating nature of twinned martensite variants. Upon removal of the stress the superelastic strain recovers at a lower stress level than at which it is induced, i.e. along a hysteresis curve (Figure 1).

1.3 Microstructural texture

As described, Nitinol derives its unique mechanical behavior from the coordinated atomic movements manifesting in a phase transformations from cubic austenite to monoclinic martensite. Therefore, any significant alignment of the atomic planes resulting from crystallographic texture in the polycrystalline material can have a marked influence on the mechanical response. It has been experimentally shown that the crystallographic texture has a significant effect on crack trajectories in NiTi tube specimens subjected to uni-axial cyclic loading [8]. In addition, significant variations in mechanical behavior have been

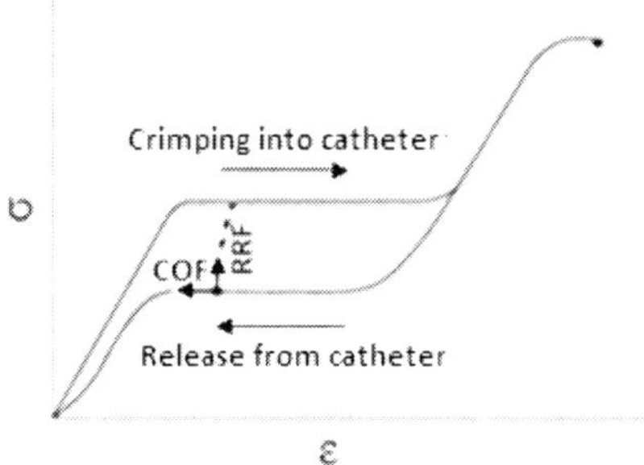

Figure 1: Superelastic stress–strain curve portraying the stent's chronic outward force (COF) on blood vessel in retaining it open from a stressed configuration, and radial resistive force (RRF) in resisting stent collapse (adapted from [3]).

observed between rolled and transverse directions of cold-drawn NiTi sheets [6, 7]; this can be attributed to the hindrance of deformed martensite structures and defects in the specimen leading to differing dominant detwinning and reorientation modes and dislocation densities. This, in turn, may result in discrete areas of the microstructure no longer being able to transform back into austenite (stabilized SIM); on a microstructural scale the stress fields of the dislocations are high enough to prevent unloading, even if the material is macroscopically without load. Thus, the material locally behaves as a standard metallic material accumulating plasticity during cyclic mechanical loading. Clarification into the effect of this process with regard to fatigue performance, however, still remains unexplored in literature.

Initial processing techniques used to form Nitinol specimens have also been investigated to determine the correlation between differing product forms and resulting microstructures. Results confirm that the crystallographic texture in Nitinol can be significantly different depending on product form selected. Specifically, thin-section flattened tube materials (0.2–0.4mm) Nitinol specimens showed higher fatigue thresholds of $\Delta K_{th} > 2.5$ MPa√m, as compared to values of $\Delta K_{th} < 2$ MPa√m in thick-section bar material (9–10mm) [9]. It was observed that crystallographic texture anisotropies are present when specimens were compressed in different crystallographic directions [10]; varying recovery strains and strength levels were displayed. Namely, the compression strength of polycrystalline NiTi is stronger when tested in the longitudinal direction, which has a [111] texture, as compared with NiTi material that was tested in the transverse direction with a [110] crystallographic texture. These results confirm there are underlying microstructural mechanisms taking place with varying textures and this, in turn, can lead to profound effects on the overall macroscopic material behavior.

1.4 Fatigue of Nitinol

For the past four decades, researchers have been eagerly attempting to fully understand the unique behavior of this remarkable superelastic material. Melton and Mercier [11] were the first to extensively study the S/N fatigue behavior of Nitinol. While these initial studies provide tremendous insight into baseline fatigue properties, much of their work was under stress control conditions, which are difficult to adapt to strain life; that being relevant for stent fatigue analyzes. This difficulty can be attributed to the near constant stress plateau associated with the phase transformation upon loading and unloading (Figure 1). In addition, Nitinol research utilizing *in situ* loading in conjunction with microscopy is seldom found in literature; yet these experiments can offer a rich understanding of Nitinol's fundamental material response as they address both microstructural and macro-mechanical response simultaneously.

Nevertheless, standard and thermal cameras have been successfully employed to capture the macro-transformation thermal signatures of Nitinol upon loading [12, 13]. With evidence of the latent heat release, which occurs during the first-order transformation, the thermal camera supports the hypothesis that SIMT is occurring. Similarly, high-speed and infra-red cameras have been used to observe the complete transformation process, from nucleation to growth and to the eventual vanishing of SIM, in a tube specimen under displacement-controlled uni-axial tension [14]. A similar evaluation of Nitinol's surface in strain-free, strained and unloaded states was carried out using scanning electron microscopy (SEM) [15]. In all cases mentioned however, attention is simply focused on a single macroscopic deformation domain; no further investigation for possible effects of microstructural deformation evolution on the transformation process is performed.

In a study by Brinson [16] the microstructural and macroscopic transformation behavior of polycrystalline NiTi specimens was investigated under low level cyclic loading. In this study, focus is again placed on the macroscopic behavior of the material rather than delving into the underlying microscopic phenomena taking place. However, one key result is presented which offers a correlation between the microscopic and macroscopic behaviors; localized plastic deformation is observed in the NiTi specimen after as few as 10 cycles (potentially being caused by the presence of stabilized SIM). The plastic deformation is observed to appear with cycling in the vicinity of the forming martensitic plates (Figure 2). Due to the increased localized deformation, it is

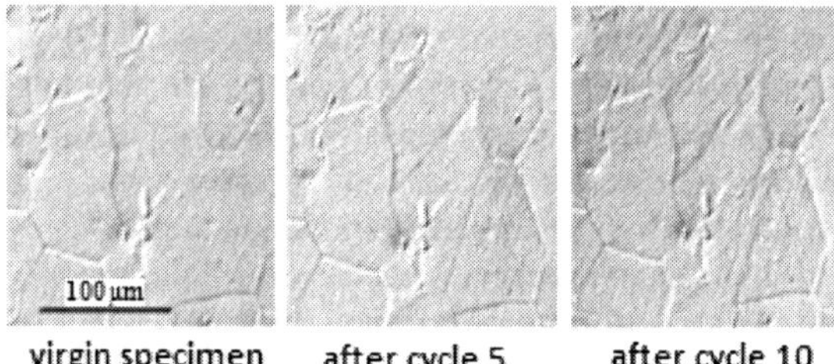

Figure 2: Accumulation of localized plastic deformation within the grains upon loading cycling for 0, 5 and 10 cycles. Specimen fully unloaded following each cycle [16].

observed that additional variants are formed at each cycle, providing a microscopic explanation for the macroscopically observed strain hardening.

As an exciting step forward such studies begin to correlate micromechanical behavior with the overall macroscopic material response of Nitinol, however further work in this research field is still required. The aim of the work presented in this paper was to establish a correlation between microscopic behavior, in particular the volume fraction of SIM, and the overall macroscopic material response employing Finite Element Analysis. In addition, using Finite Element models incorporating a granular structure, identification of possible effects of crystallographic texture on the SIMT was also investigated.

2 Material characterization

2.1 Material imagining

To create a realistic Finite Element Analysis (FEA) model of polycrystalline superelastic Nitinol, microstructural geometries of suitable specimens were required. Nitinol wire samples underwent a recrystallization anneal (750°C for 1hr) and the transformational behavior was subsequently analyzed using the bend-free recovery test; the austenite finish temperature (A_f) was found to be 1.5±0.5°C. Specimens were then set in an epoxy resin for ease for handling. Specimens were manually polished to a finish of 0.06μm and chemically etched using a $1HF:4HNO_3:5H_2O$ solution. Finally, specimens were viewed under the optical microscope; grain distribution and average grain size were specifically identified (Figure 3(a)). Utilizing the obtained micrographs, realistic FEA models were created which incorporated the complex microstructural nature of the superelastic austenitic material (Figure 3(b)); these models will be discussed further in Section 3.2.

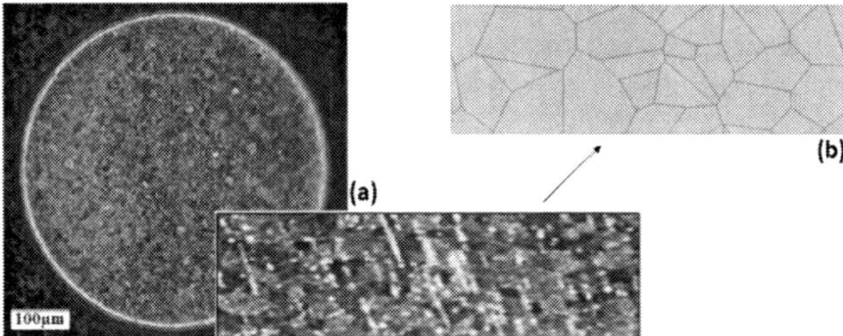

Figure 3: (a) Experimental micrograph and, (b) Finite Element representation of the microstructure of superelastic austenite wire.

2.2 Test specimens

Nitinol 'v-strut' specimens were custom fabricated for this study. The 'v-strut' is representative of a single strut of a Nitinol self-expanding stent; in this way the mechanical behavior of the individual stent components can be identified and analyzed clearly. The manufacturing and processing techniques used on the

specimens are consistent with those preformed on commercially available endovascular stents. The specimen's A_f temperature was found to be 11.2±2°C using the bend-free recovery method, thus confirming its suitability for use as a self-expanding stent as it will exhibit superelastic behavior at human body temperature. To accurately characterize Nitinol's asymmetric behavior for input into a user defined UMAT/Nitinol, uni-axial tensile and compressive experimental testing were performed on linear support struts (Figure 4); as will be discussed further in Section 2.2.1 and Section 2.2.2. Excess materials at both ends of the specimens, along with the support struts, were included in the design to provide precise alignment, structural stability and secure gripping during testing.

Figure 4: SEM of 'v-strut' specimen used, showing vertical support struts.

2.2.1 Experimental set-up

As Nitinol's mechanical properties are highly temperature dependent, all investigations were carried out at 37°C, i.e. body temperature, to accurately represent the material's behavior *in vivo*. This was achieved using the EnduraTEC ELF/3200 in conjunction with an environmental chamber with air-heating fan (Figure 5). The exothermic martensite transformation, and endothermic reverse transformation, has a considerable influence on the stress-strain response of Nitinol due to the release of latent heat due to the exothermic nature of the transformation process. Therefore, to ensure accuracy of data the test temperature was strictly monitored and controlled. Improved heat exchange

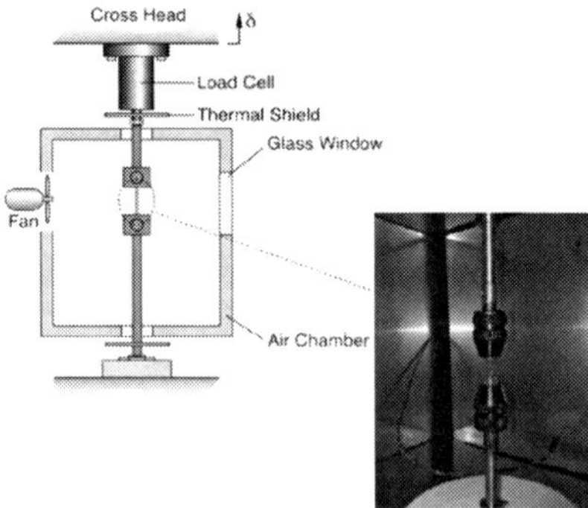

Figure 5: Experimental set-up using in this study for uni-axial tensile and compressive material testing procedures.

results in more accurate and repeatable results. In addition, it allows higher strain rates to be used, due to the faster dissipation of heat from the sample, therefore reducing testing time.

2.2.2 Experimental results

To accurately characterize Nitinol's material behavior, uni-axial tensile and compressive experimental testing were performed on the support struts of the specimens (as discussed in Section 2.2). From the stress-strain curves (Figure 6(a)), key material properties are extracted for input into the user defined UMAT/Nitinol. Validation of the UMAT was completed through comparison of the simulated stress-strain response of a simple cubic FEA model to the experimental stress-strain data; ensuring the model was accurately predicting the complex superelastic material behavior of the specimen (Figure 7).

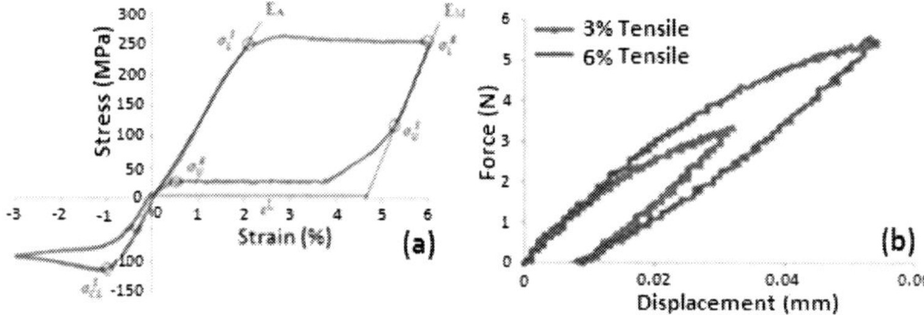

Figure 6: (a) Stress-strain behaviour of Nitinol to 3% compressive strain and 6% tensile, (b) Load-deflection curve for 'v-strut' specimens at 3% and 6% tensile strain.

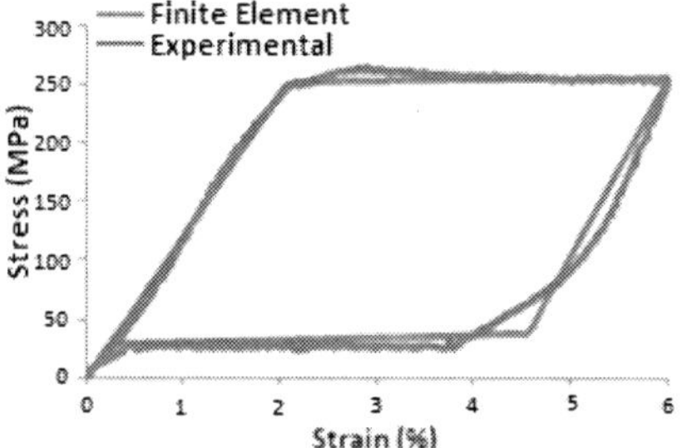

Figure 7: Validation of FEA model employing the UMAT/Nitinol.

3 Finite element model development

3.1 Correlation of SIMT to fatigue behavior

To investigate the effect of stress-induced martensite (SIM) on the fatigue behavior of a Nitinol stent device, a 3-dimensional homogenous FEA model of a 'v-strut' stent subcomponent was created. Employing a user defined material subroutine (UMAT/Nitinol), the model was capable of predicting Nitinol's

complex transformational behavior on a macro-scale level. To accurately represent Nitinol's macroscopic material behavior in the FEA model, data from the uni-axial tensile and compressive experimental tests (as described in the Section 2.2.2) were inputted into the UMAT. Following validation (Figure 7), the influence of SIM on the overall macroscopic stress-strain behavior of the 'v-strut' specimens was established for various strain levels. Three strains spanning the experimentally established superelastic plateau of the specimen are targeted for analysis (namely, 2.5%, 4% and 6%) to investigate various physiologically relevant conditions experienced by a stent; variations of mean strain experienced by a stent *in vivo* are attributed to the vessel anatomy in which they are placed and extent of stent over-sizing.

Native vessels undergo diameter changes of approximately 0.16–0.34% when subjected to a typical 100mmHg pulse pressures in a healthy vessel [1], these changes being potentially lower in a stenosed vessel. This results in the endovascular stent experiencing strain amplitudes *in vivo*. The mean strain (mid-pulse), however, has the most significance when discussing the fatigue life of self-expanding Nitinol stents. Typical engineering materials will experience a decrease in fatigue life with increasing mean strain, however, Nitinol has been shown to exhibit improved fatigue performance in the region of approximately 1.5–6% mean strain [1, 2]. Therefore, employing FEA, the aim of this work is to investigate the macroscopic behavior of Nitinol 'v-strut' stent subcomponents at these various mean strains and evaluate the potential influence of SIMT on the observed macroscopic material response; through identification of the volume fraction of SIM at critical locations. The hypothesis presented in this paper states that, at sufficiently high mean strains, the resulting stress levels induce and stabilize the SIM and this is what gives rise to the increased fatigue performance.

3.1.1 Simulated loading modes

The use of detailed FEA simulations allows evaluation of the stress-strain behavior of individual Nitinol stent components at various displacements; these being representative of various levels of physiologically relevant mean strains. Through systemic analysis, identification of the corresponding simulated displacements to induce the desired strains at critical locations in the FEA model (peak tensile strain on outer apex of 'v-strut') is achieved. It is also of significant importance, to ensure accuracy of results, that the chosen simulated loading modes represent physiological conditions experienced by the stent. As displayed in the contour plots in Figure 8, there is a considerable variation in the specimen response when loaded in compression (Figure 8(a)) or tension (Figure 8(b)). *In vivo*, a stent experiences compressive loading due to the pulsatile loading of the diastolic-systolic motion and thus compressive loading procedures are therefore performed in this study.

3.2 Effect of texture on fatigue behavior

To evaluate the effect of microstructural texture on Nitinol's macroscopic response, a 2-dimensional rectangular model was created using Voronoi Tessellation based on the obtained micrographs (Figure 3(b)). As the images had

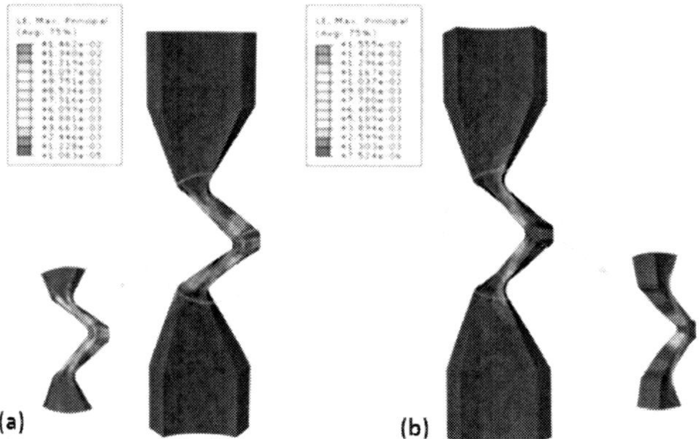

Figure 8: FEA models of 'v-strut' at 1.5% (a) compressive and (b) tensile strain.

an aspect ratio close to 1, it was determined the complex realistic models were properties were obtained from literature [17] for similarly processed NiTi not necessary; idealized hexagonal units were instead used in which each hexagonal unit represented an individual grain (Figure 9(b)). In this way, the constitutive behavior of Nitinol on a micro-scale was represented. Single crystal specimens (as those used in the material characterization of Section 2.1) and were assigned to the individual grains. The effects of individual grain orientation on SIMT were therefore investigated for randomly generated crystallographic textures.

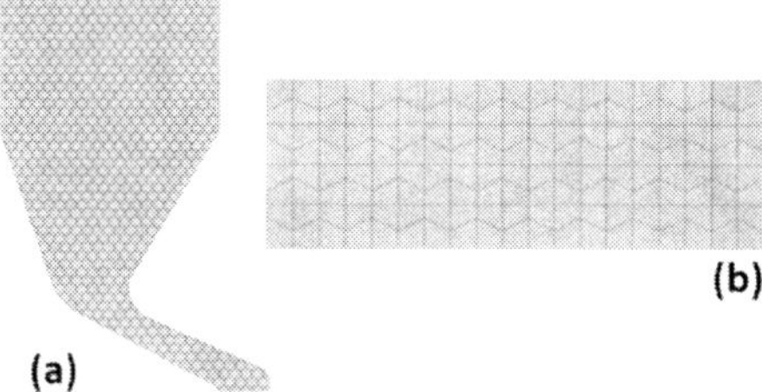

Figure 9: (a) 2-dimensional 'v-strut and, (b) rectangular FEA models with idealized hexagonal grain structures used to investigate the effect of texture.

In order to investigate the influence of crystallographic texture on the behavior of a stent material, a similar 2-dimenstional 'v-strut' model with an idealized hexagonal grain structure was created (Figure 9(a)). It should be mentioned, common microstructures in Nitinol stents feature an ultrafine-grained microstructure in the order of 40–100 nm. In this study, however, larger grain geometries are employed; annealed NiTi single crystal properties are used in this study which would represent an average grain size of approximately 20μm [17]. Nonetheless, employing FEA, the effect of a randomly generated microstructure was represented in the complex 'v-strut' geometry. The overall macroscopic

stress-strain behavior of the specimen was analyzed, in addition to identification and analyzes of critical locations of peak tensile stresses on the outer apex of the 'v-strut' model which may represent a potential fatigue failure initiation site.

4 Results and discussion

4.1 Correlation of SIMT to fatigue behaviour

The highly advantageous trait exhibited by Nitinol of increased fatigue performance with increasing mean strain still remains unexplained in literature, in particular in the correlation between microstructure and the observed macroscopic phenomenon. In this paper, it is hypothesized that the increased fatigue life is attributed to the phase transformation (SIMT) experienced by the superelastic material. The volume fraction of SIM is shown to increase with increasing strain at critical locations (Figure 10); namely at the regions of peak stresses under the compressive loading at the inner apex of the 'v-strut'. Under strain-controlled conditions, superior fatigue lives are shown in specimens that experience increased strain levels [1, 2]; it is therefore suggested there is a strong connection relating SIM, in particular at strain levels high enough to stabilize SIM, and the experimentally observed increased fatigue performance.

However, the most significant location for analysis is that of peak tensile strain (initiation site of fatigue failure). The volume fraction of SIM at this critical location is found to follow a similar trend to that of the experimentally established constant life diagram for Nitinol [1, 2]; see Figure 11. Furthermore,

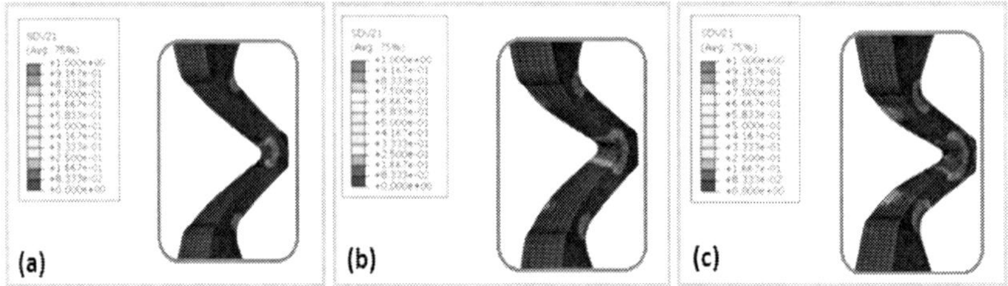

Figure 10: The volume fraction of SIM in 'v-strut' under (a) 2.5%, (b) 4% and (c) 6% compressive strain.

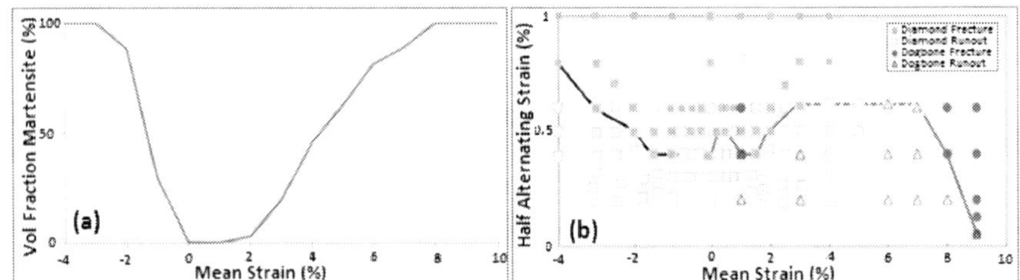

Figure 11: Comparing (a) the volume fraction of SIM present in the FEA model for various mean strains, and (b) the experimentally established constant life diagram for Nitinol [1, 2].

the observed on-set of increasing fatigue performance correlates with the onset of SIM at approximately 1.5% mean strain level which presents further evidence for the proposed hypothesis. It is proposed, at such strains the SIM is stabilized (through methods discussed in Section 1.3) and a strain hardening effect occurs, thus resulting in a strengthening effect on the Nitinol specimens. This, in conjunction with the self-accommodating nature of the twinned martensite, presents itself as a possible explanation to the observed increase in fatigue performance.

4.2 Effect of texture on fatigue behavior

This study successfully demonstrates the profound effect crystallographic texture (grain orientation distribution) has on overall macroscopic material response of Nitinol. In applying a load-unload procedure on the idealized hexagonal rectangular models described in Section 3.2, the effect of the randomly generated textures on the stress-strain response becomes apparent. It is found, grains orientated in (110) crystallographic direction inhibit, while grains in (111) orientations promote the SIMT [17]. This becomes particularly evident when comparing Figures 12(a) and (b); in including a higher volume of (111) orientated grains in model (b), it can be seen notably lower stress values are reached along the upper plateau due to the ease at which SIMT occurs.

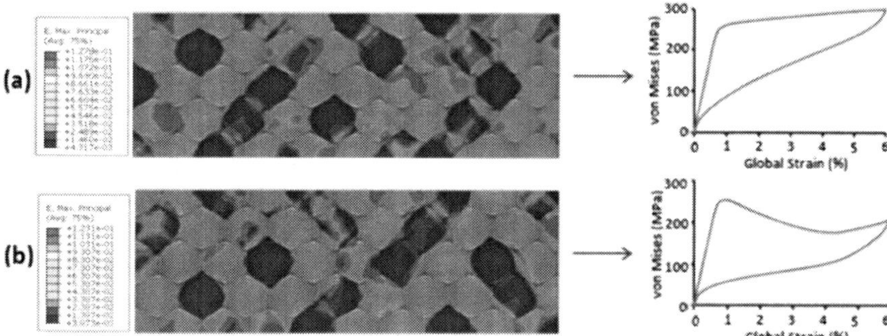

Figure 12: Maximum principal strains in idealized hexagonal model with two random texture deformed (a) and (b) to 6% global strain.

The difference in transformation behavior in the individual grains results in localized stress gradients over grain boundaries; these peak stresses become particularly important when discussing fatigue behavior. To further explore this, the influence of texture was investigated in the 'v-strut' stent subcomponent; a 2-dimenstional idealized hexagonal model was created for analysis (Figure 13). Due to the presence of (110) orientated grains at critical locations in the 'v-strut' geometry (Figure 13(a)), SIMT has been hindered and therefore localized peak strains have been induced. Consequently, the peak tensile stresses induced on the outer apex of the 'v-strut' become significant as fatigue failure initiation sites in the 'v-strut' stent subcomponent.

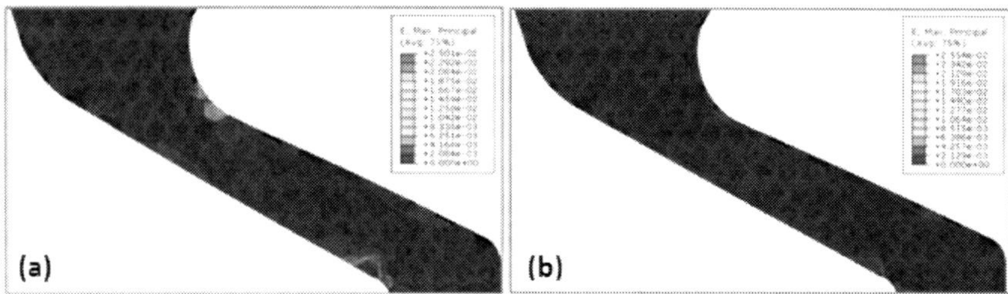

Figure 13: Maximum strains in 'v-strut' with two varying textures (a) and (b). Peak strains in texture (a) due to presence of (110) orientated grains.

As discussed in Section 3.2, the granular FEA models created in this study contain larger average grain sizes (20μm) than those typically found in Nitinol endovascular stent devices (40–100nm). Therefore models created in this work demonstrate a more severe impact of crystallographic texture on macroscopic material response; due to the larger sized grains. However, it can be stated, in understanding the impact of local microstructural effects on global mechanical response, it can lead to a much fuller understanding of the causes of deviation of the mechanical response from predictions and unforeseen fracture in Nitinol biomedical devices.

5 Concluding remarks

Finite Element Analysis (FEA) models capable of predicting the complex constitutive behavior of Nitinol were created to determine the influence of stress-induced martensite (SIM) on the fatigue behavior of Nitinol 'v-strut' stent subcomponents. Utilizing FEA, the volume fraction of SIM was established in the specimens for various levels of mean strain. As the strain levels in the specimen increase, the volume fraction of SIM present correspondingly increases. Furthermore, the percentage of SIM at critical locations of fatigue failure initiation follows a similar trend to that of the constant life diagram experimentally established for Nitinol [1, 2]. Therefore it can be suggested, there exists a strong correlation between SIM and the observed increase in fatigue performance due to possible combined effects of strain hardening (from stabilized SIM) and the self-accommodating nature of the twinned martensite itself.

This study also successfully demonstrates the profound effect texture has on stress-induced martensite transformation (SIMT); it was demonstrated that grains orientated in (110) crystallographic direction inhibit, while grains in (111) orientations promote the stress-induced martensite transformation progression [17]. This conflict between individual granular behavior has been shown to result in localized stress and strain gradients over grain boundaries which may previously have been neglected in simplistic homogenous FEA models disregarding influence of individual grain anisotropies. However, due to the mechanical property variation with texture, these induced peak tensile stresses become particularly important when discussing fatigue behavior in complex stent geometries and may present an explanation for unforeseen fatigue failures in stent devices.

Acknowledgements

This study was funded by Science Foundation Ireland (SFI) Research Frontiers Programme. The authors would also like to acknowledge Veryan Medical for producing the test specimens used in this study.

References

[1] Pelton, A.R., Schroder, V., Mitchell, M.R., Gong, X.Y., Barney, M., Robertson, S.W., Fatigue and Durability of Nitinol Stents. *Journal of Mechanical Behavior of Biomedical Materials (1)*, pp. 153-164, 2008.

[2] Pelton, A.R., Gong, X.-Y., Duerig, T., Fatigue Testing of Diamond-Shaped Specimens. *Proceedings of the Materials and Processes for Medical Devices Conference*, pp. 199-204, 2003.

[3] O'Brien, B., Bruzzi, M., Shape Memory Alloys for Use in Medicine. *Comprehensive Biomaterials (1)*, pp. 49-72, 2011.

[4] Allie, D.E., Hebert, C.J., Walker, G.M., Nitinol Stent Fractures in the SFA. *Endovascular Today (Jul/Aug)*, pp. 1-8, 2004.

[5] Van Humbeeck, J., Stalmans, R., Characteristics of Shape Memory Alloys. *In: Otsuka K & Wayman CM (eds) Shape Memory Materials*, pp. 149-183, 1998.

[6] Liu, Y *et al.* Effect of Texture Orientation on the Martensite Deformation of NiTi Shape Memory Alloy Sheet. *Acta Materialia 47(2)*, pp. 645-660, 1999.

[7] Lin, H.C., Wu, S.K., Lin, J.C., The Martensitic Transformation in Ti-rich TiNi Shape Memory Alloys. *Materials Chemistry and Physics 37(2)*, pp. 184-190, 1994.

[8] Robertson, S.W., Stankiewicz, J., Gong, X.Y., Ritchie, R.O., Cyclic Fatigue of Nitinol. *Proceedings of the International Conference of Shape Memory and Superelastic Technologies*, pp. 79-88, 2004.

[9] Stankiewicz J.M., Robertson S.W., Ritchie R.O., Fatigue-crack Growth Properties of Thin-walled Superelastic Austenitic Nitinol Tube for Endovascular Stents. *Journal of Biomedical Materials Research A*, pp. 685-691, 2006.

[10] Fonte, M., Saigal, A., Effects of Crystallographic Texture Directionality on the Compressive Stress–strain Response of Shape Recovered Polycrystalline Nitinol. *Scripta Materialia 63(3),* pp. 320-323, 2010.

[11] Melton, K.N., Mercier, O., Fatigue of NiTi Thermoelastic Martensites. *Acta Metallurgica (27)*, pp. 137-144, 1979.

[12] Shaw, J., Thermomechanical Aspects of NiTi. *Journal of the Mechanics and Physics of Solids* 43(8), pp. 1243-1281, 1995.

[13] Shaw, J., Kyriakides, S., Initiation and Propagation of Localized Deformation in Elasto-plastic Strips under Uniaxial Tension. *International Journal of Plasticity* 13(10), pp. 837-871, 1997.
[14] Feng, P., Sun, Q., Experimental Investigation on Macroscopic Domain Formation and Evolution in Polycrystalline NiTi Microtubing under Mechanical Force. *Journal of the Mechanics and Physics of Solids 54(8)*, pp. 1568-1603, 2006.
[15] Shabalovskaya, S., Anderegg, J., Vanhumbeeck, J., Recent Observations of Particulates in Nitinol. *Materials Science and Engineering A (481-482)*, pp. 431-436, 2008.
[16] Brinson, L., Stress-induced Transformation Behavior of a Polycrystalline NiTi Shape Memory Alloy: Micro and Macromechanical Investigations via in-situ Optical Microscopy. *Journal of the Mechanics and Physics of Solids 52(7)*, pp. 1549-1571, 2004.
[17] Gall, K., Sehitoglu, H., Chumlyyakov, Y.I., Kireeva, I.V., Tension-Compression Asymmetry of the Stress-Strain Response in Aged Single Crystal and Polycrystalline NiTi. *Acta Metallurgia (47)*, pp. 1203–1217, 1998.

Nanostructured ceria synthesized by detonation method and its ultraviolet absorption performance

Z. W. Han[1], L. F. Xie[1], Y. C. Han[2], O. Q. Ni[1], J. Y. Chen[1] & Y. C. Xie[1]
[1]School of Chemical Engineering, Nanjing University of Science and Technology, China
[2]State Key Laboratory of Fire Science, University of Science and Technology of China, China

Abstract

Nanostructured ceria particles have been successfully synthesized by detonation method with emulsion explosives. The effects of mass content of $Ce(NO_3)_3{\cdot}6H_2O$ in emulsion explosives on particle size of CeO_2 were investigated. X-ray diffraction (XRD) and Transmission Electron Microscope (TEM) were carried out to characterize as-synthesized powders. The ultraviolet absorption performance of nanostructured ceria was tested by ultraviolet spectrophotometer. The results indicated that the mass content of $Ce(NO_3)_3{\cdot}6H_2O$ in emulsion explosives affected the particle size of ceria mostly. When the mass content of $Ce(NO_3)_3{\cdot}6H_2O$ was 53.2%, the particle size of ceria was even and on the small side. The UV absorption performance of as-synthesized nanostructured ceria was excellent, especially for ultraviolet B.
Keywords: nanostructured ceria, detonation synthesis, UV absorption, emulsion explosives, UVB.

1 Introduction

Excessive exposure to ultraviolet rays (UV) in sunlight induces coloring and deterioration in paper, fabrics and resins. It also causes harmful damage to the human body such as premature aging of skin, skin cancer and cataracts. Among all the wavelengths of UV, UVB (290–320nm) is the most harmful. It is the chief

WIT Transactions on Engineering Sciences, Vol 77, © 2013 WIT Press
www.witpress.com, ISSN 1743-3533 (on-line)
doi:10.2495/MC130031

cause of skin reddening, sunburn and cancer. Nowadays, most of inorganic UV-blocking filters are based on titanium dioxide (TiO_2) and zinc oxide (ZnO). However, a number of reports exist indicating the possibility of brain cells [1], blood lymphocytes [2] and lymphoblastic cells [3] damaged by titania nanoparticles. Moreover, nanoparticles of zinc and titanium oxides possess enormous photocatalytic activity [4], thus inducing the increase in their toxicity upon irradiation. Reactive oxygen species (ROS) forming during photocatalytic processes decompose not only components of cosmetics but even skincells [5, 6]. It was demonstrated that under UV-irradiation TiO_2 and ZnO nanoparticles being the part of sunscreen cosmetics generate hydroxyl radicals [7, 8] damaging DNA of skin cells [9]. Phototoxicity of titania against fibroblasts has been also confirmed [10].

Ceria(CeO_2) is one of the today's most promising nanomaterials [11]. Due to the high concentration of oxygen defects in ceria lattice the recombination of free charge carriers (electrons and holes) forming upon UV-irradiation of ceria proceeds very rapidly. It should be also noted that UV-extinction coefficient of ceria is rather high. Therefore this compound is considered as a promising UV-filter in sunscreen cosmetics [12].

Thus this paper focused on a novel method that never has been reported to prepare nanostructured ceria. And the UV absorption performance of nanostructured ceria was tested. We are trying to investigate the application of nanostructured ceria synthesized by detonation method on UV protection.

2 Experimental

2.1 Preparation of emulsion explosives

As well as the ordinary emulsion explosives which were used for demolishing mines or mountains, this special emulsion explosive for preparing nanostructured ceria was also composed of oxidant, incendiary agents and emulsifiers. In order to obtain nanostructured ceria, cerium nitrate hexahydrate ($Ce(NO_3)_3 \cdot 6H_2O$, AP) was regarded as the main oxidant with the mass fraction from 13.3% to 66.5%. Five kinds of emulsion explosives were prepared with mass fraction of cerium nitrate hexahydrate by 13.3%, 26.6%, 39.9%, 53.2% and 66.5% respectively. Ammonium nitrate (NH_4NO_3, AP) was regarded as the auxiliary oxidant which played the role of oxygen provider. Compound oil which was a mixture of paraffin wax and machine oil with a certain proportion was regarded as the incendiary agent and SP-80 as the emulsifier. Resin microballs were regarded as the sensitizer. The main components of the emulsion explosives are listed in Table 1.

The aqueous phase (disperse phase) and the oil phase (continuous phase) were mixed with an emulsion machine under a proper oil-bath temperature and some emulsion matrix was obtained. The mixing process lasted for 5 minutes at a mixing speed of 2000 rpm with the emulsion machine. Then the emulsion matrix was sensitized with resin microballs. We got the special emulsion explosives ready for ceria synthesis.

Figure 1 is a SEM picture of emulsion matrix. In the picture, we can see some droplets of aqueous phase, which look like bulb. The diameters of the droplets were about several micrometers. It was oil phase among the droplets. The aqueous phase and the oil phase composed a stable system called emulsion matrix.

Table 1: Main ingredients of the emulsion explosive.

Phase	Chemical name	Chemical formula	Mass content (wt. %)
Disperse phase	Cerium nitrate hexahydrate	$Ce(NO_3)_3 \cdot 6H_2O$	13.3–66.5
	Ammonium nitrate	NH_4NO_3	25–75
	water	H_2O	0–10
Continuous phase	Paraffin wax	$C_{18}H_{38}$	6
	Machine oil	$C_{12}H_{26}$	
Emulsifier	SP-80	$C_{24}H_{44}O_6$	3
Sensitizer	Resin microballs	$C_{15}H_{16}O_2$	1

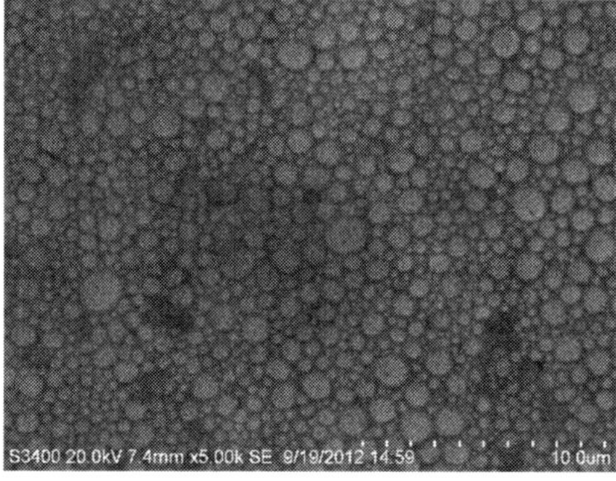

Figure 1: SEM picture of emulsion matrix.

2.2 Synthesis

The emulsion explosives were charged in oiled paper tubes which are 30mm in inner diameter. Then the experiments were carried out in an explosion vessel. The sketch of explosion vessel is shown in Figure 2. The illustration of each part is noted in the sketch. The vessel consists of two main parts which are the cover plate and the body of explosion vessel. Those two parts are connected by several bolts. In the side wall of the body of explosion vessel, there is a relief valve to release pressure before collecting detonation soot. Because the synthesis process of ceria needs a high pressure and high temperature atmosphere, the explosion vessel must be sealed before detonating. All parts of the vessel must be fabricated precisely. The internal dimensions are $\Phi 200 \times 300$mm. The metal used to fabricate the vessel is strong enough to endure the explosion of emulsion explosives. The designed TNT equivalence of the vessel is 100g. After the detonation of emulsion explosives, some detonation soot was collected from the bottom of the detonation vessel. The soot needed to be filtered and then dried at 80°C. Some off-white powders were obtained.

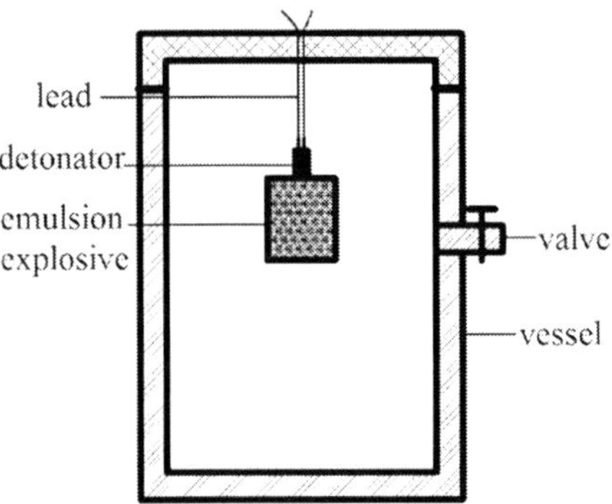

Figure 2: Schematic diagram of explosion vessel.

2.3 Characterizations

The powders were assessed with a Bruker D8 X-ray diffractometer (XRD) at room temperature using copper anode as X-ray source (λ=0.154056 nm) at a scanning rate of 4°/min in the 2θ range from 20° to 70°. The morphology of powders was investigated by using a JEM200 transmission electron microscope (TEM). The ultraviolet absorption performance was detected by an Evolution 220 ultraviolet spectrophotometer. The range of UV wavelength was from 240nm to 420nm.

3 Results and discussion

3.1 Feasibility of detonation synthesis of ceria

Detonation velocity of emulsion explosives with different mass fraction of cerium nitrate hexahydrate was detected by ion probe method. Before the test was carried out, a pair of probes were inserted into the emulsion explosives charged in oiled paper tubes. The distance between the probes was measured, marked with L. Due to the ionization of detonation wave, an equipment could record the time (marked with Δt) when the detonation wave passing by the probes. Then detonation velocity (marked with D) of the emulsion explosives could be computed according to formula (1).

$$D=L/\Delta t \tag{1}$$

The relation curve between detonation velocity and mass content of $Ce(NO_3)_3{\cdot}6H_2O$ in emulsion explosives is shown in Figure 3. From Figure 3 we can see that the detonation velocity was decreasing when the mass content of $Ce(NO_3)_3{\cdot}6H_2O$ in emulsion explosives increasing. As the fraction was up to

66.5%, the emulsion explosives could not be initiated by a detonator because of its high water content and density [13]. Thus it was impossible to use that formula to synthesize ceria by detonation method. In this case, we chose the rest four kinds of emulsion explosives with mass content of $Ce(NO_3)_3 \cdot 6H_2O$ 13.3%, 26.6%, 39.9%, 53.2% respectively to synthesize nanostructured ceria.

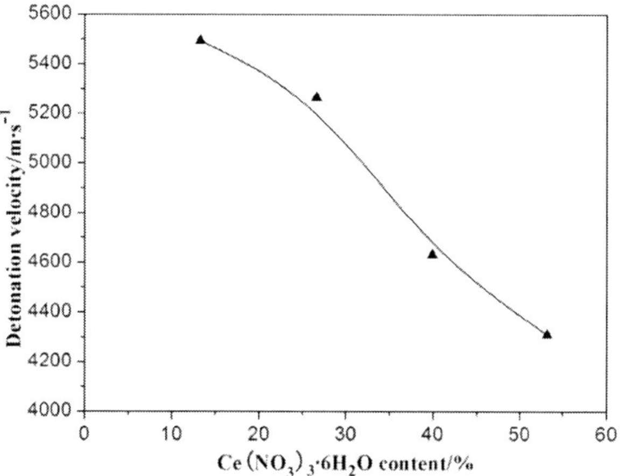

Figure 3: Relationship between detonation velocity and $Ce(NO_3)_3 \cdot 6H_2O$ content.

3.2 Effects of cerium nitrate hexahydrate content in emulsion explosives

SEM images of emulsion matrix with different formulas are shown in Figure 4. Seen from Figure 4, as a result of the high speed stirring paddles of the emulsion machine, the aqueous phase was dispersed into an oil phase as lots of water solution droplets. The mean diameter of water solution droplets tended to be smaller when the content of $Ce(NO_3)_3 \cdot 6H_2O$ increasing. Meanwhile, the dimensional homogeneity of the droplets was better with higher mass content of $Ce(NO_3)_3 \cdot 6H_2O$. During the preparation process of different emulsion matrix (emulsion process), we found that the viscosity was growing as the mass content of $Ce(NO_3)_3 \cdot 6H_2O$ increasing. Because the operational temperature of emulsion process was above 110°C and the mass content of aqueous phase was more than 90%, we have an assumption that the emulsion matrix was Newton fluid. Newton law of viscosity was shown in equation (2).

$$\tau = \mu \frac{du}{dy} \tag{2}$$

where τ was shearing strength, μ was viscosity of fluid and $\frac{du}{dy}$ was velocity gradient. In equation (2), $\frac{du}{dy}$ was constant due to a specified stirring paddle and its corresponding rotation speed. We found that higher mass content of $Ce(NO_3)_3{\cdot}6H_2O$ in emulsion matrix generally meant higher viscosity (μ) of emulsion matrix during the preparation process of emulsion matrix. According to equation (2), we know that shearing strength (τ) was proportional to viscosity (μ). So we concluded that it was prone to obtain homogeneous aqueous phase droplets by increasing the viscosity of fluid and shearing strength.

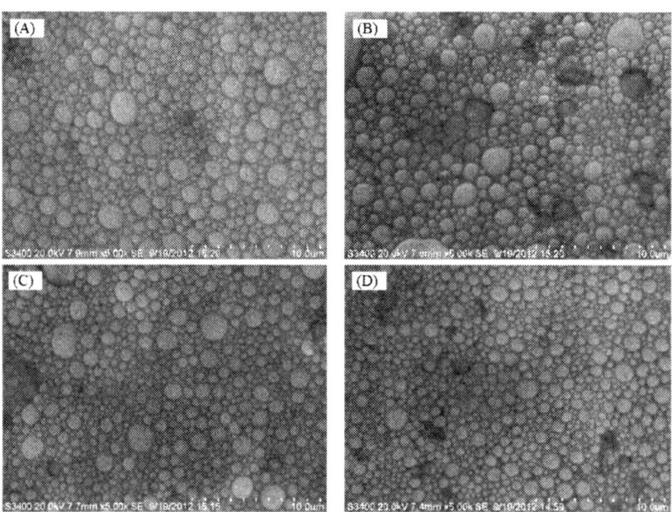

Figure 4: SEM images for emulsion matrix of differfent formulas: (A) 13.3% $Ce(NO_3)_3{\cdot}6H_2O$, (B) 26.6% $Ce(NO_3)_3{\cdot}6H_2O$, (C) 39.9% $Ce(NO_3)_3{\cdot}6H_2O$, (D) 53.2% $Ce(NO_3)_3{\cdot}6H_2O$,.

3.3 Characterization of as-synthesized powder

The crystal properties of four samples that synthesized by four kinds of formulas were studied by using XRD (X-Ray Diffractometer). Figure 5 is the XRD patterns of four samples. The specification of different samples are shown in Figure 5. Seen from the patterns, we can see that there are six typical diffraction peaks for the four samples, followed successively by 28.59°, 33.12°, 47.53°, 56.38°, 59.14° and 69.46°. We know that all the four samples were cubic ceria. Actually, all the six diffraction peaks have a right shift about 0.05°. That is because the detonation process of emulsion explosives is so short that the atmosphere with high pressure and high temperature caused by the detonation could only last for about several microseconds or even shorter. During the extremely short time, the crystallites could hardly grow toward preferred orientation, which makes the crystallites have some lattice defects. That is why all four diffraction peaks have a right shift.

The crystallite size of samples was estimated by applying full-width-half-maximum (FWHM) of characteristic peak (111) to Scherrer equation:

$$D = \frac{k\lambda}{\beta \cos\theta} \tag{3}$$

where D was the crystallite size, k was a constant (0.89 in this study), λ was wavelength of copper target (0.15406 nm in this study), β was FWHM of characteristic peak (111) and θ was the diffraction angle for the (111) plane. Seen from the XRD patterns, the FWHM of characteristic peak (111) was increasing as the mass content of $Ce(NO_3)_3 \cdot 6H_2O$ in emulsion matrix was growing. According to equation (3), higher FWHM means smaller crystallite size. That means granules prepared by emulsion explosives with higher mass content of $Ce(NO_3)_3 \cdot 6H_2O$ were more prone to grow into smaller particles. It was proved by the results of the TEM images shown in Figure 6. The powders obtained from the emulsion explosives with higher mass content of $Ce(NO_3)_3 \cdot 6H_2O$ had relatively smaller dimensions (sample (d)).

Seen from Figure 6, sample (d) had relatively uniform particle size. That means an increase in mass content of $Ce(NO_3)_3 \cdot 6H_2O$ in emulsion explosions leads to smaller particle size and better dimensional homogeneity of nanostructured ceria. The results can be attributed to the preparation process of emulsion matrix. There was a certain relationship between the dimensional parameters of aqueous phase droplets in emulsion matrix and that of its corresponding ceria particles. Uniform ceria particles with smaller size could be obtained by improving the uniformity of the aqueous phase droplets in emulsion matrix. Base on the results and discussion above, sample (d) was prepared for its ultraviolet absorption performance test.

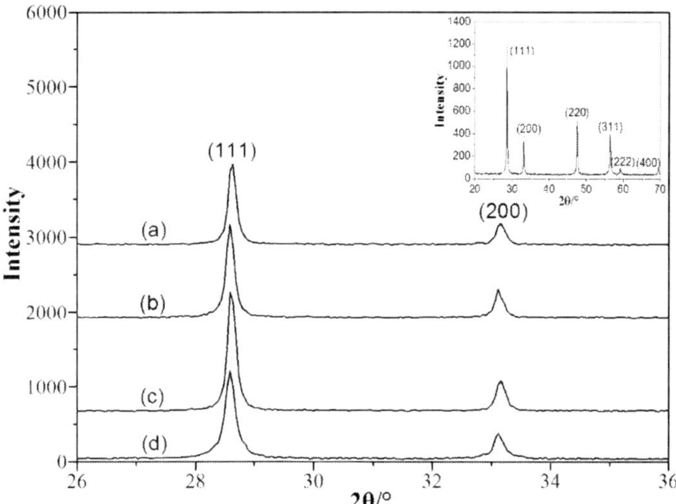

Figure 5: XRD patterns of as-synthesized powder at different conditions: (a) 13.3% $Ce(NO_3)_3 \cdot 6H_2O$, (b) 26.6% $Ce(NO_3)_3 \cdot 6H_2O$, (c) 39.9% $Ce(NO_3)_3 \cdot 6H_2O$, (d) 53.2% $Ce(NO_3)_3 \cdot 6H_2O$.

www.witpress.com, ISSN 1743-3533 (on-line)

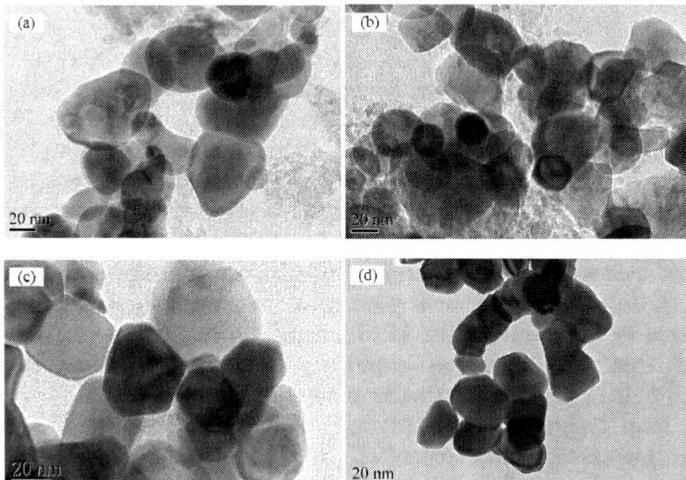

Figure 6: TEM images of nanostructured ceria synthesized at different conditions: (a) 13.3% $Ce(NO_3)_3 \cdot 6H_2O$, (b) 26.6% $Ce(NO_3)_3 \cdot 6H_2O$, (c) 39.9% $Ce(NO_3)_3 \cdot 6H_2O$, (d) 53.2% $Ce(NO_3)_3 \cdot 6H_2O$.

3.4 UV absorption performance of nanostructured ceria

The ultraviolet absorption spectrum of sample (d) was shown in Figure 7. Seen from Figure 7, the samples (d) had an excellent performance to filter ultraviolet radiation, whose wavelengths are between 240-360nm. More precisely, both the ultraviolet type B (UVB, 290–320nm) and the short ultraviolet type A (short UVA, 320–340nm) could be filtered by samples (d). Actually, the ultraviolet

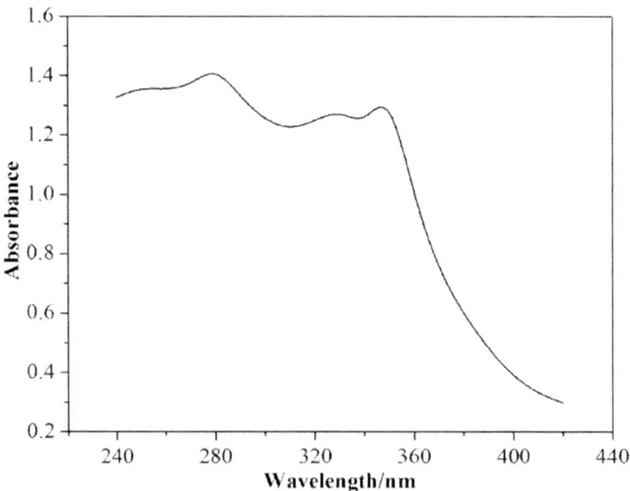

Figure 7: Ultraviolet absorption spectrum of sample (d).

type A radiation is divided into two domains. One is called short UVA mentioned above. The other is called long UVA, whose wavelengths are from 340nm to 400nm. However, compared with long UVA, short UVA comprises the most energetic and thus it is the most harmful type of UVA radiation. Short UVA is indeed implicated in skin cancers [14]. In this case, sample (d) will be used as potential materials for ultraviolet filtration. It is preferred to be used in sunscreen cosmetic products.

4 Conclusions

In the present study we have prepared nanostructured ceria by the detonation method and demonstrated the relationship between the dimensional parameters of nanostructured ceria particles and mass content of $Ce(NO_3)_3{\cdot}6H_2O$ in emulsion explosives. From the results, the particle size and its dimensional uniformity of prepared nanostructured ceria are strongly dependent on the mass content of $Ce(NO_3)_3{\cdot}6H_2O$ in emulsion explosives. Nanostructured ceria with smaller size and better dimensional uniformity was obtained by increasing the mass content of $Ce(NO_3)_3{\cdot}6H_2O$ to 53.2%. This kind of nanostructured ceria had excellent UV filtration performance according to its UV absorption spectrum, especially for harmful UVB. It was a kind of potential material for UV filtration in sunscreen cosmetic products.

Acknowledgement

The present work was financially supported by Natural Science Foundation of China (Project 11102091).

References

[1] Long T.C., Tajuba J., Sama P., *et al.*, Nanosize titanium dioxide stimulates reactive oxygen species in brain microglia and damages neurons in vitro. Environmental Health Perspectives, 115(11), pp. 1631–1637, 2007.
[2] Kang S.J., Kim B.M., Lee Y.J., *et al.*, Titanium dioxide nanoparticles trigger p53-mediated damage response in peripheral blood lymphocytes. Environmental Molecular Mutagenesis, 49(5), pp. 399–405, 2008.
[3] Wang J.J., Sanderson B.J. and Wang H., Cyto- and genotoxicity of ultrafine TiO_2 particles in cultured human lymphoblastoid cells. Mutation Research, 628(2), pp. 99–106, 2007.
[4] Herrmann J.M., Heterogeneous photocatalysis: fundamentals and applications to the removal of various types of aqueous pollutants. Catalysis Today, 53(1), pp. 115–129, 1999.
[5] Serpone N., Dondi D. and Albini A., Inorganic and organic UV filters: Their role and efficacy in sunscreens and suncare products. Inorganica Chimica Acta, 360(3), pp. 794–802, 2007.

[6] Brezova V., Gabcova S., Dvoranova D., *et al.*, Reactive oxygen species produced upon photoexcitation of sunscreens containing titanium dioxide (an EPR study), Journal of Photochemistry and Photobiology B: Biology, 79(2), pp. 121–134, 2005.

[7] Uchino T., Tokunaga H., Ando M., *et al.*, Quantitative determination of OH radical generation and its cytotoxicity induced by TiO_2-UVA treatment. Toxicology in Vitro, 16(5), pp. 629–635, 2002.

[8] Sayes C.M., Wahi R., Kurian P.A., *et al.*, Correlating nanoscale titania structure with toxicity: A cytotoxicity and inflammatory response study with human dermal fibroblasts and human lung epithelial cells, Toxicological Sciences, 92(1), pp. 174–185, 2006.

[9] Hidaka H., Kobayashi H., Koike T., *et al.*, DNA damage photoinduced by cosmetic pigments and sunscreen agents under solar exposure and artificial UV illumination. Journal of Oleo Science, 55(5), pp. 249–261, 2006.

[10] Wamer W.G., Yin J.J. and Wei R.R., Oxidative damage to nucleic acids photosensitized by titanium dioxide. Free Radical Biology and Medicine, 23(6), pp. 851–858, 1997.

[11] Zholobak N.M., Ivanov V.K., Shcherbakov A.B., *et al.*, UV-shielding property photocatalytic activity and photocytotoxicity of ceria colloid solutions. Journal of Photochemistry and Photobiology B: Biology, 102(1), pp. 32–38, 2011.

[12] Truffault L., Ta M.T., Devers T., *et al.*, Application of nanostructured Ca doped CeO_2 for ultraviolet filtration. Materials Research Bulletin, 45(5), pp. 527–535, 2010.

[13] Yunoshev A.S., Plastinin A.V. and Sil'vestrov V.V., Effect of the density of an emulsion explosive on the reaction zone width. Combustion, Explosion, and Shock Waves, 48(3), pp. 319–327, 2012.

[14] Matsumura Y. and Ananthaswamy H.N., Toxic effects of ultraviolet radiation on the skin. Toxicology and Applied Pharmacology, 195(3), pp. 298–380, 2004.

A microstructural study on the high temperature oxidation, carburation and sulfidation of HK 40 and Incoloy 802

I. Caminha[1], C. Barbosa[1,2], I. Abud[1], S. Santana de Carvalho[1,2], F. C. de Souza Coelho dos Santos[2] & M. de Jesus Monteiro[1]
[1]Laboratory of Characterization of Mechanical and Microstructural Properties, National Institute of Technology (INT), Rio de Janeiro, Brazil
[2]Center of Characterization on Nanotechnology, National Institute of Technology (INT), Rio de Janeiro, Brazil

Abstract

The main purpose of this article is an analysis of the microstructural changes in stainless steels and other special alloys due to high temperature phenomena, such as oxidation, carburization and sulfidation. These conditions arise in a plant during the production of polymeric materials, when the raw material, coke originating from petroleum, rich in carbon and sulfur, is in direct contact with tubes fabricated with metallic materials at elevated temperatures during the pyrolysis process. These changes can damage the characteristics and properties of the alloys, thus leading to premature failure in certain regions of the pipeline. In this work, techniques such as optical microscopy, field emission scanning electron microscopy (FESEM) equipped with an EDS microanalysis system and hardness tests were employed to analyze HK 40 stainless steel and Incoloy 802 used in pyrolysis tubes for polymeric material production. The presence of many undesirable oxide, carbide and sulfide particles and chromium depletion was identified mainly in the inner surface of the tubes in direct contact with the coke at elevated temperatures (around 1000°C). Variations of the corrosive attack were observed along different positions in the pipeline, depending on the more or less extensive exposure to high temperatures and corrosive agents. It can be concluded that the degradation which led to the premature failure of the pipeline

WIT Transactions on Engineering Sciences, Vol 77, © 2013 WIT Press
www.witpress.com, ISSN 1743-3533 (on-line)
doi:10.2495/MC130041

was caused by these microstructural changes in the alloys subjected to high temperature and contact with carbon and sulfur rich coke.
Keywords: microstructure, microscopy, stainless steels, special alloys, high temperature degradation.

1 Introduction

Heat resistant special alloys must present requirements of corrosion resistance and mechanical strength at high temperatures applications. However, when components fabricated with such materials work in very aggressive environments, the presence of harmful agents can degrade the microstructural characteristics and consequently their properties, favored by the acceleration and enhancement of diffusion of these deleterious atoms in such high temperatures [1–6].

In the production of polymeric materials, petroleum rich in carbon and sulfur is firstly converted in coke, where these elements remain and this raw material is subjected to high temperature conditions, around 1000°C, and transported through a pipeline usually fabricated with materials that present very high corrosion resistance, such as HK 40 stainless steel and Incoloy 802. In spite of these very good microstructural characteristics and chemical and mechanical properties, these materials can be attacked in very aggressive conditions. At this high temperature atomic diffusion is accelerated in such manner that oxygen atoms from the atmosphere and carbon and sulfur atoms from the coke can migrate into the metallic material, where it forms harmful compound such as oxides, carbides and sulfides, which cause many damages: loss of mass, degradation of toughness and corrosion resistance and eventually the collapse of the pipeline structure. Despite this catastrophic aspect, the degradation degree varies along the length of the pipeline: in some regions, where the conditions are still more aggressive (higher temperatures and/or more exposure to the aggressive elements, for instance), whereas in other regions these conditions are not as severe, and as consequence it is more intense in some regions than in other areas of the pipeline.

In this work, samples of pipelines fabricated with HK 40 stainless steel and Incoloy 802 nickel based superalloy were analyzed by optical microscopy (OM), field emission scanning electron microscopy (FESEM) and Vickers hardness tests. The results showed the presence of many undesirable oxide, carbide and sulfide particles, and chromium depletion was also identified, mainly in the inner surface of the tubes in direct contact with the coke at elevated temperatures (around 1000°C). Variations of the corrosive attack were observed along different positions in the pipeline, depending on the more or less extensive exposure to high temperatures and corrosive agents. Probably the premature failure of the pipeline was caused by these microstructural changes in the alloys subjected to high temperature and contact with carbon and sulfur rich coke.

2 Materials and methodology

Samples of stainless steel and superalloy used in the pyrolysis stage of polymeric materials fabrication are presented in figure 1. In this image sample 81 (mounted: left) and a sample of the inner surface of the tube (right) with greenish stains are displayed. Several similar samples were received and analyzed. Two types of metallic materials were used in the fabrication of the tubes: HK 40 stainless steels (25% Cr and 35% Ni) at the input of the system and Incoloy 802 (35% Cr and 45%Ni) at the output (higher temperatures: 35% Cr and 45%Ni) of the system.

Figure 1: Samples: after metallographic preparation (left) and as received (right).

Failure occurred in different circumstances: while samples 74 and 76 presented normal lifetime (about 6 years), other samples (66, 70, 81, 82 and inner surface with greenish stains) were removed from tubes which showed premature failure (shorter life time, around 3 years). One of the samples (73) was removed from a nearly new tube, only as a reference for comparison. Sample 70 presented similar results to sample 66, as well as sample 76 with the same results of sample 74.

The chemical composition of the materials is presented in table 1.

Table 1: Nominal chemical composition: HK 40 stainless steel [7].

Element	C	Cr	Ni	Si	Fe
Mass %	0.35-0.45	23.0-27.0	19.0-22.0	1.75 max.	Balance

Table 2: Nominal chemical composition: Incoloy 802 [7].

Element	C	Cr	Ni	Si	Mn	Fe
Mass %	0.4	21.5	32.5	0.4	0.8	46.0

www.witpress.com, ISSN 1743-3533 (on-line)

2.1 Metallographic analysis

The samples for metallographic analysis were subjected to the standard preparation method [8–9]. The HK 40 stainless steel samples were etched with a solution of 10 g oxalic acid in 100ml of distilled water at 6 V for 45 to 60 s, while the Incoloy 802 samples were etched with a solution of 70 ml H3PO4 with 30 ml of distilled water, 5 to 10 V and 5 to 60 s.

2.2 Mechanical properties

Vickers hardness tests were performed with a 5 kgf (around 49 N) load according to NBR ISO 188-1:99 standard [10].

2.3 Field Emission Scanning Electron Microscopy (FESEM)

The same samples of the metallographic analysis and hardness tests were subjected to Field Emission Scanning Electron Microscopy (FESM). This equipment was operated at 20 kV and in the same equipment EDS (energy dispersion spectroscopy) was performed with the purpose of identifying different elements present in these alloys, although in a semi quantitative basis.

3 Results

3.1 Optical Microcopy (OM)

Figure 2 presents the microstructure of the sample 81 observed in an optical microscope, while figure 3 reveals details of the aspect of greenish stains in the inner surface of the pyrolysis tubes.

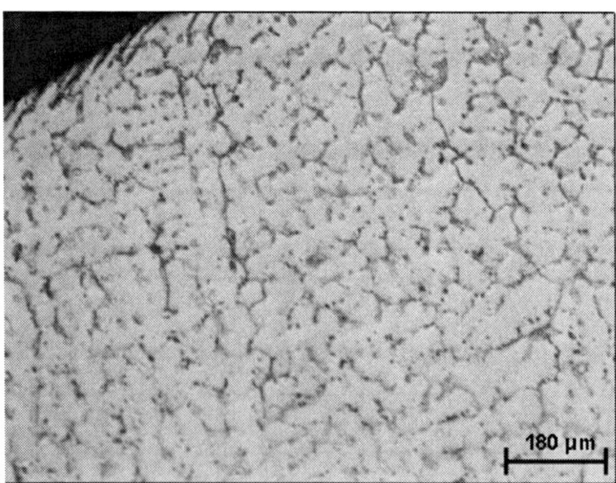

Figure 2: Microstructure of sample 81 (inner surface) – optical microscope.

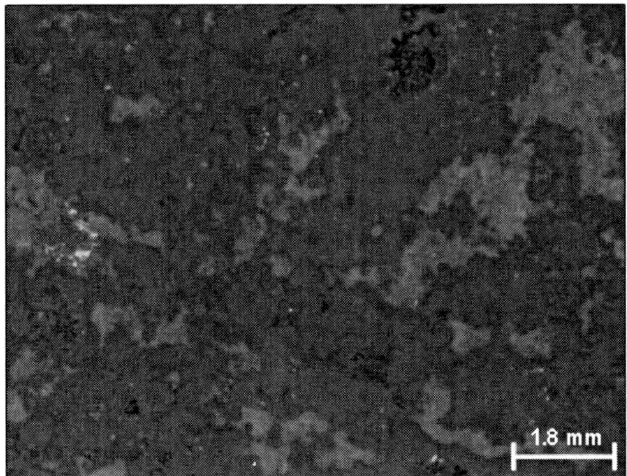

Figure 3: Inner surface of sample 81 with greenish stains: stereomicroscope.

3.2 Field Emission Scanning Electron Microscopy (FESEM)

Figure 4 shows the aspect of the microstructure of sample 66 observed in a field emission scanning electron microscope (FESEM) with backscattering electrons. Figure 5 presents the regions where EDS analysis was performed and the EDS spectrum corresponding to region 6.

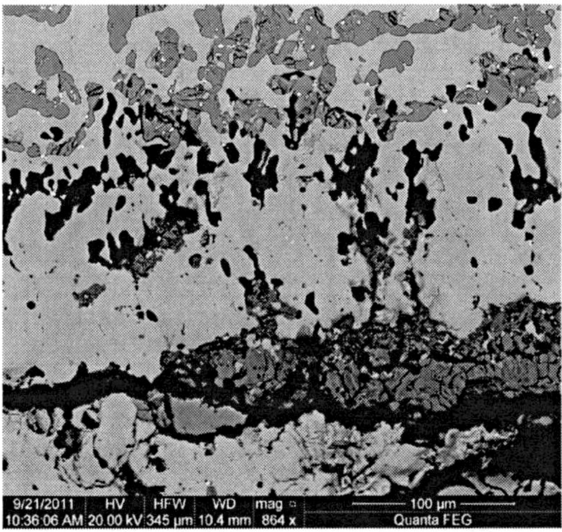

Figure 4: SEM BSE image sample 66 inner surface, output.

While Figures 4 and 5 refer to sample 66, the next images (Figures 6 and 7) show equivalent results for sample 82, and the following ones (Figures 8 and 9) for the inner surface of the same sample.

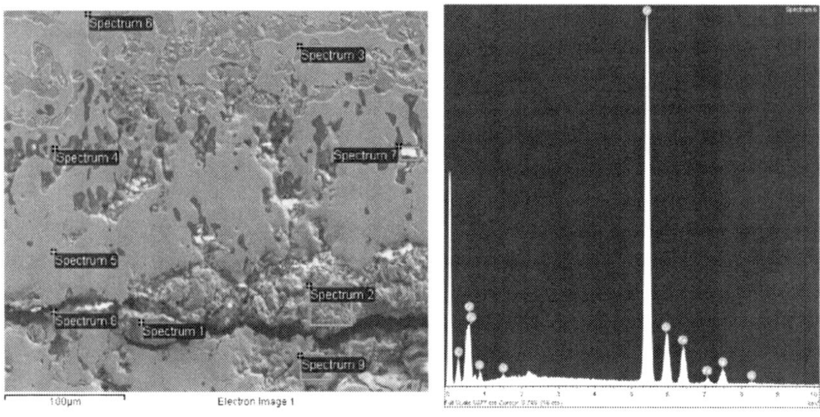

Figure 5: Regions selected for EDS analysis: sample 66 inner surface, output (left side) and EDS spectrum obtained from region 6 (right side).

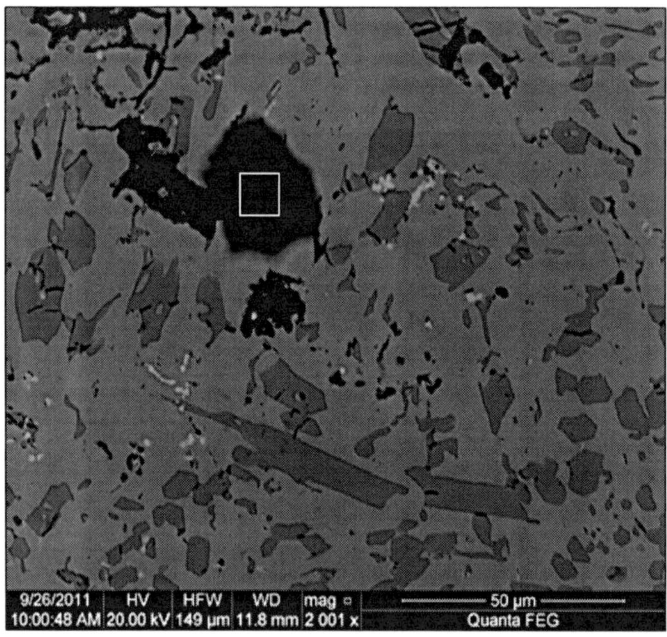

Figure 6: SEM backscattering electrons image of region with S rich particles. Region of sample 82 subjected to EDS analysis: indicated by a rectangle (inside).

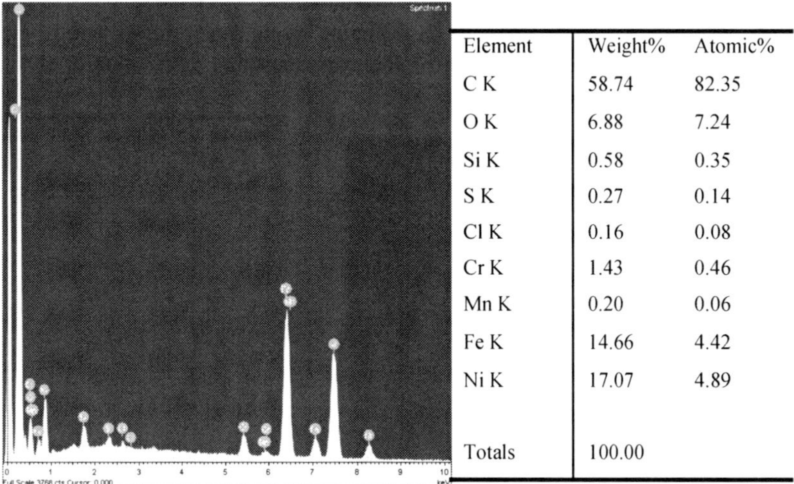

Element	Weight%	Atomic%
C K	58.74	82.35
O K	6.88	7.24
Si K	0.58	0.35
S K	0.27	0.14
Cl K	0.16	0.08
Cr K	1.43	0.46
Mn K	0.20	0.06
Fe K	14.66	4.42
Ni K	17.07	4.89
Totals	100.00	

Figure 7: EDS spectrum analysis: S rich particles in the region indicated by rectangle Figure 6.

Figure 8: Sulfur rich particle in another region: inner surface of sample 82 with greenish stains.

Figure 9 shows the corresponding EDS microanalysis spectrum.

Figure 10 presents a SEM image with different layers forms the surface to the inner region of the sample and figure 11 the corresponding EDS microanalysis with C and Cr contents in each region.

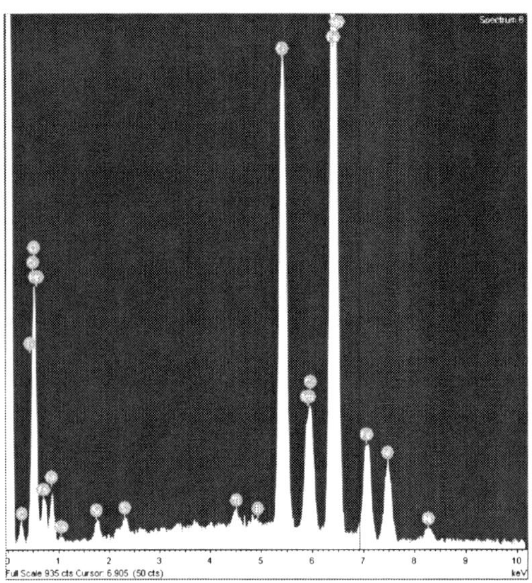

Element	Weight%	Atomic%
C K	3.27	10.83
O K	10.39	25.85
Na K	0.36	0.63
Si K	0.44	0.63
S K	0.28	0.35
Ti K	0.32	0.27
Cr K	22.01	16.84
Mn K	4.71	3.41
Fe K	49.74	35.45
Ni K	8.47	5.74
Totals	100.00	

Figure 9: EDS spectrum obtained from region 6 (previous image in Figure 8): sample 82 surface with greenish stains.

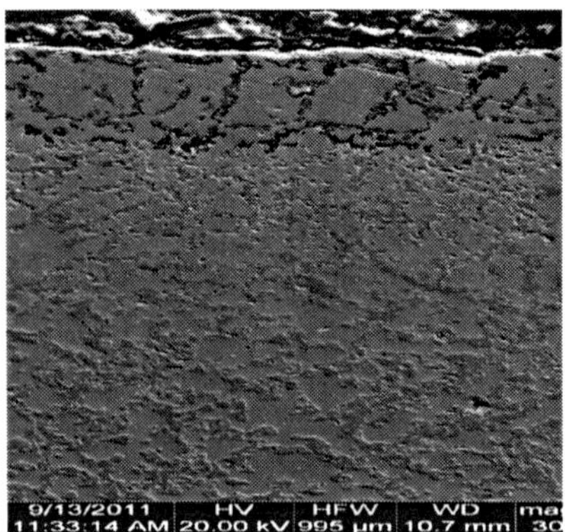

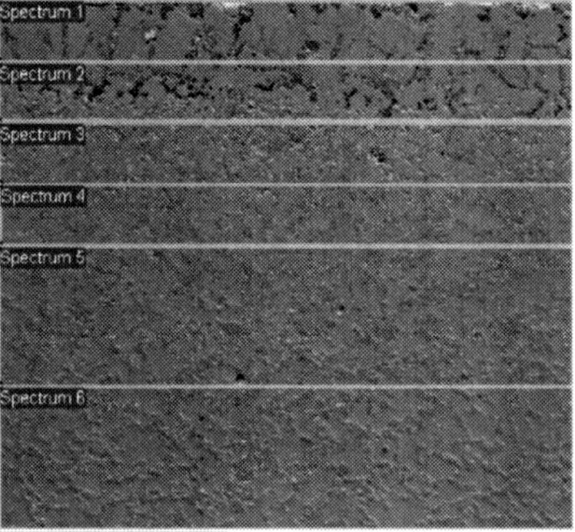

Figure 10: SEM image with different layers from the surface to the inner region of the sample 73 inner surface.

www.witpress.com, ISSN 1743-3533 (on-line)

Spectrum	C	O	Al	Si	Cr	Mn	Fe	Ni	Nb	Total
1	4.80	12.46	0.36	4.32	18.28	0.69	15.86	43.22		100.00
2	7.08	8.54	0.29	3.88	21.77		15.86	42.59		100.00
3	9.94		0.31	0.59	34.22		14.85	38.06	2.02	100.00
4	9.86		0.25	0.85	34.39	0.66	14.43	38.09	1.47	100.00
5	10.61		0.23	1.03	36.74	0.73	13.75	35.64	1.28	100.00
6				1.44	39.96	1.00	15.50	40.61	1.49	100.00
Max.	10.61	12.46	0.36	4.32	39.96	1.00	15.86	43.22	2.02	
Min.	4.80	8.54	0.23	0.59	18.28	0.66	13.75	35.64	1.28	

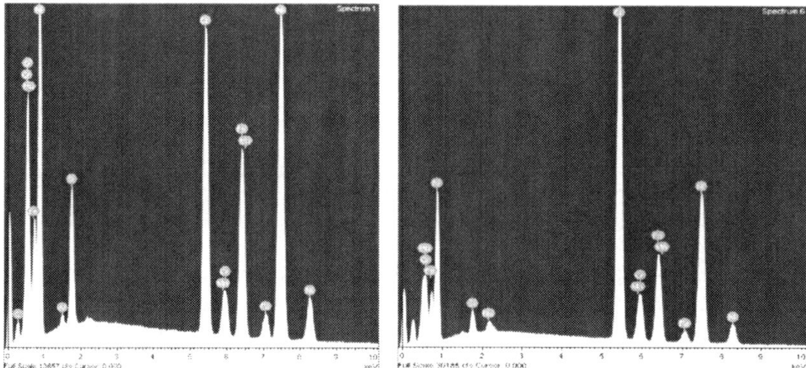

Figure 11: EDS microanalysis spectra, layers 1 (surface) and 6 (inner layer) from the sample 73-11-1P inner surface.

3.3 Hardness tests

Table 3 presents the hardness values of sample 73 inner surface (Vickers with 1 kgf or 9.8 N load: HV1), revealing the variation from the outer layer (1) to the inner layer (6). Five indentations were carried out on each layer.

Table 3: Vickers hardness (HV5) values of sample 73.

HV5	Points					
Layer (Spectrum)	1	2	3	4	5	Average
1	274.73	253.99	184.16	236.82	442.61	278.46
2	437.45	477.26	470.48	407.51	464.50	451.44
3	455.75	499.60	435.66	402.16	409.69	440.57
4	464.83	459.93	430.03	470.82	381.04	441.33
5	353.75	369.10	379.82	347.07	366.78	363.31
6	255.86	271.91	253.06	249.01	224.62	250.89

Based on standard deviation values (about 95% confidence) the variation of hardness with the layers is plotted in figure 8.

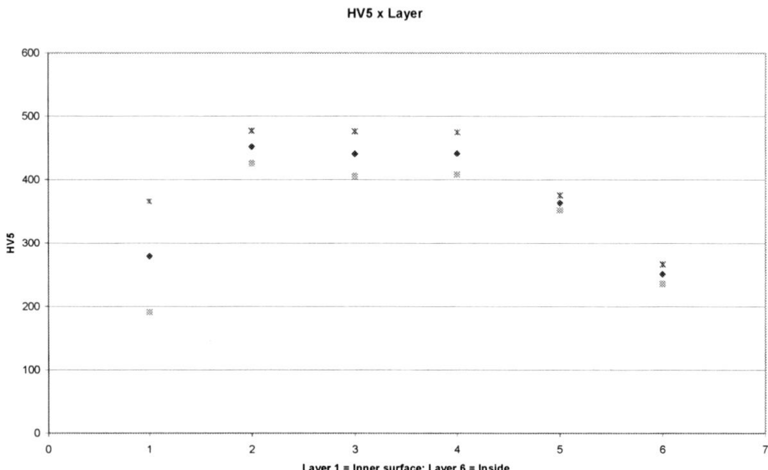

Figure 12: Hardness (HV5) variation along the 6 layers: 1 inner surface, 6: inside the tube.

This plot shows a clear tendency: the hardness of the inner surface (layer 1) is relatively low and when the distance to this surface increases the hardness value also increases. Then it reaches a steady level (layers 2 to 4) until at some distance it starts to diminish (layers 5 and 6), reaching on layer 6 a slightly harder value than layer 1.

4 Discussion

The several results obtained in this work confirm that the problem in the tubes was associated to the formation of brittle particles and to the chromium depletion, since this element is responsible for the formation of a passive Cr2O3 scale on the surface of the material, which provides corrosion resistance, mainly on the inner surface, subjected to contact with the coke formed during the pyrolysis process, earlier stage of the production of polymeric materials fabricated with petroleum as raw material.

In nearly all analyzed samples the chromium depletion was observed, from the middle region of the sample to its surface. As an example, in sample 73 (figure 11) the chromium content diminishes from 39.38% in the middle region to 18.28% in the surface, while in sample 74 it varied from 25.67 to 12.58% in the respective equivalent regions.

Several particles were observed and analyzed by an EDS device installed in the FESEM: mainly oxides, carbides and, in just some instances, sulfides (samples 66, 82 and inner surface with greenish stains, as shown in figures 5, 7 and 9, respectively). In these samples, and also in sample 81 (not shown in this paper), the presence of sodium (Na) was also identified.

Several carbide forming elements were identified, such as chromium, iron, niobium, titanium and silicon. Oxide forming elements were also found, such as chromium, iron, manganese and even aluminum. Sulfide particles are scarcer, as already mentioned, but in the regions where sulfur was identified, sulfide forming elements were also found, such as chromium, nickel, manganese, iron and sodium, but their presence cannot be considered negligible. Other less common elements were observed in some regions or particles, such as phosphor (P), brome (Br) and chlorine (Cl).

The technical literature reports the degradation of this type of material at high temperatures (around 1000°C), associated to the formation of carbides in carbon rich environments, which absorb mainly chromium (but also some other elements), which was present in the matrix and in the passive oxide scale, which rendered good corrosion resistance, thus favoring the progress of the corrosive process at high temperatures [1–2].

Hardness variation across the different layers of the material can be interpreted in this way: in the surface (layer 1) there are several cracks that affect the indentations. According to the literature [11], the propagation of cracks leads to energy release, thus lowering stress levels, which can contribute to hardness decrease. Therefore, it is not unexpected that hardness values are lower in this region than in other layers. In the following layers (2 to 4), placed more distant to the surface, hardness is kept in a higher level that nearly does no vary along these layers. It seems that the carbon enrichment in these layers can lead to the formation of carbides that can harden these layers, where cracks were not seen. In the fifth layer hardness starts to decrease. Probably in this region the distance to the surface is large enough to diminish the carbon enrichment and this tendency is still more evident in the sixth layer, the most distant from the surface, where the hardness is still lower, probably due to still lower carbon content.

5 Conclusion

- Degradation and premature failure of the tubes is associated to the formation of brittle particles and to the chromium depletion. This is the main element responsible for the formation of the passive scale which provides good corrosion resistance in these metallic materials. The chromium depletion is still more marked in the layers which are nearer to the surface, mainly in the inner surface which is in direct contact with coke formed during pyrolysis process.
- In all samples, which were analyzed in this study, the chromium depletion was observed to be stronger in the surface than in the inner regions of the sample.
- Among several particles found in these metallic materials oxides, carbides (of chromium, silicon, iron and manganese) and, in some samples, sulfides (of chromium, nickel, manganese, iron and sodium).
- In some regions the presence of sodium (Na), phosphor (P), bromine (BR) and chlorine (Cl) was identified.

References

[1] Luton, R.P.; Ramanarayan, T.A.; Mixed Oxidant Attack of High Temperature Alloys in carbon and Oxygen Containing Environments, Oxidation of Metals, 34, (5/6), 1990, 381-400.

[2] Shi, S.; Lippold J.C.; Microstructure evolution during service exposure of two cast heat resisting stainless steels HP-Nb modified and 20-32 Nb, Materials Characterization, 59, 2008, 1029-1040.

[3] Tjokro, K.; Young, D.J.; Johansson, R.E.; Ivarsson, B.G.; High temperature sulfidation oxidation of stainless steels, Journal de Physique IV, 3, 1993, 357-364.

[4] Jakobi, D. (Schmidt + Clemens Group), S + C Spun casting, Failure Mechanisms for Radiant Tubes (Attack by Contaminants, e.g. Sodium), www.schmidt-clemens, 3 p.

[5] Specs: 309/309S, 310/310S, Sandmeyer Steel Company, Philadelphia, PA, USA, www.sandmeyersteel.com, 7 p.

[6] Sims, C.T.; Hagel, W.C.; "The Superalloys", John Wiley & Sons, New York, 1972, p. 1- 565.

[7] ASM Handbook, volume1, properties and selection: Irons, Steels and High performance Alloys, ASM International, Materials park, Ohio, USA, 1990 (2001), p.910-915.

[8] ASTM E 3 Standard: "Standard Guide for Preparation of Metallographic Specimens", 2011.

[9] ASTM E 407 Standard: 407: "Standard Practice for Microetching Metals and Alloys", 2007 (2011).

[10] ISO 6507-1 Standard: "Metallic materials – Vickers hardness test – Part 1: Test method", 2005.

[11] Broek, D.; Elementary Engineering Fracture Mechanics, Kluwer Academic Publishers, 4th edition, Dordrecht, Netherlands, 1986 (2002), p.150-178.

Effect of pre-treatment procedure on micropore system characteristics of subbituminous coal

T. Zelenka
Department of Chemistry, University of Ostrava, Czech Republic

Abstract

Various pre-treatment procedures like hot oven drying under and without vacuum, purging with an inert gas at elevated temperatures, lyophilisation and long-term drying in a desiccator were applied on a sample of a subbituminous coal. Variously dried samples with known residual moisture contents were interpreted according to the microporous textural parameters. Adsorption isotherms of CO_2 at 30°C were determined using a volumetric sorption analyser. Adsorption data were evaluated according to Dubinin–Radushkevich and Medek's equation in order to determine micropores' volume and micropores' surface and modus of the micropores' radii.

It was found that the drying method has an influence on the microporous textural properties. Samples which were not exposed to oxygen during drying (vacuum drying, inert gas purging) have better micropore characteristics; likely due to the lower contents of the residual moisture in the order of a few tenths wt.%. If a sample is exposed to an oxidizing atmosphere at a comparable temperature and drying time, the reduction of the textural parameters was observed. It may be explained by a higher residual moisture content of this sample (1.4 wt. %). The effect of the drying temperature has not been clearly proven despite the fact that mean S_{mi} and V_{mi} parameters have an increasing tendency with a decreasing drying temperature. It was also found that 2 hours is an effective time for the vacuum drying process at 60°C; even despite the difference in residual moisture contents of these samples.

Keywords: subbituminous coal, pre-treatment procedure, surface area, micropores.

www.witpress.com, ISSN 1743-3533 (on-line)
doi:10.2495/MC130051

1 Introduction

Subbituminous coal is considered to be a naturally microporous carbon material which contains a significant amount of moisture in its structure [1, 2]. Studies concerning the influence of moisture on the CO_2 sorption capacity generally state that the presence of moisture leads to a decrease in sorption capacity [3–5]. For instance, Švábová *et al.* [6] examined the effect of the moisture on three coal samples which differ in the degree of coalification. They found out that in all of the cases, the moisturized coal showed lower sorption capacity of the carbon dioxide than the dry coal.

It is obvious that proper moisture removal from the structure of the coal is very important for achieving the maximum sorption capacity due to the good accessibility of adsorbed gas. Brennan *et al.* [7] found out that the mechanism that leads to a reduction in CO_2 adsorption is the formation of clusters which can effectively block the entire pore space. Day *et al.* [3] examined that in low rank coals, moisture had a greater effect on the sorption capacity than in high rank coals. Probably because of the greater proportion of polar sites which are preferentially occupied by water at the expense of CO_2 and also CH_4.

Moisture content of the coal is connected with oxygen containing functional groups. These functional groups act as primary adsorption centres [6]. At any time during the pre-treatment process (including not only the drying procedure, but also storing of the coal) when the sample is in contact with oxygen, physical and chemical changes can occur [1, 2]. This natural process is known as low-temperature oxidation of the coal and it progresses at a normal temperature up to 100°C [2]. Some authors consider higher temperatures up to 200°C [2, 8]. The low-temperature oxidation leads to formation and/or elimination of polar functional groups, especially hydroxyl groups but also carboxyl, methoxyl and carbonyl groups [1, 2]. There is an expectation that the presence of a non-oxidizing atmosphere during the pre-treating process, such as vacuum or inert, will repress oxidation of the sample. However, the oxidation process of the coal is expected during drying in an oxidizing atmosphere [2].

Various drying processes have been developed since the 1920s. Among them were the superheated and pressurized steam drying, hot and supercritical water drying and hydro thermal-mechanical compression drying processes [9, 10]. On the other hand, the use of high-pressure equipment provides crucial economical disadvantages in the process.

The micropore properties of coals with respect to its moisture content have already been studied [3, 6]. Nevertheless, the influence of various pre-treatment methods according to assessment of microporous properties of (subbituminous) coal has not been published yet.

For this purpose, various pre-treatment procedures were chosen and applied to the samples of the subbituminous coal. The aim of this paper was to describe a different final state of the subbituminous coal obtained after various pre-treatment procedures. This final state has been characterized with respect to their microporous textural parameters.

2 Experimental part

2.1 Sample

A sample of subbituminous coal taken from the Czech part of the North Bohemian Coal Basin was chosen for the study. The sample was crushed in a ball mill to obtain the grain size of 0.06 - 0.15 mm. The thus obtained material was stored in closable plastic container in a refrigerator at 3°C. The basic quantitative parameters of the coal are given in table 1.

Table 1: Basic characteristics of the studied coal sample (wt.%).

Ultimate and proximate analysis						
C^{daf}	H^{daf}	N^{daf}	O_{dif}^{daf}	S_t^d	A^d	W_t
74.4	6.5	1.0	16.8	1.3	8.0	18.0
Ash analysis						
CaO	SiO_2	Al_2O_3	Fe_2O_3	MnO	MgO	TiO_2
4.0	51.2	27.5	6.4	0.01	0.8	3.2

A – ash, *d* – dry basic, *daf* – dry and ash-free basic, *t* – total, *dif* – by difference, *W* – moisture content.

2.2 Pre-treatment methods

Five pre-treatment procedures were studied on this sample with regard to its textural parameters.

(i) Hot oven drying under vacuum is intended to prevent oxidation of the sample. The effect of the various drying temperatures and time was investigated. Monitoring the impact of the vacuum drying at various elevated temperatures (60, 80 and 105°C) was carried out at a constant drying time of 5 hours. In the case of different pre-treatment times under vacuum (2, 5 and 8 hours), the constant drying temperature of 60°C was chosen. These experiments were performed in the hot oven (mfr. Javoz) using the vacuum pump (mfr. Laboratorní přístroje Praha; limit of pressure around 100 Pa).

(ii) Purging of the sample at elevated temperature with an inert nitrogen atmosphere is another method that protects the adsorbent to its oxidation. This procedure was carried out in the same (above mentioned) oven at 60°C for 5 hours at nitrogen flow 50 cm^3 min^{-1}. The sample was placed on the bottom of the glass U-tube during the purging procedure.

(iii) Another vacuum drying procedure used to prevent oxidation of the adsorbent is lyophilisation (mfr. Labconco), which unlike the method (i) takes place at a low temperature. This non-evaporate method differs from others in the context of the phase transition (sublimation occurs). Before the lyophilisation process, the sample in a glass lyophilisation flask was frozen by liquid nitrogen. The actual sample temperature during this procedure could not be monitored, but the temperature in the condenser was -87°C during the entire process. This type of pre-treatment procedure lasted for 70 hours at pressure 50 Pa.

(iv) Convex hot air oven pre-treating is a process performed under elevated temperature. It is one of two methods in which oxidation is to be expected. The pre-treatment process was carried out at 60°C for 5 hours (mfr. Chirana).

(v) Finally, the ambient temperature (at ~25°C) and ambient pressure drying process took place in a desiccator with a silica gel. The sample was exposed to air on a long-term basis (216 hours) therefore oxidation of the coal sample is to be expected.

The effects of the pre-treatment temperature and time as well as the effect of the residual moisture content were evaluated with regard to all the above-mentioned pre-treated methods. All the pre-treatment parameters are presented in table 2.

Table 2: Parameters of the pre-treatment procedures.

Sample	Pre-treatment procedure	Temperature [°C]	Time [hour]
1	vacuum	60	5
2		80	
3		105	
4		60	2
5			8
6	inert gas		5
7	hot air oven		
8	lyophilisation	-87	70
9	desiccator	~ 25	216

The treated sample was separated into two portions. The first portion was used for carbon dioxide adsorption analysis.

The second part of the sample was dried at 105°C in hot air oven (mfr. Chirana) to constant weight in order to determine the amount of the residual moisture according to the norms ČSN ISO 5069-1 and 5069-2. There is an expectation that the residual moisture content will vary depending on the used drying method, which should be reflected in the textural parameters of the sample.

2.3 Volumetric sorption analysis

Carbon dioxide isotherms were measured by means of a Setaram PCTPro E&E volumetric analyser. All the measurements were carried out at 30°C and over the relative pressure range up to 0.014 p/p_s (0–1 bar of absolute pressure expression). The weight of the sample was 0.7 g. A dead volume of the sample cell was measured first by the helium expansion technique. Hy-Data E&E software (Setaram) was used to specify different adsorption parameters. In the course of the sorption process, the sample temperature was automatically controlled.

2.4 Interpretation of carbon dioxide isotherms

Obtained adsorption data were interpreted according to Dubinin theory, based on the effect of overlapping adsorption potentials of opposite walls in micropores. Here, an amplified adsorption field is formed where volume filling occurs at pressures much lower than the saturation pressure and the rest of the surface is negligibly covered [11–13]. This sorption mechanism is described by the following Dubinin–Radushkevich (DR) eqn (1):

$$\ln V = \ln V_0 - \frac{R^2T^2}{E^2}\ln^2\left(\frac{p_s}{p}\right) \tag{1}$$

where V is the volume of micropores occupied by adsorbed carbon dioxide at equilibrium pressure p, V_0 is a constant representing limiting (total) volume of micropores (corresponds with V_{mi}), R is the gas constant, E is the characteristic energy and p_s is the saturated pressure of CO_2 vapour at temperature T.

This DR equation manifests a linear plot in coordinates $ln\ V$ versus $ln^2(p_s/p)$ (see fig. 1). From the $ln\ V$ intercept of the plot, the limiting value of the micropores volume V_0 was evaluated. In addition, from the slope of the DR plots, characteristic energy E was determined which was further used for calculation of the micropores surface area S_{mi} by means of Medek's eqn (2) [12]:

$$S_{mi} = 1.86 \cdot V_0 \cdot \left(\frac{E}{k}\right)^{\frac{1}{3}} \tag{2}$$

where k is the constant characterizing interactions between carbon dioxide and the carbonaceous surface [11, 12, 14]. Values of all necessary parameters applied to evaluate adsorption isotherms of CO_2 at 30°C according to DR eqn (1) and Medek's eqn (2) are given in table 3.

Table 3: Properties of CO_2 at 30°C.

Molar density of a liquid phase	73.595 $cm^3\ mol^{-1}$	[15]
Saturated pressure, p_s	72.763 bar	[15]
Interaction constant with carbonaceous surface, k	3.145 kJ $nm^3\ mol^{-1}$	[11, 12, 14]

The distribution curve of micropores was calculated using following eqn (3) on the basis of Medek's equation [12]:

$$\frac{dV}{dr} = 6 \cdot V_0 \cdot \left(\frac{k}{E}\right)^2 \cdot r^{-7} \cdot e^{-\left(\frac{k}{E}\right)^2 \cdot r^{-6}} \tag{3}$$

The modus of the micropores radii r_{mode} (eqn (4)) was derived from the micropores size distribution [12]:

$$r_{mode} = \left[\left(\frac{6}{7}\right)^{\frac{1}{2}} \cdot \frac{k}{E}\right]^{\frac{1}{3}} \tag{4}$$

3 Results and discussion

3.1 Error of the measurements

According to the rectified Dubinin-Radushkevich isotherms and the high determination coefficients R^2 better than 0.99, it is obvious that this subbituminous coal sample complies very well with the dependence in the Dubinin coordinates $ln\ V$ versus $ln^2(p_s/p)$ (see fig. 1).

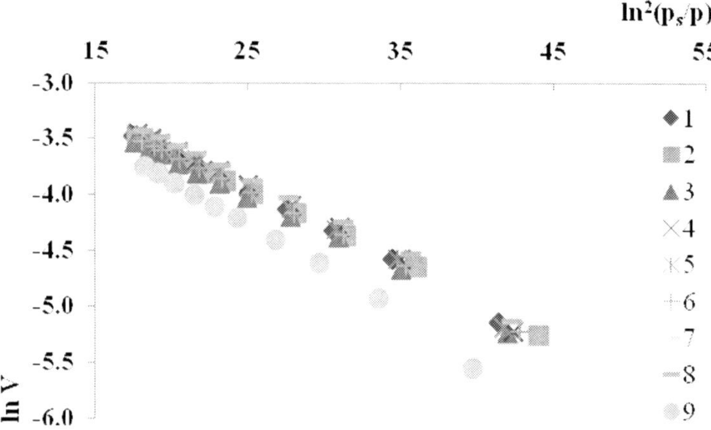

Figure 1: Dubinin–Radushkevich plots of the adsorption data of CO_2 at samples of the coal. The numbers correspond to different ways of drying from table 2. All samples exhibit high determination coefficients better than 0.99.

Each of the nine DR isotherms consisting of 10 – 12 points were statistically evaluated by linear regressions according to the 95% confidence intervals.

Table 4 summarizes textural parameters of all the samples. For example, sample 1 exhibits 95% confidence interval 0.109±0.009 $cm^3\ g^{-1}$ for the V_{mi} and 294±26 $m^2\ g^{-1}$ for the S_{mi} as well as E = 9.64±0.22 kJ mol^{-1}. It is obvious that the 95% confidence intervals of most samples are overlapping. These samples can be considered as equivalent according to the monitored parameters.

3.2 Effect of the oxidizing atmosphere

It was found that the pre-treatment process has a significant influence on the final state of the sample. As presented in table 4, there is no pre-treatment procedure which would lead to a completely dry coal sample. However, with regard to the accuracy of the used analytical method of determination of the

residual moisture content, values ≤ 0.2 wt. % can be considered as dry states of the coal samples.

The effect of the oxidizing atmosphere was studied on the samples 1, 6 and 7, which were pre-treated at 60°C for 5 hours. The samples 1 and 6, which were not exposed to the oxidizing processes, were dried under vacuum and purged with an inert gas, respectively.

Table 4: Textural parameters of the microporous subbituminous coal dried by various methods.

	Dubinin-Radushkevich		Medek		
Sample	V_{mi} [cm^3 g^{-1}]	E [kJ mol^{-1}]	S_{mi} [m^2 g^{-1}]	r_{mode} [nm]	W_{res} [wt. %]
1	0.109±0.009	9.64±0.22	294±26	0.67	≤0.2
2	0.099±0.004	9.83±0.12	268±12	0.67	≤0.2
3	0.097±0.006	9.69±0.17	262±19	0.67	≤0.2
4	0.113±0.010	9.55±0.23	303±30	0.67	0.6
5	0.111±0.009	9.54±0.22	297±28	0.67	≤0.2
6	0.100±0.006	9.75±0.16	272±17	0.67	0.4
7	0.093±0.007	9.10±0.15	247±19	0.68	1.4
8	0.115±0.013	9.66±0.29	311±38	0.67	0.5
9	0.106±0.012	8.84±0.24	278±35	0.69	2.1

V_{mi} – volume of micropores, E – characteristic energy, S_{mi} – surface of micropores, r_{mode} – modus of micropore radii, W_{res} – residual moisture content.

On the other hand, drying of the sample 7 was carried out under oxidizing conditions. All the microporous parameters are presented in table 4 and fig. 2. Note that the Y axis corresponding with the S_{mi} has its minimum at 200 m^2 g^{-1} and S_{mi} of the samples 1, 6 (and also 7) corresponds with the drying temperature 60°C (corresponding points are slightly shifted for readability of the error bars). As can be seen, the mean values of the 95% confidence intervals of the samples dried at non-oxidizing atmosphere correspond with the micropore surfaces 294 and 272 m^2 g^{-1}. Their micropore volumes are 0.109 and 0.100 cm^3 g^{-1}. In contrast, one air dried sample exhibits S_{mi} 247 m^2 g^{-1} and V_{mi} 0.093 cm^3 g^{-1}.

As expected, the micropore characteristics of the samples which were not exposed to oxygen during drying are better (according to their mean values). This fact may be explained by the presence of the residual moisture, which is higher for the hot air dried sample (1.4 wt. %). Higher moisture content may also be responsible for lower characteristics energy 9.10±0.15 kJ mol^{-1} and higher values of the r_{mode} 0.68 nm. Samples pre-treated under a non-oxidizing atmosphere exhibit W_{res} ≤0.4 wt. %. This is in agreement with the claim of Švábová *et al.* [6]. They confirmed that drying under oxidizing atmosphere may be a consequence of reduction of the textural parameters because of filling of micropores by moisture likely occurs. However, this claim should relate to only samples 1 and 7 because their 95% confidence intervals (for such statistical evaluation) do not overlap (see fig 2 or table 4).

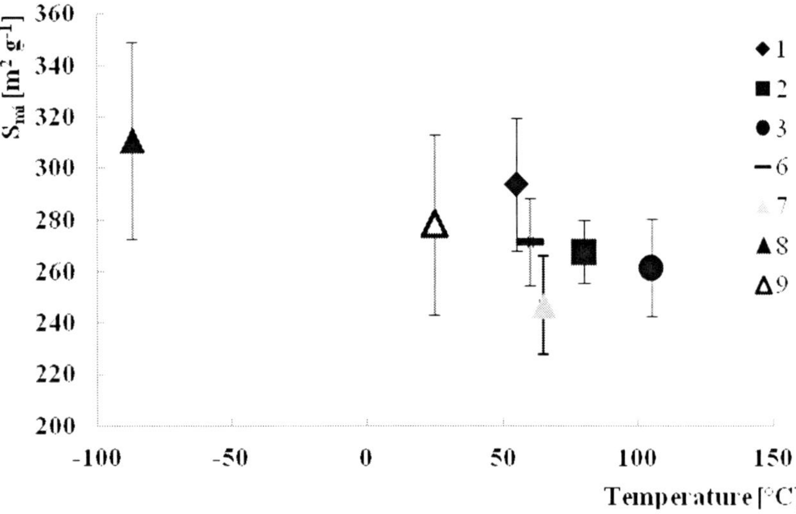

Figure 2: Effect of the drying temperature (across the various pre-treatment methods) on S_{mi}. A micropore surface area of the samples 1, 6 and 7 corresponds with the temperature 60°C. Each error bar represents the error of the measurement.

3.3 Effect of drying temperature

In the case of samples which were not exposed to oxidation (specif. 1, 2, 3), the drying process was carried out under vacuum for 5 hours. The effect of temperature drying was monitored at 60, 80 and 105°C.

The micropore parameters S_{mi} and V_{mi} exhibit the range 262–294 $m^2\ g^{-1}$ and 0.097–0.109 $cm^3\ g^{-1}$, respectively (in favour of the sample dried at a lower temperature) and constant modus of micropore radii 0.67 nm. This drying procedure running under non-oxidizing atmosphere, where the oxidation process is repressed, produces basically dry adsorbents ($W_{res} \leq 0.2$ wt. %) irrespective of the drying temperature. The effect of the temperature could also be assessed with respect to the sample 8, which was dried using lyophilisation at low temperature. The oxidizing processes are also repressed according to this method. However, the drying time is far longer (70 hours) in comparison with the above discussed samples. As can be seen, the long drying time is not a guarantee of getting a dry sample (residual moisture content was 0.5 wt. %). This drying method performed under -87°C exhibits the same (r_{mode} = 0.67 nm) and slightly higher mean values of the V_{mi} = 0.115 $cm^3\ g^{-1}$ and S_{mi} = 311 $m^2\ g^{-1}$ than samples 1 - 3. It may seem that S_{mi} and V_{mi} have increasing tendency with decreasing drying temperature (from 105 to -87°C), but all the micropore parameters lie within the 95%

confidence interval, consequently this fact cannot be clearly confirmed from these data. The small scatter of values of the E (from 9.64 to 9.83 kJ mol^{-1}) and constancy of the r_{mode} (0.67 nm) indicates the same adsorption potential standpoint of all the samples.

The influence of temperature is also related with the influence of various phase transitions when drying under vacuum. While drying the samples at elevated temperatures leads to removal of moisture through evaporation, in the case of the lyophilisation process the moisture leaves through the sublimation.

Capillary effects are suppressed and destruction of the microporous coal during the lyophilisation procedure should not occur. However, based on established textural parameters, the lyophilisation process did not significantly improve the textural properties of the sample. This is in line with the claim of Evans *et al.* [16], which found out that collapse of the coal structure through capillary effect is relatively small. Nevertheless, Deevi *et al.* [17] found out that shrinkage (collapse) of the coal during drying occurs (only) at the level of macropores.

The effect of the temperature of pre-treating in an oxidizing atmosphere was studied by means of hot air oven drying (sample 7) and using of a desiccator (sample 9) for 5 and 216 hours, respectively (note that drying times are not equal). It was found that samples exhibit mean micropore volumes of 0.093 and 0.106 cm^3 g^{-1}, respectively and mean micropore surfaces of 247 and 278 m^2 g^{-1}, respectively. In such a way pre-treated samples also have lower characteristic energy values (9.10 and 8.84 kJ mol^{-1}, respectively) and higher values of the r_{mode} (0.68 and 0.69 nm, respectively). These procedures were carried out under ambient pressure conditions in the presence of air. Therefore, the low-temperature oxidation of the coal samples likely occurred. The residual contents of the moisture correspond with the 1.4 and 2.1 wt. % for hot air oven drying (sample 7) and desiccator drying (sample 9), respectively. Using of these methods cannot be expected to result in a completely dry state of the subbituminous coal.

3.4 Effect of drying time

The influence of the pre-treating time was mainly studied according to the vacuum pumping method. Samples 4, 1 and 5 were dried for 2, 5 and 8 hours, respectively at temperature 60°C. It can be seen (table 4) that samples 4 and 5 exhibits basically the same microporous parameters (S_{mi} ~ 300 m^2 g^{-1}, V_{mi} ~ 0.11 cm^3 g^{-1}, r_{mode} = 0.67 nm) despite the difference in residual moisture content.

From this one can conclude that two hours is sufficient time (at 60°C) for the vacuum pre-treatment method for this subbituminous coal. Incomparably longer drying time is probably needed in the case of lyophilisation (70 hours) and the desiccator drying procedure (216 hours). On the other hand, kinetic data from these procedures were not collected. Further experiments have to be realized for in-depth examination of this issue.

4 Conclusions

Experimental adsorption data were successfully fitted by the modified DR model. The micropore parameters obtained from this fitting show the following findings. It was found that the drying method of the subbituminous coal influences its final state.

(i) It was found that samples that were not exposed to oxygen during drying have better micropore characteristics, likely due to the lower contents of the residual moisture (≤0.4 wt. %). However, the reduction of the textural parameters of non-oxidizing atmosphere is likely related with filling of micropores by residual moisture, which is higher for a hot air dried sample (1.4 wt. %).

(ii) It was found that there are not significant differences according to microporous textural parameters (statistically evaluated in such way) with respect to a comparison of the minimal and maximal residual water content values. On the other hand, the residual moisture content may be responsible for the shifting of the mode of the micropores radii to the higher values with respect to air dried samples of the coal. The effect of the drying at elevated temperatures under vacuum was not proved, nor was a significant effect of temperature pre-treatment regardless of the used method. However, 60°C is sufficient temperature for vacuum drying.

(iii) Monitoring of the effect of the pre-treatment time proved that 2 hours is sufficient time for a clean surface using vacuum pumping at 60°C. The obtained microporous parameters of samples dried for 2 and 8 hours were negligible (even despite the difference in residual moisture contents of these samples).

Acknowledgements

The article has been drawn up in connection with the project of the Institute of Environmental Technologies, reg. No. CZ.1.05/2.1.00/03.0100 supported by the Research and Development for Innovations Operational Programme financed by Structural Funds of the European Union and from the means of the state budget of the Czech Republic. The paper was also supported by SGS, reg. No. SGS02/PřF/2013.

References

[1] Yu, J., Tahmasebi, A., Han, Y., Yin, F., Li, X., A review on water in low rank coals: The existence, interaction with coal structure and effects on coal utilization. *Fuel Processing Technology*, **106**, pp. 9-20, 2013.

[2] Taraba, B., *Nízkoteplotní oxidace a samovzněcování uhelné hmoty,* Ostravská univerzita: Ostrava, 2003 (in Czech).

[3] Day, S., Sakurovs, R., Weir, S., Supercritical gas sorption on moist coals. *International Journal of Coal Geology*, **74**, pp. 203-214, 2008.

[4] Siemons, N., Busch, A., Measurement and interpretation of supercritical CO_2 sorption on various coals. *International Journal of Coal Geology*, **69**, pp. 229-242, 2007.

[5] Clarkson, C.R., Bustin, R.M., Binary gas adsorption/desorption isotherms: effect of moisture and coal composition upon carbon dioxide selectivity over methane. *International Journal of Coal Geology*, **42**, pp. 241-271, 2000.

[6] Švábová, M., Weishauptová, Z., Přibyl, O., The effect of moisture on the sorption process of CO_2 on coal. *Fuel*, **92**, pp. 187-196, 2012.

[7] Brennan, J.K., Thomson, K.T., Gubbins, K.E., Adsorption of water in activated carbons: effect of pore blocking and connectivity. *Langmuir*, **18**, 5438–5447, 2002.

[8] Pisupati, S.V., Scaroni, A.W., Effects of natural weathering and low-temperature oxidation on some aspects of the combustion behaviour of bituminous coals. *Fuel*, **72**, pp. 779-785, 1993.

[9] Fei, Y, Artanto, Y, Giroux, L., Marshall, M., Jackson, W.R., MacPhee, J.A., Charland, J-P., Chaffee, A.L., Allardice, D.J., Comparison of some physico–chemical properties of Victorian lignite dewatered under non-evaporative conditions. *Fuel*, **85**, pp. 1987-1991, 2006.

[10] Bergins, Ch., Hulston, J., Strauss, K., Chaffee, A.L., Mechanical/thermal dewatering of lignite. Part 3: Physical properties and pore structure of MTE product coals. *Fuel*, **86**, pp. 3-16, 2007.

[11] Medek, J., Possibility of micropore analysis of coal and coke from the carbon dioxide isotherm. *Fuel*, **56**, pp.131-133, 1977.

[12] Medek, J., Weishauptová, Z., Vliv mikropórovitosti koksů na jejich vlastnosti, 1. část – Mikropórovitost jako charakteristická složka porézní struktury koksů. *Uhlí*, **35**, pp. 170-174, 1987 *(in Czech)*.

[13] Lowell, S., Shields, J.E., Thomas, M.A., Thommes, M., *Characterisation of porous solids and powders: Surface area, pore size and density*, Springer: Dordrecht, pp. 29-33, 2004.

[14] Medek, J., Weishauptová, Z., The microporous phase of carbonaceous substances and its fractal dimension. *Fuel*, **79**, pp. 1621-1626, 2000.

[15] Taraba, B. Personal communication, 8 October 2012, Professor at the Department of Chemistry, University of Ostrava.

[16] Evans, D.G., The brown-coal/water system: Part 4. Shrinkage on drying. *Fuel*, **52**, pp. 186-190, 1973.

[17] Deevi, S.C., Suuberg, E.M., Physical changes accompanying drying of western US lignites. *Fuel*, **66**, pp. 454–460, 1987.

www.witpress.com, ISSN 1743-3533 (on-line)

Cellulose acetate-based carbon xerogels and cryogels

J. Štefelová, M. Mucha & T. Zelenka
Department of Chemistry, Faculty of Science,
University of Ostrava, Czech Republic

Abstract

Low density carbonaceous materials based on biosources were prepared from cellulose acetate with a catalyst and a crosslinking agent. Standard preparation of these materials included dissolution, gelation and subsequent drying. In this study, cellulose acetate xerogels or cryogels were prepared, and the influence of the different drying techniques applied was investigated. The internal surface area of cellulose xerogels and cryogels varied from 700 to 850 m^2 g^{-1}, the volume of pores was 0.3 ml g^{-1}. Other properties such as thermal decomposition or surface chemistry were gained and both materials were analysed by adsorption experiments of selected heavy metal and organic dye from aqueous phase.
Keywords: cellulose acetate, xerogel, cryogel, freeze-drying.

1 Introduction

Carbon aerogels are fascinating, unique materials with an extensive porous network, high values of specific surface areas and low densities due to the fact that more than 90% of their volume is air. Inorganic aerogels made from silica or metal oxides such as zinc, aluminium or titanium are the most common materials. In most cases, organic aerogels are made from resorcinol and formaldehyde or from their derivates by sol-gel synthesis. Cellulose aerogels were firstly prepared by S. Kirstler [1, 2]. Cellulose is a renewable material and the most common source of biomass; it is the most abundant biopolymer and has really good properties such as biocompatibility, biodegradability as well as thermal and chemical stability [3]. Cellulose is a fibrous polysaccharide consisting of β-D-glucopyranosyl units joined by 1, 4-glycosidic bonds. It is

WIT Transactions on Engineering Sciences, Vol 77, © 2013 WIT Press
www.witpress.com, ISSN 1743-3533 (on-line)
doi:10.2495/MC130061

almost insoluble and represents the major structural component of cell walls of plants [4].

Generally, hydrogels are made by condensation of small polymeric particles. Aggregation of these particles takes place by means of sol-gel process, after that, the sponge-like, three-dimensioned porous network with liquid-filled pores is obtained. The liquid is mostly water or alcohol; therefore the resulting wet gels are named hydrogels or alcogels. When the water or alcohol in pores is replaced by means of supercritical or freeze-drying, the products are aerogels and cryogels respectively. When drying under ambient condition is used, xerogels are produced, but these materials are characterised by large shrinkage during drying and have a collapsed pores network. This fact is caused by capillary forces, which destroy the pores during evaporation of the liquid [1].

The aim of this study is preparation of carbon material from cellulose acetate with a catalyst and a crosslinking agent using two different types of drying, and characterisation of prepared materials.

2 Experimental

2.1 Synthesis of cellulose acetate gel

The cellulose acetate-based hydrogel was synthesized according to the procedure reported in literature [5] with minor modifications. 20 g of cellulose acetate was dissolved in 140 g of acetone at room temperature with continuous stirring. After 24 hours, a transparent solution was obtained. 0.05 g of a DABCO catalyst (33% solution of triethylenediamine in propylene glycol) and 4 g of a poly [(phenyl isocyanate)–co–formaldehyde] crosslinking agent were separately dissolved in 20 g of acetone and added to the acetate solution. The resulting mixture was poured into a bottle and the gel was aged for 1 week at room temperature.

2.2 Drying

Normal drying was performed by leaving the gel for several days at room temperature. Small, yellow, hard xerogels were obtained and named CA-N. Freeze-drying was performed using the FreeZone Freeze dry system device by Labconco. The pieces of gels were leached in distilled water and then frozen by immersing into liquid nitrogen. The frozen pieces were freeze-dried at ~52 Pa for 48 hours to obtain cryogels named CA-FD.

2.3 Pyrolysis

The obtained xerogels or cryogels were pyrolysed in a flow of nitrogen (150 ml min^{-1}). The temperature program consisted of an initial isothermal step at 100°C (30 min) followed by heating to 500°C with a heating rate of 10 K min^{-1}. This maximum temperature was then kept for 1 hour. The samples after pyrolysis were marked with an X (CA-NX, CA-FDX).

2.4 Sample analysis

2.4.1 Surface analysis

The S_{BET} values for porous characterisation of both samples after pyrolysis were obtained by dynamic nitrogen desorption at -196°C using the CHROM4 device; TiO_2 standards were used during the measurement. The microporous volume was calculated by the Dubinin–Radushkevich equation according to [6–8] from the data obtained by CO_2 sorption. The carbon dioxide isotherms were measured using the Setaram PCTPro E&E volumetric analyser at 30°C and at absolute pressure range from 0 to 1 bar. Microporous surface characterisations were acquired by the Medek's equation.

2.4.2 FT-IR analysis

The surface chemistry of the carbon materials was characterised by means of IR spectroscopy (Nicolet 6700). The samples (CA-N, CA-NX, CA-FD and CA-FDX) were analysed in an attenuated total reflectance mode (ATR) with a single bounce diamond crystal. Each spectrum was obtained under 4 cm^{-1} resolution and a total of 64 scans were performed.

2.4.3 Thermal analysis

A thermal analysis (Setsys Evolution, Setaram) with an analysis of evolved gasses by a mass spectrometer (QMG 700, Pffeifer, coupled by Supersonic system, Setaram) was carried out from 20 to 1300°C with an isothermal step at 100°C with a heating rate of 10 K min^{-1} in an argon flow. Mass spectroscopy was used for observation of selected ions, which were generated during the thermoanalytical measurement.

2.4.4 Adsorption analysis

To determine the maximal adsorbed capacity, the sorption of methylene blue and Cu(II) from aqueous phase were performed as described in [9] with minor modifications. The adsorption experiments were carried out in small amount where 10 mg of the sample after pyrolysis was placed in contact with 10 ml of methylene blue or Cu(II) solution at various concentrations for 24 hours at room temperature with occasional shaking. In the case of methylene blue, the used initial concentrations were 1000, 750, 500, 250, 100, 50, 25 and 10 mg l^{-1}, concentrations after adsorption were analysed using the UV/VIS spectrometer (Varian Cary 50 Conc) at 645 nm. In the case of Cu(II), the initial concentrations were 1.0, 0.8, 0.6, 0.4, 0.2 and 0.1 mmol l^{-1}. The adsorption data was gained by the atomic absorption spectrometer (Varian AA240FS). The adsorbed amount was calculated by the following equation:

$$a = \frac{(c - c_e)}{100.m} \tag{1}$$

where a [mg g^{-1} or mmol g^{-1}] is the adsorbed amount, c [mg l^{-1} or mmol l^{-1}] is the concentration before sorption, c_e [mg l^{-1} or mmol l^{-1}] is the equilibrium concentration after 24 hours, m [g] is the mass of the adsorbent (xerogel or cryogel).

3 Results and discussion

3.1 Pore texture

The porous characteristics of the analysed carbon materials are reported in Table 1. The specific surface areas measured by nitrogen desorption were 700 m^2 g^{-1} for CA-NX and 770 m^2 g^{-1} for CA-FDX. The microporous volume was 0.3 ml g^{-1} for both samples. The microporous surface area characterised by the Medek's equation was 820 m^2 g^{-1} in the case of CA-NX and 860 m^2 g^{-1} for CA-FDX.

Table 1: Porous characteristics for cellulose-acetate based materials.

Sample	S_{BET} [m^2 g^{-1}]	V_{DUB} [ml g^{-1}]	S_{MICRO} [m^2 g^{-1}]
CA-N	< 10	–	–
CA-FD	< 20	–	–
CA-NX	700	0.3	820
CA-FDX	770	0.3	860

The resulting data implies that the pyrolysis has a major influence on the surface properties of the final materials. Xerogel and cryogel have very low values of surface areas while S_{BET} values of pyrolysed CA-NX and CA-FDX are in the hundreds of m^2 g^{-1}. It is apparent that the major reaction or reactions forming the resulting porous material take place during the pyrolysis.

3.2 Thermal analysis and mass spectroscopy

Figure 1 shows the resulting thermogravimetric (TG) and derivative thermogravimetric (DTG) curves for the pyrolysis of xerogel and cryogel. From the graphs it is evident that there is no significant difference between the analysed samples. The main mass loss occurs in the temperature range from 200 to 400°C but above this temperature, there is a next, less significant pyrolytic step. As mentioned above, the pyrolysis has a big influence on the porous characteristics of these samples and therefore, the temperature used for pyrolysis could be crucial. Regarding the gradual final step of the pyrolysis, it would be desirable to use a temperature of at least 600°C during this treatment.

The gases evolved during the pyrolysis of xerogel as well as cryogel were analysed by mass spectroscopy to better understanding what happens in the sample. For both materials, a water loss was observed at a temperature about 100°C (residual moisture) and another loss of water occurred in the range from 200 to 500°C (Figures 2 and 3).

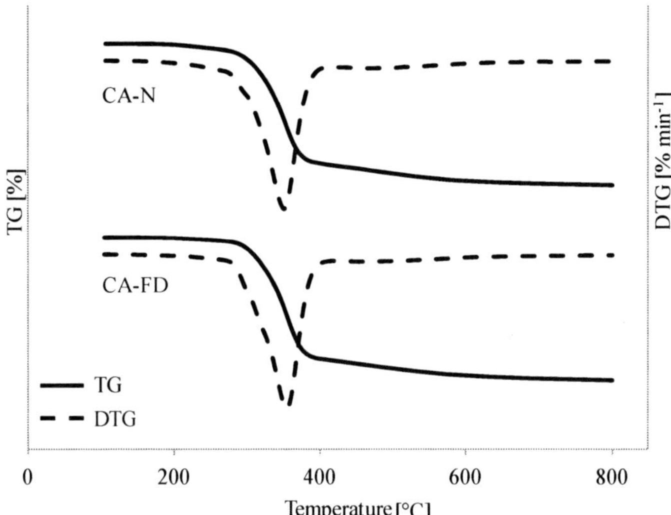

Figure 1: The TG (a) and DTG (b) curves of CA-N and CA-FD pyrolysis.

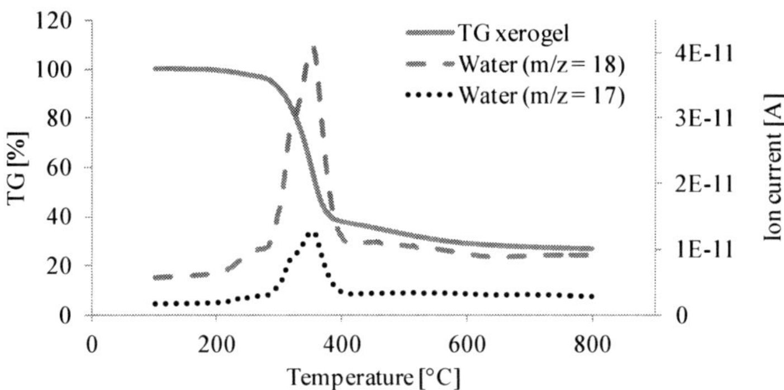

Figure 2: Water loss during xerogel CA-N pyrolysis.

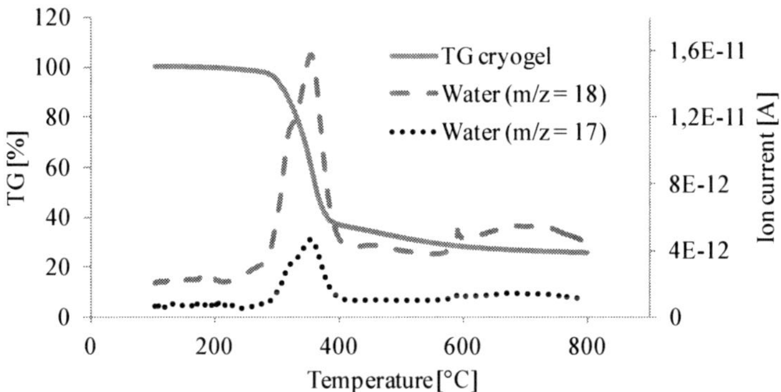

Figure 3: Water loss during cryogel CA-FD pyrolysis.

www.witpress.com, ISSN 1743-3533 (on-line)

During the preparation of CA, carbon dioxide was generated as a result of the reaction between water and PMDI. CO_2 was created in the form of little bubbles immediately after pouring the PMDI solution into dissolved cellulose acetate. As can been seen on the MS spectra, which are the same for both analysed gels, the residual water and carbon dioxide were released during the main thermal decomposition step (Figures 4 and 5). Maybe, not all water is reacted with PMDI and this reaction could have taken place at a higher temperature or some generated CO_2 could have been bonded in the structure of xero or cryogel and released during the pyrolysis. It is clear that some CO_2 left the structure, because the shape of the samples was totally different after the pyrolysis – from small, hard and almost non-porous material, porous, black carbon xerogel or cryogel was made.

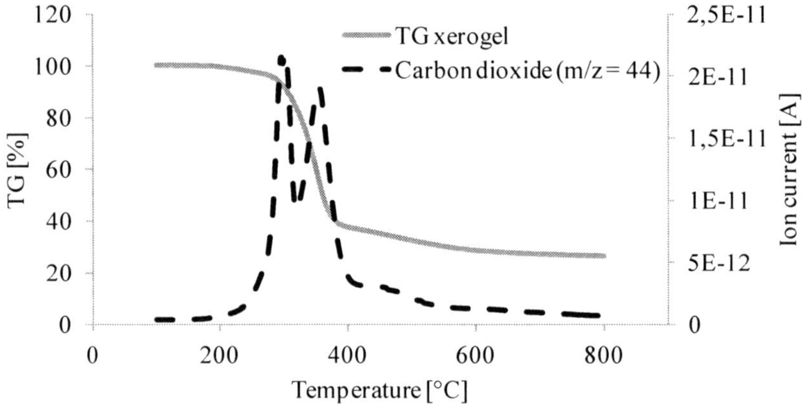

Figure 4: Release of carbon dioxide during xerogel CA-N pyrolysis.

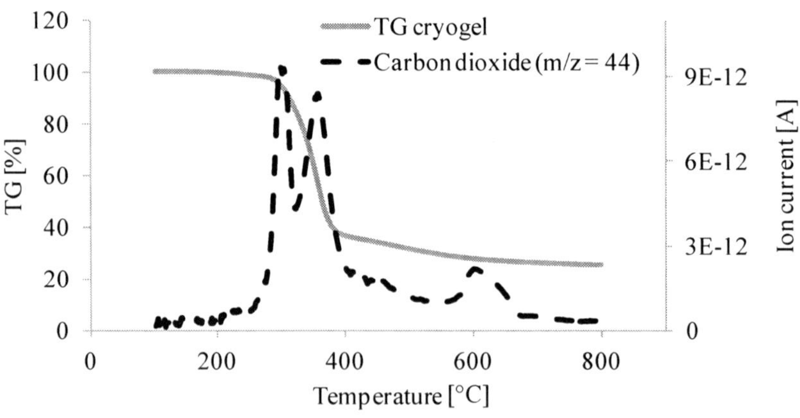

Figure 5: Release of carbon dioxide during cryogel CA-FD pyrolysis.

3.3 Infrared spectroscopy

Xerogel and cryogel as well as their pyrolysed forms were analysed by FT-IR to determine their surface chemistry (Figures 6 and 7). The spectra of both cryogels and xerogels are very similar and therefore, a more detailed description is provided only for xerogel. The distinction between the materials before (CA-N, CA-FD) and after (CA-NX, CA-FDX) the pyrolysis is evident from the FT-IR record. For CA-N, the adsorption bands at 2923 and 2854 cm^{-1}correspond to the C–H aliphatic bonds and the value of 1367 cm^{-1} can be attributed to the deformation mode of the C–H bond. The peaks at 1512 and 1537cm^{-1} can be associated with the urethane bonds, which was generated during the preparation of the acetate gel. The wide peak at 3479 cm^{-1} can be probably attributed to the NH group. The peaks at 1736, 1217 and 1034 cm-1 can indicate the presence of the C=O and C–O structures. For CA-NX, it is clear, that the spectrum is poorer, the sharp peaks disappeared and instead of them, wider and less identifiable peaks are present. Some small peaks at 3400 and 3055 cm^{-1} indicate that the structure is aromatic. The peak at 1587 and 1417 cm^{-1} can be attributed to the C=C and C=N. Three peaks at 876, 802 and 746 cm^{-1} may be associated with aromatic rings. The pyrolysis is likely to cause the degradation of oxygen functionalities which is confirmed by the disappearance of the peaks at 1736, 1217 and 1034 cm^{-1}.

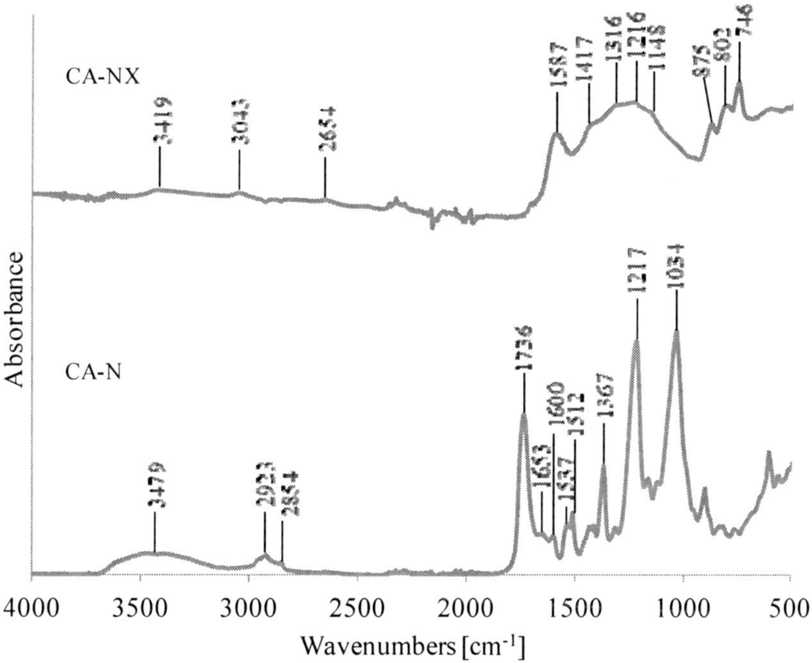

Figure 6: The FT-IR analysis of CA-N and CA-NX.

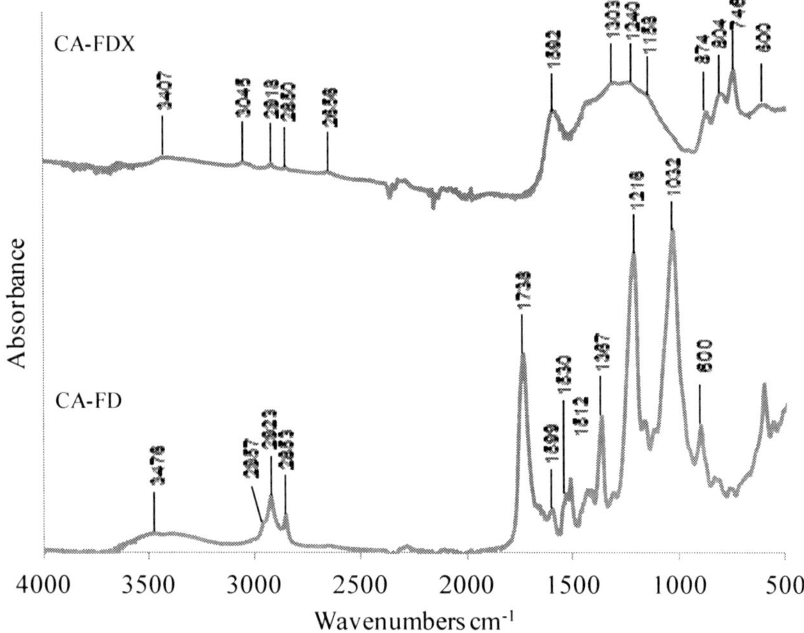

Figure 7: FT-IR analysis of CA-FD and CA-FDX.

3.4 Sorption experiments

The obtained data of concentration and adsorbed amount were plotted to get the maximal adsorption capacity which is defined as the maximum amount of adsorbed substance related to 1 g of sorbent. This experiment was repeated three times to obtain three adsorption curves. The maximal adsorption capacity was about 100 mg g^{-1} for CA-NX and about 200 mg g^{-1} for CA-FDX (Figure 8). There is a significant difference between the used sorbents and better sorption ability is probably a results of the freeze-drying technique applied during the preparation, which preserved an open porous structure with less damages.

In addition, the sorption of copper ions was performed. The experiment was carried out as mentioned above in the case of methylene blue in the same ratio of 10 mg of sorbent (xerogel or cryogel) and 10 ml of copper solution. In the case of Cu(II), sorption was also repeated to obtain three curves. The maximal adsorption capacity was calculated according to equation (1) and it was about 0.12 mmol.l^{-1} for xerogel and about 0.25 mmol l^{-1} for cryogel (Figure 9). The difference between the used adsorbents is apparent; cryogel has a higher adsorption capacity, especially at higher concentrations.

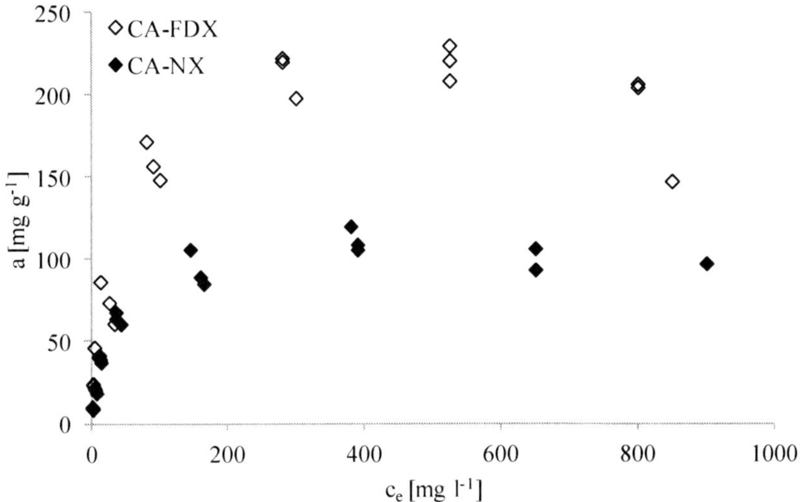

Figure 8: The adsorption isotherms of methylene blue on CA-NX and CA-FDX.

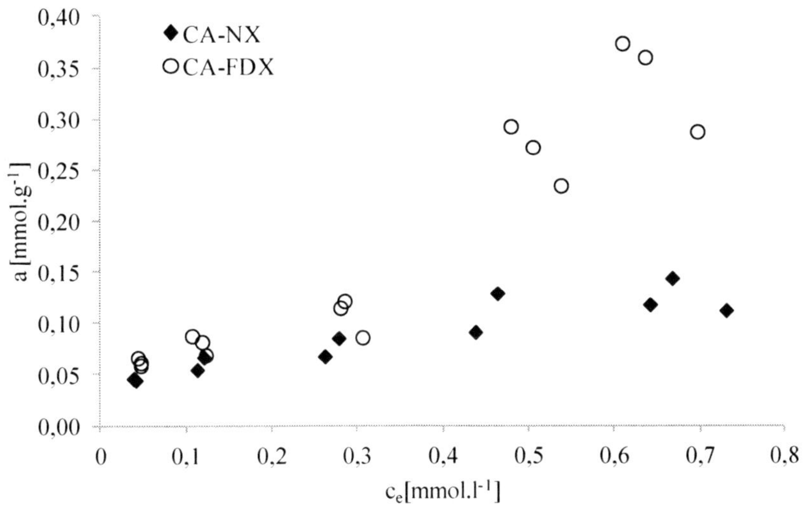

Figure 9: The adsorption isotherm of Cu(II) on samples CA-NX and CA-FDX.

4 Conclusion

The present paper describes the preparation of carbon xerogels and cryogels from cellulose acetate. The resultant materials and their pyrolysed forms were analysed by various techniques to obtain complex information of their properties. From a chemical point of view, xerogel and cryogel are very similar; they have almost identical surface chemistry, thermal behaviour and porous structures. The volume of pores was 0.3 ml g^{-1}, S_{BET} values were from 700 to 770 m^2 g^{-1} and S_{micro} was about 850 m^2 g^{-1}. During the adsorption experiments, it was observed that freeze-dried materials have better adsorption abilities in the adsorption of organic dye and Cu(II) from aqueous phase. The maximal adsorption capacity was almost twice as high compared to xerogel. This suggests that during preparation, freeze-drying is a sensitive method which prevents the nanoporous network of gels from collapsing and preserves an open pores structure in contrast to normal drying. Their high surface area, adsorption ability and the fact that xerogels and cryogels are made from a biodegradable resource make these materials ideal candidates for various applications.

Acknowledgements

This study was performed in connection with a project Institute of Environmental Technologies, reg. No. CZ.1.05/2.1.00/03.0100 supported by the Research and Development for Innovations Operational Programme financed by the Structural Funds of the European Union and from the means of state budget of the Czech Republic. The University of Ostrava supported this research by means of SGS reg. No. SGS02/PřF/2013.

References

[1] Hüsing, N. and Schubert, U., Aerogels–airy materials: chemistry, structure, and properties. *Angew. Chem. Int. Ed.*, 37, pp. 22–45, 1998.

[2] Innerlohinger, J., Weber, H. K. and Kraft, G., Aerocellulose: Aerogels and aerogel-like materials made from cellulose. *Macromol. Symp.*, 244, pp. 126–135, 2006.

[3] Deng, M., Zhou, Q., Du, A., van Kasteren, J. and Wang, Y., Preparation of nanoporous cellulose foams from cellulose-ionic liquid solutions. *Materials Letters*, 63, pp. 1851–1854, 2009.

[4] Paljevac, M., Primožič, M., Habulin, M., Novak, Z. and Knez, Ž., Hydrolysis of carboxymethyl cellulose catalyzed by cellulase immobilized on silica gels at low and high pressures. *J. of Supercritical Fluids,* 43, pp. 74–80, 2007.

[5] Grzyb, B., Hildenbrand, C., Berthon-Fabry, S., Bégin, D., Job, N., Rigacci, A. and Achard P., Functionalisation and chemical characterisation of cellulose-derived carbon aerogels. *Carbon*, 47, pp. 2297–2307, 2010.

[6] Medek, J., Possibility of micropore analysis of coal and coke from the carbon dioxide isotherm. *Fuel*, 56, pp. 131–133, 1977.

[7] Medek, J. and Weishauptová, Z., Vliv mikropórovitosti koksů na jejich vlastnosti, 1. část – Mikropórovitost jako charakteristická složka porézní struktury koksů. *Uhlí*, 35, pp. 170–174, 1987 (in Czech).
[8] Lowell, S., Shields, J.E., Thomas, M.A. and Thommes, M., Characterisation of porous solids and powders: Surface area, pore size and density, *Springer: Dordrecht*, pp. 29–33, 2004.
[9] Nunes, C. A. and Guerreiro, M. C., Estimation of surface area and pore volume of activated carbons by methylene blue and iodine numbers. *Quim. Nova,* 34(3), pp. 472–476, 2011.

Statistical structure-to-property relationships for fuel cell materials

K. Artyushkova[1], A. Patel[1], P. Atanassov[1], V. Colbow[2], M. Dutta[2], D. Harvey[2] & S. Wessel[2]
[1]*Chemical and Nuclear Engineering, University of New Mexico, USA*
[2]*Ballard Power Systems, Burnaby, BC, Canada*

Abstract

Development and optimization of components for fuel cells is hindered by the complex nature of the materials, a partial understanding of the reaction mechanisms and precise chemistry of active site or sites and relationship between performance and morphological properties such as size of particles, surface area, roughness, porosity, etc. XPS is one of the most widely utilized surface spectroscopic techniques for the analysis of material structure. Surface morphology is one of the most important factors affecting functional performance of fuel cell components. Correlation of structure, morphology and performance of Pt catalysts within MEA subjected to different accelerated test protocols (AST) is achieved through the application of principal component analysis (PCA) to curve-fits of high-resolution XPS spectra combined with morphological parameters extracted from SEM images and results of electrochemical measurements.
Keywords: structure-to-property correlation, X-ray photoelectron spectroscopy, SEM, morphology, digital image processing, durability, fuel cells, electrocatalysts, principal component analysis.

1 Introduction

Polymer electrolyte membrane fuel cells (PEMFCs) are a promising alternative energy source, capable of being deployed as successful replacements in many stationary and mobile applications. Durability of catalyst layer (CL) is of key importance in the large-scale deployment of PEMFCs.

WIT Transactions on Engineering Sciences, Vol 77, © 2013 WIT Press
www.witpress.com, ISSN 1743-3533 (on-line)
doi:10.2495/MC130071

Accelerated stress tests (ASTs) have become the standard for PEMFC aging protocols [1, 2]. Using ASTs researchers are capable of aging the cells at more rapid rates than they would typically achieve in standard operation.

For Pt supported on carbon catalysts, CL degradation is linked to Pt dissolution and agglomeration, ionomer degradation, carbon support degradation, and the changes in pore morphology and surface properties [3–5]. X-ray photoelectron spectroscopy (XPS) is a powerful technique to study the chemical changes in the catalysts layers and polymer membrane.

The quantitative information contained in SEM images is largely underutilized with qualitative analysis being a preferred approach. However, digital image processing (DIP) of SEM images is able to convert the 2-D intensity distribution into 1-D image descriptors (values) that can be utilized for quantitative morphology analysis. A major difficulty in direct correlation between morphology and functions of materials is the lack of information on the scale of surface topography and the texture that are responsible for different functions of the materials. Different phenomena and processes related to the material performance correspond to different diffusion regimes at different length scales. Therefore, it is important to access these multiscale morphological features of materials.

The objective of this study was to characterize physical and chemical changes of aged MEAs, and to correlate these changes to perceived performance losses. Multiple carbon support materials were tested as cathode catalysts for MEAs. These MEAs were cycled using AST protocols in order to rapidly degrade the cathode catalyst. Conditioned and aged catalysts coated membranes (CCMs) were then analyzed using XPS, to quantify chemical structural changes, and SEM to quantify morphological changes. These data sets were then processed using PCA in order to draw statistical correlations between the chemical, physical, and performance changes observed.

2 Experimental details

2.1 Materials

In order to understand the effect of the carbon support on catalyst degradation, Pt catalysts on different carbon supports, low surface area (LSA), mid-range surface area (MSA), high surface area (HSA) carbons and in addition, heat treated high HSA surface area carbon (HSAHT) Pt catalyst, each with 50wt% Pt, were investigated. CCMs including following catalysts with 0.4 $mgPt/cm^2$ Pt loading were studied: Pt50LSAC, Pt50MSAC, Pt50HSAC1, Pt50HSAC1-HT, Pt50HSAC2, and Pt50HSAC2-HT. The anode portion of the membrane used Pt50LSAC at a loading of $0.1 mgPt/cm^2$, and membrane was Nafion® NR-211.

2.2 MEA testing

MEAs were operated at 80°C, 100% RH with 5 psig back pressure of hydrogen and air on the anode and cathode sides, respectively. The cells were run at high stoichiometric ratios. The MEAs were conditioned overnight by being held at a current density of 1.3A/cm^2. Following the conditioning step, the BOL polarization curves for the membranes were obtained. The fuel cells were then cycled from 0.6V for 30s to 1.2V for 60s. MEAs using Pt50LSAC as the cathode, however, were subjected to additional cycling protocols which changed the upper potential limit (UPL) to 1, 1.3, 1.4V or 1.6 V (vs. RHE). Irrespective of which AST protocol was used, the cells were cycled 4700 times.

2.3 XPS analysis

Conditioned and aged CCMs were analyzed using a Kratos Axis DLD Ultra spectrometer using monochromatic Al Kα X-ray source, with emission voltage of 15 kV and emission current of 10 mA. High resolution spectra for carbon, fluorine, oxygen, and platinum were acquired to determine to what extent chemical shifts had altered the carbon support, platinum nanoparticles, or Nafion® ionomer.

2.4 SEM and digital image processing

SEM images were acquired for conditioned and degraded CCMs. The images were taken at 2.5K and 50K magnification at a voltage of 1kV. Multiple locations were chosen per sample, and the locations were chosen randomly in order to prevent user bias from influencing the results. The images were analyzed through Digital Image Processing software to quantify changes to sample morphology due to conditioning and accelerated stress testing.

3 Results and discussion

3.1 XPS results

XPS provides chemical composition from top 5–10 nm of the material. High resolution C and Pt spectra were curve-fit using individual symmetrical peaks to derive speciation information. Figure 1 shows changes in speciation as a function of UPL of AST for Pt50LSAC CCM sample. The C1s speciation obtained by cycling the cell to a maximum of 1V gave identical results to cycling the CCMs to 1.2V, however subjecting the cells to higher potentials resulted in drastically different composition. At 1.3V and 1.4V, the amount of graphitic carbon from the carbon support and fully fluorinated carbons from ionomer, CF_2, decreased, while the carbonate and carboxylate species increased. These results suggest oxidation of both carbon of the catalyst support and of the ionomer.

Moreover, cycling Pt50LSAC resulted in significantly lower PtO and Pt-C content while the metal form of platinum increased. Though the carbon speciation for the 1V and 1.2V AST protocols were identical, there were differences in the platinum speciation. These were manifested by detachment of Pt associated with C and growth of Pt particles resulting in larger amount of metallic Pt.

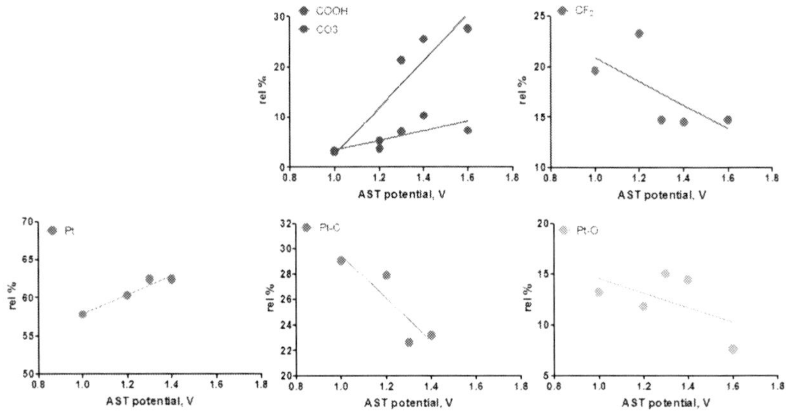

Figure 1: Changes in relative % of carbon and Pt species for Pt50LSAC sample as a function of UPL AST potential.

3.2 SEM results

Figure 2 shows SEM images for conditioned and aged CCMs at two different magnifications. Careful consideration was taken to ensure that the magnification, voltage, brightness, and contrast were at constant values for all images. From Figure 2, it is clearly visible that the structure of material has changed drastically following the cycling protocol. The corroded sample no longer shows individual carbon black particles that are clearly visible in the conditioned CCM image.

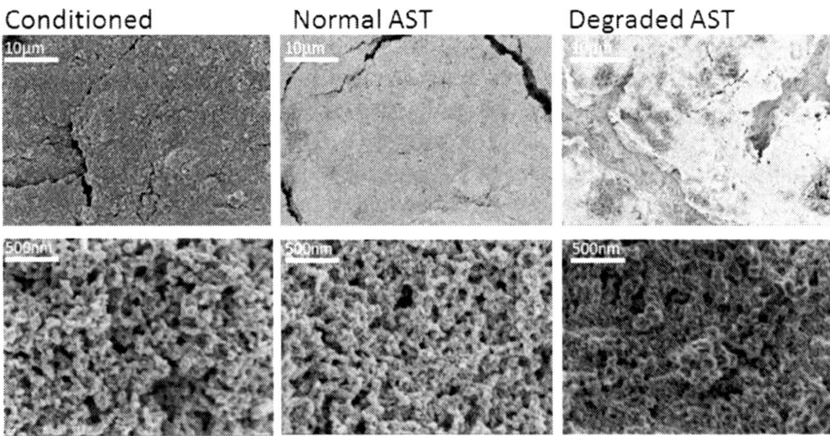

Figure 2: SEM images of conditioned, aged and corroded CCM at 2.5K (top row) and 50K (bottom row) magnifications.

Even though through visual inspection of SEM images, major changes in morphology are obvious, much more morphological details are contained within the images that are not obvious to a human eye. In order to understand impacts of AST on roughness, porosity and texture of CCMs, digital image processing of SEM images were performed [6, 7]. For this study we focused on skewness (*Rsk*) as it describes asymmetry of the roughness, pointing to the domination of pores or peaks in the image. Its value increases as amount of pores (dark values of intensity within images) increases.

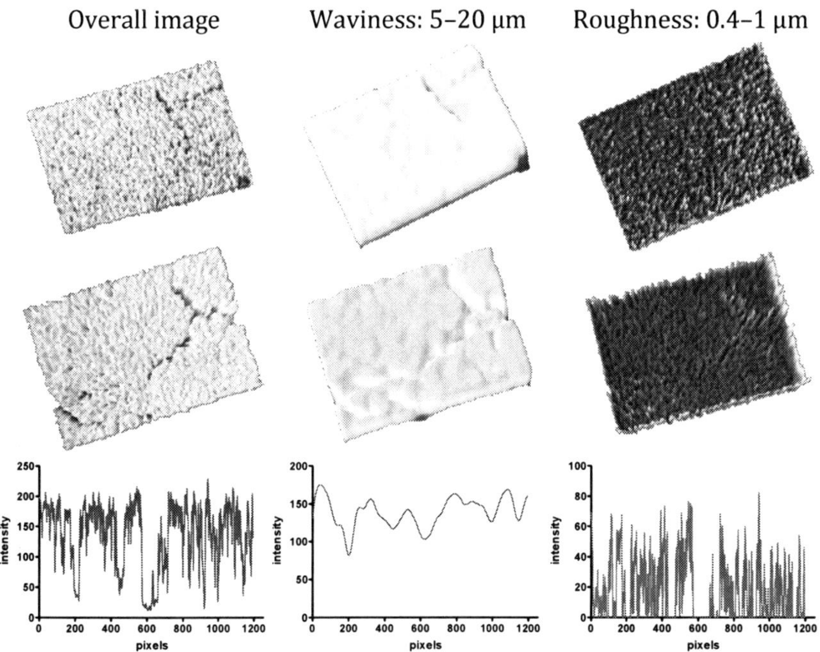

Figure 3: Sample images and line profiles for overall image and their separated waviness (low-frequency) and roughness (high-frequency) components.

A typical surface exhibits roughness superimposed over waviness. Waviness is due to the more widely spaced repetitive deviations (low-frequency component of images) and roughness is due to the finer irregularities of surface roughness (high-frequency component of images). It is important to separate and evaluate waviness and roughness parameters individually. For this purpose, we have applied high-pass filter to remove low-frequency component and low-pass filter to remove high-frequency component from images to produce roughness and waviness image components, respectively [7].

Figure 3 shows image components, and corresponding line profiles, from which skewness was calculated separated by filtering of SEM images acquired at 2.5K magnification. At this magnification, high-frequency component images

correspond to roughness in the range of 0.4–1 μm, and low-frequency component images correspond to 5–20μm. The same has been done for images at 50K magnification. At this magnification, filtering separates images into low-frequency components at 0.2–1 micron scale and high-frequency component at 20–100 nm. Figure 4 shows hierarchy of scales extracted from SEM images. Because images were acquired at 2 different magnifications, the high frequency and low frequency components porosity represent two length scales with the high frequency components at 2.5K being approximately the same length as the low frequency components at 50K magnification.

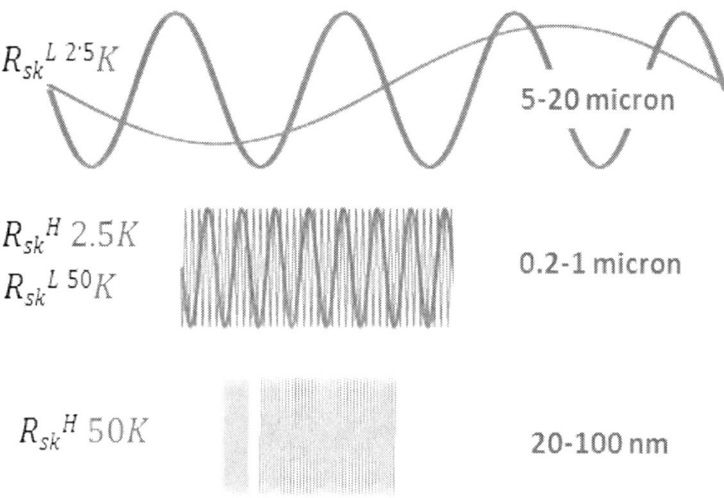

Figure 4: Profiles extracted from SEM images by filtering shown at scale.

Figure 5 shows skewness parameters extracted from image components separated by filtering. Overall skewness and skewness at scale of 5–20 microns does not change significantly with increasing AST potential. Dramatic increase in porosity as represented by skewness at scales of 0.2–1 microns and 20–100 nm is detected as UPL of cycling increases. In comparison with conditioned sample, the porosity at larger scale of 5–20 microns increases for all tested samples. Degradation of MEA, however, is linked mainly to changes at scale from tens of nanometer up to a micron. This microroughness and microporosity introduced plays a significant role in water management, potentially, causing failure of MEAs at higher UPL of cycling.

To get an additional insight into the mechanism of MEA degradation for classes of different types of carbons used as supports, multivariate analysis, particularly, Principal Component Analysis was applied to the data combining structural, morphological and durability parameters shown in Table 1.

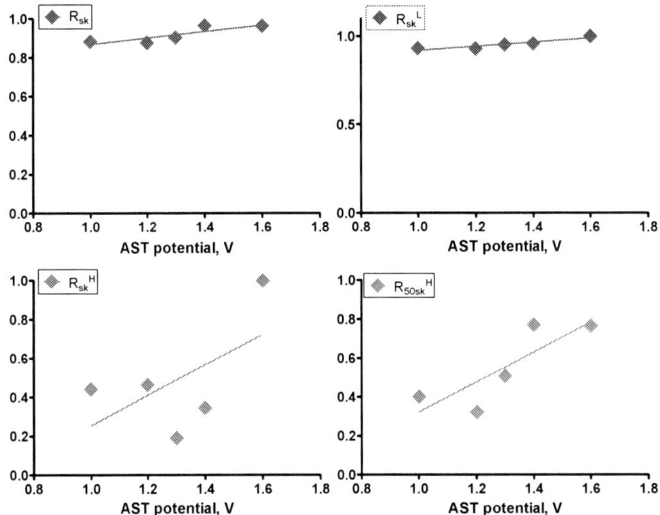

Figure 5: Skewness parameters, overall, low and high frequency from 2.5K and high frequency for 50K SEM images as a function of AST potential.

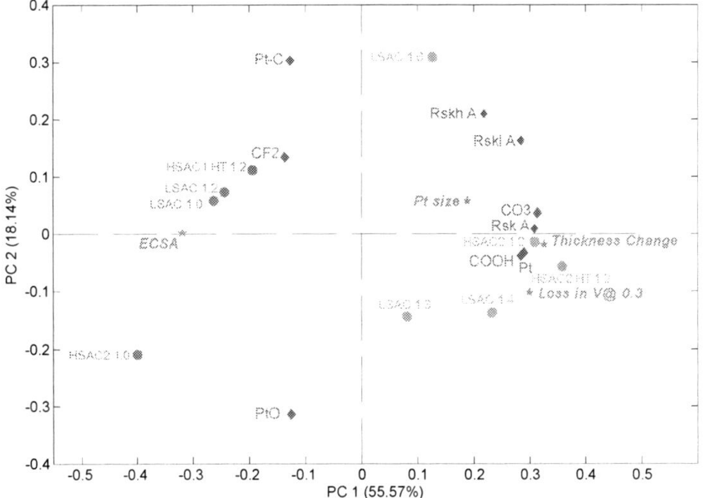

Figure 6: PCA biplot from data in Table 1.

These parameters combine relative % of carbon and Pt species, such as plotted in Figure 1, skewness parameters extracted from SEM images, such as plotted in Figure 5, Pt particle size obtained from TEM of cycled MEAs, % thickness change, loss in V at 0.3 A/cm^2 and Electrochemically Accessible

Surface Area (ECSA). The sample set include Pt50LSAC at various UPL of AST cycling and 4 high surface area carbon-based catalysts, HSAC, original and heat treated.

Table 1: All data for statistical structure-to-property correlations.

	ECSA	Pt size	Thickness Change	Loss in V: 0.3 A/cm^2	R_{sk}	R_{sk}^H	R_{sk}^L
LSAC 1.0 V	75.0	8.5	1.1	11.5	1.05	0.18	1.02
LSAC 1.2 V	73	8.1	0.0	24.0	1.04	0.19	1.02
LSAC 1.3 V	51	8.6	8.8	138.2	1.07	0.08	1.04
LSAC 1.4 V	18	7.5	9.5	161.1	1.15	0.14	1.05
LSAC 1.6V	68	7.7	8.9	57.3	1.14	0.40	1.10
HSAC2 1.0 V	132	6.5	0.7	39.4	1.07	-0.07	1.01
HSAC2 1.2 V	21	10.1	7.6	361.9	1.14	0.20	1.06
HSAC1 HT 1.2	65	9.5	1.8	21.6	1.08	0.00	1.04
HSAC2 HT 1.2	9	10.1	9.1	357.0	1.19	0.33	1.07

	COOH	CO_3	CF_2	Pt	Pt-C	PtO
LSAC 1.0 V	3.1	3.3	19.6	57.8	29.0	13.2
LSAC 1.2 V	3.7	5.3	23.3	60.3	27.9	11.8
LSAC 1.3 V	7.1	21.3	14.7	62.4	22.6	15.0
LSAC 1.4 V	10.3	25.5	14.5	62.4	23.2	14.4
LSAC 1.6V	7.3	27.6	14.7	58.0	34.3	7.6
HSAC2 1.0 V	3.3	2.4	14.0	53.7	26.2	20.2
HSAC2 1.2 V	9.8	20.6	14.6	62.7	26.6	10.7
HSAC1 HT 1.2	4.0	5.7	25.6	59.6	29.6	10.8
HSAC2 HT 1.2	5.3	20.1	18.9	64.4	21.8	13.8

Figure 6 shows PCA results. Non degraded aged MEA samples, LSAC cycled at 1.0 and 1.2 V, HSAC 2 sample cycled at 1.0 V and HSAC 1 HT cycled at 1.2 V show larger retained ECSA correlated with smaller loss of Pt associated with C and oxides. Severally degraded samples, those showing larger loss in voltage, being LSAC tested at all voltages higher than 1.2 V and HSAC2 samples original and heat-treated tested at 1.2 V have larger amounts of carboxylates and carbonates and metallic Pt correlated with larger Pt size. Thickness change is being accompanied by larger porosity at all scales as provided by DIP of SEM images.

Figure 7 shows direct correlations between the parameters highlighted by statistical structure-to-property correlations obtained via PCA. Increase of relative % of metallic Pt as determined by XPS is correlated with the size of Pt particles as determined by TEM. The relative % of Pt-O and Pt associated with carbon retained after testing is correlated with lager ECSA. Increase in surface oxides such as carbonates and carboxylates is linearly related to the loss in performance as manifested by loss in the V at constant current. And finally, change in thickness of CL is associated with total increase in porosity as calculated by DIP of SEM images.

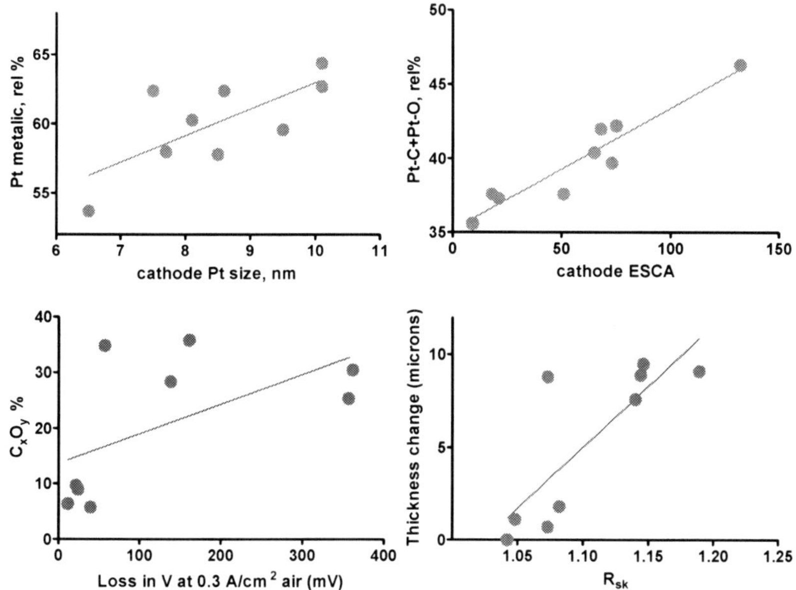

Figure 7: Correlation between structural, morphological and performance parameters for all types of catalysts.

References

[1] Y. Gu, J. St-Pierre, A. Joly, R. Goeke, A. Datye, P. Atanassov, *J Electrochem Soc* 156 (2009) B485-B492.

[2] K. Panha, M. Fowler, X.-Z. Yuan, H. Wang, *Applied Energy* 93 (2012) 90-97.

[3] V. Parry, G. Berthome, J.C. Joud, O. Lemaire, A.A. Franco, *J Power Sources* 196 (2011) 2530-2538.

[4] J. Xie, D.L. Wood, K.L. More, P. Atanassov, R.L. Borup, *J Electrochem Soc* 152 (2005) A1011-A1020.

[5] F.Y. Zhang, S.G. Advani, A.K. Prasad, M.E. Boggs, S.P. Sullivan, T.P. Beebe, *Electrochim Acta* 54 (2009) 4025-4030.
[6] K. Artyushkova, S. Pylypenko, M. Dowlapalli, P. Atanassov, *Journal of Power Sources* 214 (2012) 303-313.
[7] K. Artyushkova, S. Pylypenko, M. Dowlapalli, P. Atanassov, *Rsc Advances* 2 (2012) 4304-4310.

Section 2
Mechanical characterisation and testing

Mechanical characterisation of novel polyethylene nanocomposites by nanoindentation

A. S. Alghamdi[1], I. A. Ashcroft[1] & M. Song[2]
[1]*Mechanical, Materials & Manufacturing Engineering, University of Nottingham, UK*
[2]*Department of Materials, Loughborough University, UK*

Abstract

Ultra-high Molecular Weight Polyethylene (UHMWPE) is a high performance polymer which is currently used in lightweight body armour and total joint replacement applications because of its high toughness and wear resistance, respectively. However, the high molecular weight also makes processing difficult and expensive, which has led to research to find alternative materials with similar performance but better processability. The aim of this work is to investigate the mechanical properties of novel polyethylene nanocomposites by means of nanoindentation. The localized mechanical properties were used to evaluate the dispersion of the nanoparticle into the polymer matrix. Scratch, creep and wear resistance of the polyethylene nanocomposites were also studied. Nanoindentation tests provided information about the dispersion quality of the nanoparticles, which can be used to develop processing methods. The addition of the nanoparticles was found to improve the scratch, creep and wear resistance of the polyethylene.
Keywords: polyethylene, nanocomposite, carbon nanotube, nanoclay, nanoindentation, scratch, wear.

1 Introduction

Ultra-high molecular weight polyethylene (UHMWPE) is a high performance thermoplastic with outstanding mechanical properties, such as high wear strength, chemical resistance and high toughness, which provide not only practical benefits but also scientific interest [1–3]. However, its extremely high

WIT Transactions on Engineering Sciences, Vol 77, © 2013 WIT Press
www.witpress.com, ISSN 1743-3533 (on-line)
doi:10.2495/MC130081

molecular weight, and subsequent high viscosity, raises difficulties in processing using standard techniques, such as twin screw extrusion and compression moulding. Reducing the viscosity of UHMWPE is an effective method of avoiding these processing difficulties. Blending UHMWPE with other polymers that have lower viscosity, such as high density polyethylene (HDPE), can therefore be used to improve processability. HDPE has a similar structure to UHMWPE but with lower molecular chain length, however, it exhibits lower wear resistance, yield strength and toughness than UHMWPE [1]. This reduction in performance on adding HDPE to UHMWPE can potentially be mitigated, whilst retaining the improved processability, by the addition of nano-reinforcement, which has been shown to improve the mechanical performance of polyethylene [4–11].

Depth sensing indentation (DSI) is an advanced technique, which is capable of providing valuable information about the near-surface properties of solid polymers, such as indentation elastic modulus, indentation hardness, elastic-plastic, viscoelastic (creep) and viscoplastic behaviour [12]. Recently, this technique has become increasingly popular in the investigation of the near-surface properties of polymer nanocomposites and their correlation to the nanoparticle loading [13–15]. The effects of nanoparticles on the scratch and wear resistance of polymer nanocomposites have also been studied at micro-level [14, 15]. To date, no work has been carried out to evaluate the dispersion of the nanoparticles, the creep behaviour, or the scratch and wear resistance of polyethylene nanocomposites by means of DSI. Therefore, in this paper, the dispersion of carbon nanotube (CNT) and inorganic nanoclay in a UHMWPE/HDPE blend matrix using two different mixing methods was evaluated by DSI. The effects of the nanoparticle addition on the creep behaviour, scratch and wear resistance were also investigated.

2 Experimental methods

2.1 Materials

The materials tested in this study were UHMWPE/HDPE blended polymers with CNT and nanoclay fillers. Nascent UHMWPE powders (Sabic®UHMWPE3548) were purchased from SABIC [16] which had an average molecular weight of 3×10^6 mol/g. HDPE powders (ExxonMobil™ HDPE HMA014) were purchased from ICO Ltd [17]. Natural hectorite nanoclay was supplied by Elementis specialties [18]. Multi-wall Nanotubes (MWNT) with diameters in the range of 5 nm to 50 nm, were provided by Nanocyl [19]. Butylated hydroxytoluene and Tris(nonylphenyl) phosphate, supplied by Sigma-Aldrich [20], were used as primary and secondary antioxidants, to maintain the long term thermal stability and melt processing stability, respectively.

2.2 Processing

An in-house pre-mix technology was used to incorporate the CNT or inorganic nanoclay into the UHMWPE and HDPE powders. A twin-screw extruder was used to blend the 75wt.% UHMWPE and 25wt.% HDPE powders pre-mixed with CNT or nanoclay, to form nano-filled UHMWPE/HDPE blends with low volume fraction (0.5 wt.%). Two processing methods (M1 and M2) were used and the mixing temperature was controlled using five zones from feeding port to die, the processing parameters are shown in table 1. A blend of 75wt.% UHMWPE and 25wt.% HDPE, abbreviated to U75H25, was used as the hybrid PE matrix to accommodate the nanofillers. Compression moulding was used to mould the nanocomposite materials. The raw material was placed into a square mould (100 X 100 X 1.65 mm), and then heated to 190°C, which is higher than the melting point of the composite (approximately 135°C). Various mould pressures (154, 232, 309, and 386 MPa) were investigated to optimise the properties of the material such as hardness and crystallinity. Various holding times at maximum pressure (10, 15 and 30 minutes) were also used to identify the most appropriate moulding parameters. The optimal moulding pressure and holding time were found to be 309 MPa and 15 minutes respectively, which resulted in the highest measured values of hardness and crystallinity. After compression moulding, the mould was cooled to room temperature using water.

Table 1: Processing method parameters.

Processing Method	Extruder Speed (rpm)	Processing Temperature (°C)					
		Zone 1	Zone 2	Zone 3	Zone 4	Die	Cooling
M1	400	180	190	200	210	220	water
M2	190	220	250	260	270	280	water

2.3 Material testing and characterisation

Depth sensing indentation (DSI) experiments were performed on 10 x 10 x 1.65 mm specimens at a controlled machine chamber temperature of 24.8±0.6°C using a NanoTest 600 from Micro Materials Ltd (Wrexham, UK). A Berkovich indenter, with face angle of 65.3°, was used to make a grid of 10x10 indents using 40 mN maximum load, 600s dwell period and 2 mN/s loading and unloading rates. The results were analysed using the Oliver and Pharr method [21], and then plotted using Matlab software from MathWorks (Cambridge, UK). In this method, the initial portion of the unloading curve is described by the power low relation:

$$P = \alpha (h - h_r)^m \quad (1)$$

where P is the load, α and m are constants determined by curve fitting, h is penetration depth and h_r is the depth of the residual impression. The contact stiffness (S) can be obtained by:

$$S = \frac{dP}{dh}(h = h_{max}) = m \propto (h_{max} - h_r)^{m-1} \quad (2)$$

The contact depth (h_c) at maximum load (P_{max}) can be estimated using:

$$h_c = h_{max} - \varepsilon \frac{P_{max}}{S} \tag{3}$$

where ε is a constant related to the geometry of indenter, which is 0.75 for the Berkovich indenter, h_{max} is the maximum penetration depth. Thus, the projected contact area (A_c) is determined from (h_c) by the following relation:

$$A_c \approx 24.5\ h_c^2 \tag{4}$$

and hence the indentation hardness (H) is:

$$H = \frac{P_{max}}{A_c} = \frac{P_{max}}{24.5h_c^2} \tag{5}$$

For the materials used in this study, a bulge or (nose) effect was found during the initial portion of unloading as a result of creep, which can lead to errors in the calculation of contact depth and contact stiffness. Therefore, a dwell time of 600s was introduced at maximum load to minimize the effect of viscoelastic behaviour. In this study, the Oliver and Pharr method was used to compare the mechanical resistance of the blends and nanocomposites under identical testing conditions.

Micro-scratch and wear tests were also performed on 10 x 10 x 1.65 mm specimens at 24.8±0.6°C using the NanoTest 600.A diamond probe of ~200μm tip radius (Rockwell) was used to perform single and multi-pass scratches (wear) over tracks of 150 and 100μm, respectively. The single scratch procedure involved scans at 15μm/s velocity, 25mN load, 2.97mN/s loading rate and 100μm distance between scratches. The wear experiments were performed using a multi-pass scratch (11 scratches) technique at 10μm/s velocity, 25mN load, 2.97mN/s loading rate and 100μm distance between scratches. An initial topography (0.2mN constant load) scan was applied before starting each experiment to check the sample preparation. Five repeat tests were performed on each sample and the average result was used in data analysis.

The nanoparticle dispersion was also investigated using a Philips XL30 ESEM-FEG, Scanning Electron Microscope (SEM) from FEI (Eindhoven, The Netherlands). AJEOL 2000FX Transmission Electron Microscope (TEM) from JEOL Ltd. (Welwyn Garden, UK) was used to analyse the dispersion of nanoclay in the blend matrix.

3 Results and discussion

3.1 Dispersion of nanoparticles

Nanoparticle dispersion is a very important factor in the manufacture of nanocomposites that can affect the mechanical and rheological properties of the composite. It is difficult to achieve a uniform dispersion of nanoparticles in a polymer matrix and in this work two processing methods (M1 and M2) were used in an attempt to minimize the effects of CNT aggregation and to form exfoliated nanoclay platelets.

The scanning electron microscope (SEM) and the transmission electron microscope (TEM) were used to analyse nanoparticle dispersion in the U75H25 matrix at the micro-scale, as shown in fig. 1. It can be seen that CNT, which is indicated by black arrows in fig 1a is well dispersed in the matrix. Also, the nanoclay layers are dispersed in the blend matrix. The dark region in fig. 1b is a separate nanoclay layer. However, enlarging the measurement scale can also provide information about nanoparticle dispersion, which is not achievable using the previous techniques. Therefore, in this work a DSI method was used to investigate near-surface properties of the nanocomposites, in this case indentation hardness, over an area of approximately 1 mm^2. The variations in hardness value were used to evaluate the effect of the processing method on the blend morphology and the dispersion of the CNT and inorganic nanoclay. The results are shown in figs. 2 and 3 for processing methods M1 and M2 respectively. In fig. 2, the influence of adding HDPE to the UHMWPE can be seen to significantly increase the indentation hardness and the variation of hardness across the sample, which indicates poor mixing of the two polymer phases (UHMWPE and HDPE). In fig. 3, it can be seen that the hardness of the blend is similar to that of the UHMWPE and there is less variation in hardness than seen with M1. This indicates a better blending of the HDPE into the UHMWPE microstructure.

The addition of very low volume fractions of CNT and nanoclay (0.5 wt.%) to the blend shows a slight increase in hardness using M1, as seen in fig. 2. However, the variation in hardness by the addition of 0.5 wt.% CNT indicates the formation of some aggregations of nanoparticles. In fig. 3, the CNT and nanoclay were uniformly dispersed throughout the U75H25 matrix. However, at very low volume fraction of CNT, hardness was increased significantly, which indicates a strong influence of the CNT addition on the near surface properties of U75H25.

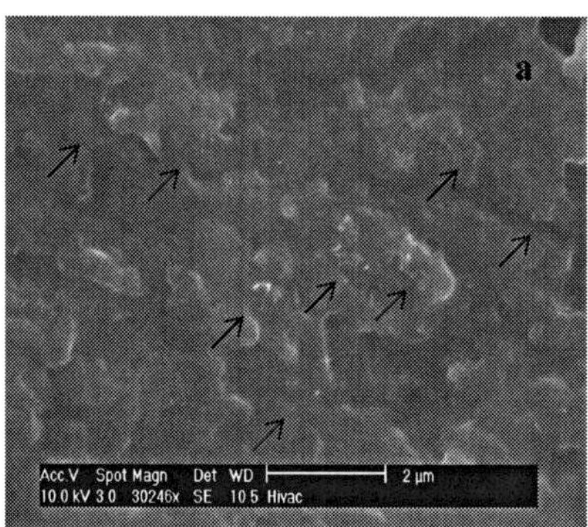

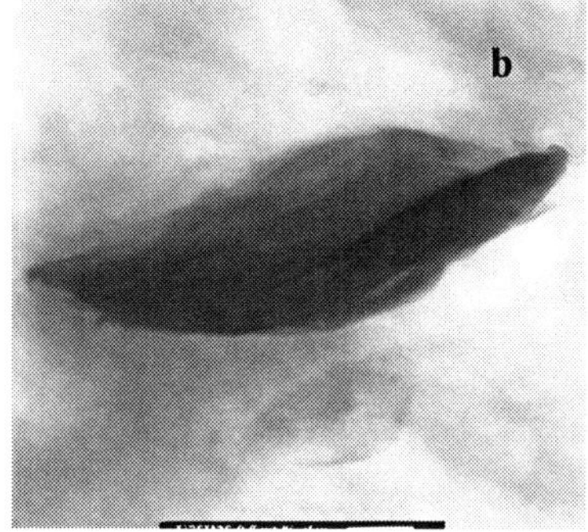

Figure 1: Dispersion of nanoparticle in the U75H25 matrix: a) SEM image for the dispersion of CNT and b) TEM image for the dispersion of nanoclay.

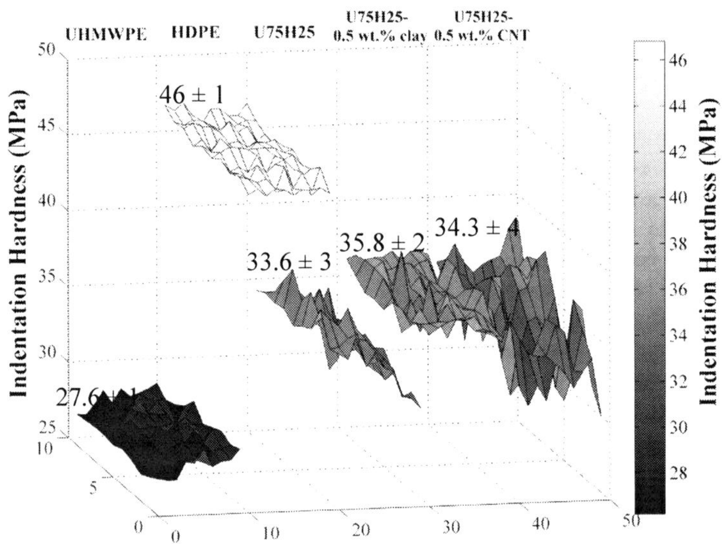

Figure 2: Indentation hardness for polyethylene blend and nanocomposites using processing method M1, including mean and standard deviation values.

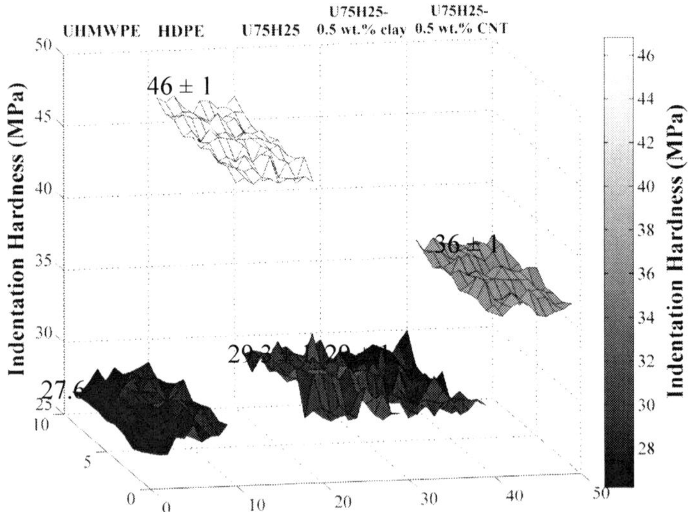

Figure 3: Indentation hardness for polyethylene blend and nanocomposites using processing method M2, including mean and standard deviation values.

3.2 Depth sensing indentation results

Depth sensing indentation (or nanoindentation) can be used to investigate the near-surface mechanical properties of materials. This technique is useful and directly applicable for elastic-plastic materials. Polymers, however, can exhibit time dependent behaviour which affects the initial portion of unloading, and consequently the contact depth and stiffness. Creep on unloading can increase indentation depth resulting in a higher or even a negative slope, which affects the calculated modulus value. Fig. 4 shows the load-displacement curves of UHMWPE, U75H25 and the two nanocomposites. A 600s holding time at the maximum load (40mN)was used to minimise the effect of creep on unloading. It can be seen that the addition of HDPE to the UHMWPE increases its stiffness. Further increases in stiffness are seen by the addition of the CNT and nanoclay fillers to the blend.

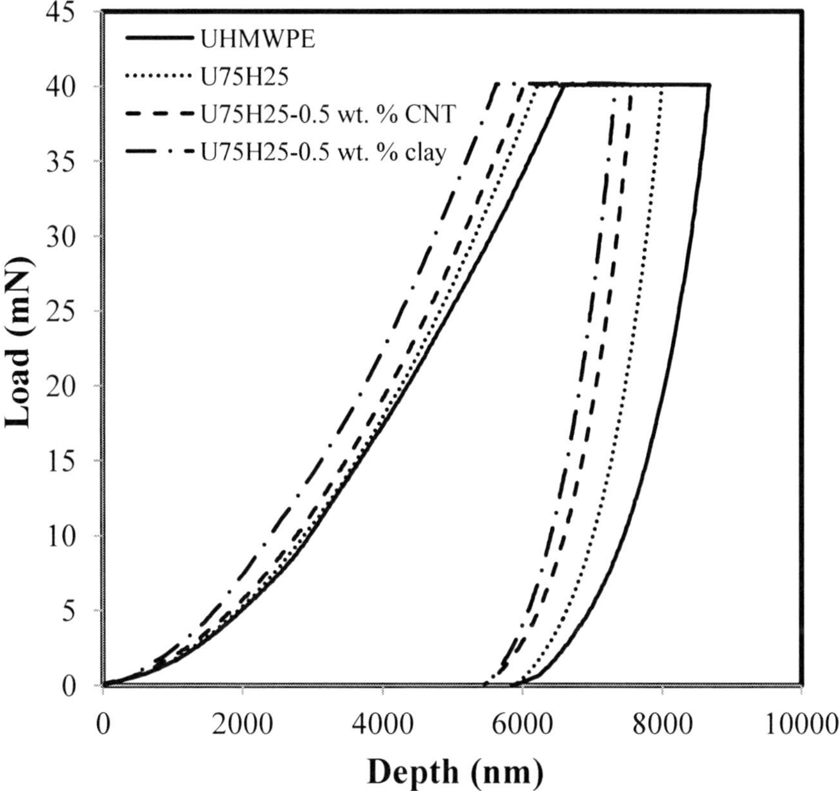

Figure 4: Comparison of the effect of nanoparticle on the nanoindentation behaviour.

3.3 Nanoindentation creep results

Fig. 5 shows the effect of CNT and nanoclay fillers on the creep behaviour of U75H25 and compares them with UHMWPE. It can be seen that the addition of 25 wt.% HDPE increases the creep resistance of UHMWPE. A further increase in creep resistance can be realised by the addition of CNT or nanoclay. The presence of CNT and nanoclay platelets can inhibit the creep deformation of the blend chains, when subjected to indentation pressure. This also indicate a good load transfer between the blend matrix and the nanoparticles. The improvement in the creep resistance for the nanocomposite materials can also be attributed to the large aspect ratio of the nanoparticle and the interfacial area between nanoparticle and U75H25 matrix, which enhance load transfer and restrict the mobility of the chains.

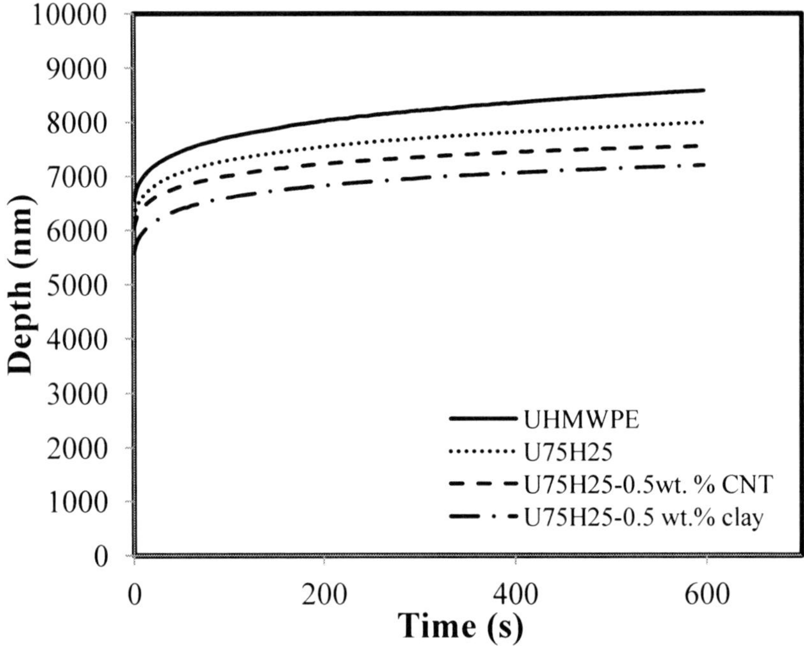

Figure 5: Effect of nanoparticle on the creep resistance of the polyethylene blend.

3.4 Scratch and wear resistance

Typical scratch depth profiles from the single and multi-pass scratch tests can be seen in figs 6 and 7, respectively. It is clear that the addition of 0.5 wt.%

nanoclay significantly increases the scratch resistance of the blend material. This can be seen from the reduction in the scratch depth over the scan distance. The addition of CNT shows a slight improvement in the scratch resistance at higher load. The addition of nanoclay induces better wear resistance, which can be evaluated from the plastic depth after 11 scratches, as shown in fig. 7. It can be seen that the scratch and wear resistance are closely related, the nanocomposite with higher scratch resistance also showing improvement in the wear resistance. Different factors can affect the performance of nanocomposites conducted to scratch or wear tests. These include the dispersion (the aspect ratio), the interfacial area, the interaction bonding between nanoparticle and polymer matrix and the exfoliation or intercalation in case of nanoclay platelets.

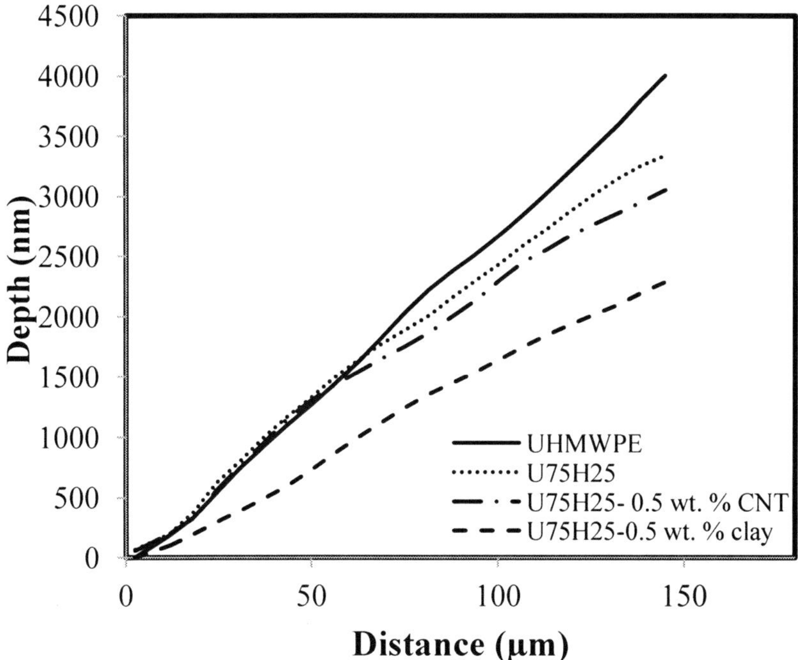

Figure 6: Scratch depth profiles for UHMWPE, blend and nanocomposites at 15μm/s scan velocity.

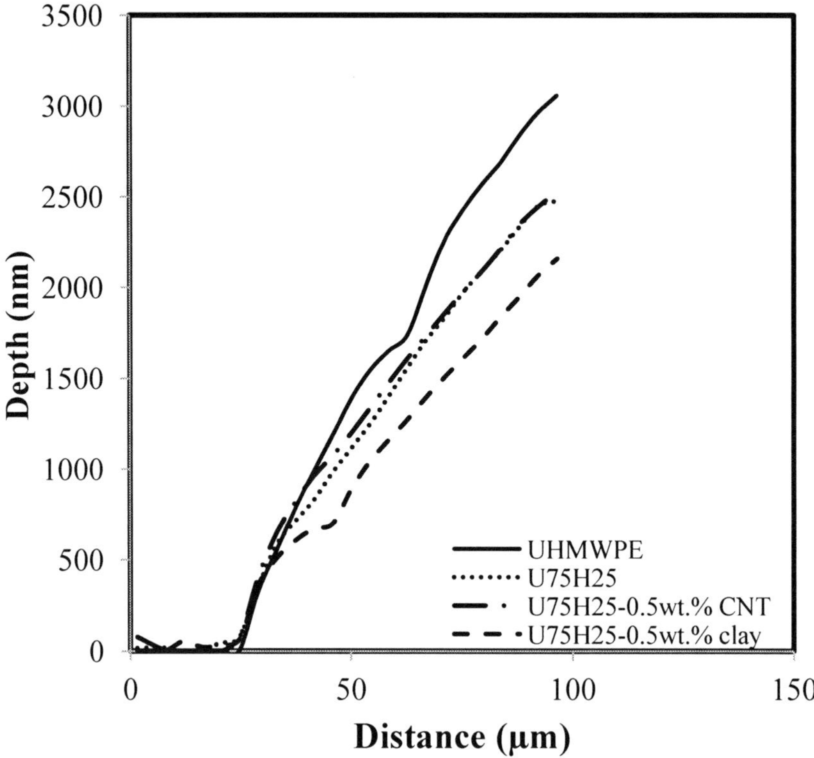

Figure 7: Wear depth profile for UHMWPE, blend and nanocomposites at 10µm/s scan velocity.

4 Conclusions

In this work, the near surface mechanical properties of UHMWPE, U75H25 and nanocomposites with CNT and nanoclay were investigated by nanoindentation. Indentation hardness values were used to evaluate the dispersion of nanoparticle into the U75H25 matrix manufactured using two different methods. Creep, scratch and wear tests were carried out to study the effect of nanoparticle addition. It was found that the addition of a low volume fraction of nanoclay (0.5 wt.%) can significantly increase creep, scratch and wear resistance of the polyethylene nanocomposites. This was attributed to the improvement in the dispersion and exfoliation of nanoclay and also the interaction between the platelets and the U75H25 matrix.

www.witpress.com, ISSN 1743-3533 (on-line)

References

[1] Kelly, J.M., Ultra-high molecular weight polyethylene. *Journal of Macromolecular Science, Part C: Polymer Reviews*, 42(3), pp. 355-371, 2002.

[2] Lucas, A.A., Ambrósio, J.D., Otaguro, H., Casta, L.C. and Agnelli, J.A.M., Abrasive wear of HDPE/UHMWPE blends. *Wear*, 270, pp. 576-583, 2011.

[3] Lim, K.L.K., Mohd-Ishak, Z.A., Ishiaku, U.S., Fuad, A.M.Y., Yusof, A.H., Czigany, T., Pukanszky, B. and Ogunniyi, D.S., High-density polyethylene/ultrahigh-molecular-weight polyethylene blend. I. The processing, thermal, and mechanical properties. *Journal of Applied Polymer Science* 97, pp. 413-425, 2005.

[4] Chen, Y., Qi, Y., Tai, Z., Yan, X., Zhu, F. and Xue, Q., Preparation, mechanical properties and biocompatibility of graphene oxide/ultrahigh molecular weight polyethylene composites. *European Polymer Journal,*48, pp. 1026-1033, 2012.

[5] Ren, P.G., Di, Y.Y., Zhang, Q., Li, L., Pang, H. and Li, Z.M., Composites of Ultrahigh-Molecular-Weight Polyethylene with Graphene Sheets and/or MWCNTs with Segregated Network Structure: Preparation and Properties *Macromolecular Materials and Engineering*, 297, pp. 437-443, 2012.

[6] Sui, G., Shong, W.H., Ren, X., Qang, X.Q. and Yang, X.P., Structure, mechanical properties and friction behavior of UHMWPE/HDPE/carbon nanofibers. *Materials Chemistry and Physics*, 115, pp. 404-412, 2009.

[7] Kontou, E. and Niaounakis, Thermo-mechanical properties of LLDPE/SiO_2 nanocomposites. *Polymer*, 47, pp. 1267-1280, 2006.

[8] Zoo, Y.S., An, J.W., Lim, D.P. and Lim, D.S., Effect of Carbon Nanotube Addition on Tribological Behavior of UHMWPE *Tribology Letters*, 16(4), pp. 305-309, 2004.

[9] Zhenhua, L. and Yunxuan, L., Mechanical and Tribological Behaviour of UHMWPE/HDPE Blends Reinforced with SBS. *Polymer-Plastic Tech. & Eng*., 51, pp. 750-753, 2012.

[10] Xue, Y., Wu, W., Jacobs, O. and Schädel, Tribological behaviour of UHMWPE/HDPE blends reinforced with multi-wall carbon nanotubes *Polymer Testing*, 25, pp. 221-229, 2006.

[11] Stoeffler, K., Lafleur, P.G., Florence, P.S., Bureau, M.N. and Denault, Micro-mechanisms of deformation in polyethylene/clay micro- and nanocomposites *J. Composites: part A*, 42, pp. 916-927, 2011.

[12] Fischer-Cripps, A.C., *Nanoindentation*, Springer-Verlag, NY, 2002.

[13] Aldousiri, B., Dhakal, H.N., Onuh, S., Zhang, Z.Y. and Bennett, N., Nanoindentation behaviour of layered silicate filled spent polyamide-12 nanocomposites. *Polymer Testing*, 30, pp. 688-692, 2011.

[14] Yusoh, K., Jin, J. and Song, M, Subsurface mechanical properties of polyurethane/organoclay nanocomposite thin films studied by nanoindentation. *Progress in Organic Coatings,*67, pp. 220–224, 2010.

[15] Wang, Z.Z., Gua, P. and Zhang, Z., Indentation and scratch behavior of nano-SiO_2/polycarbonate composite coating at the micro/nano-scale. *Wear,* 269, pp. 21–25, 2010.
[16] Sabic, Saudi Arabia, http://www.sabic.com
[17] ExxonMobil Chemical Europe, Belgium, http://www.exxonmobil chemical.com
[18] Elementis Specialties, USA, http://www.elementis-specialties.com.
[19] Nanocyl, Belgium, http://www.nanocyl.com
[20] Sigma-Aldrich, UK, www.sigmaaldrich.com
[21] Oliver, W. C. and Pharr, G. M., An improved technique for determining hardness and elastic modulus using load and displacement sensing indentation experiments. *J. Mater. Res*, 7:6, pp. 1564-1583, 1992.

Determination of frequency and temperature dependent mechanical material properties by means of an Inverse Method

J. Ilg, S. J. Rupitsch & R. Lerch
Chair of Sensor Technology, Friedrich-Alexander-University, Erlangen-Nuremberg, Germany

Abstract

We present a method to determine the frequency as well as the temperature dependence of mechanical material parameters. In general, we apply a so-called Inverse Method that adapts simulation results to match the best possible measurements. The variable quantities within the simulation are the sought-after material parameters, namely the elasticity modulus, Poisson's ratio, and a damping factor. Measurements are carried out by applying forced mechanical vibrations over a wide frequency range from quasi static up to more than 5 kHz. In particular, an electromechanical shaker harmonically excites tensile or bending vibrations of clamped plates and cylindrically shaped specimens. Two Laser Doppler vibrometers measure the vibration right next to the clamping and at the free end of the specimen, yielding a frequency dependent transfer function by relating the two measurands. Since the experimental setup is built up within an environmental chamber, temperatures from -40 to 150°C can be applied. Based on the experiment and the actual measuring points, a finite element (FE) analysis is performed to simulate the transfer function. Finally, the Inverse Method iteratively adapts the sought-after material parameters in such a convenient way that the simulated transfer function matches the measured one. The presented method was utilized to investigate the frequency and temperature dependence of different material classes: silicon rubber, plastics, metals, ceramics, and glass fibre reinforced plastics. In order to quantify and compare the dynamic material behaviour, functional relations of elasticity modulus/damping factor versus temperature are determined.
Keywords: elasticity modulus, damping factor, forced vibration testing, frequency dependence, temperature dependence, mechanical material properties.

www.witpress.com, ISSN 1743-3533 (on-line)
doi:10.2495/MC130091

1 Introduction

The development and design of every technical device demands for reliable material parameters of all involved materials. First and foremost, the parameters are used for the calculation of the structural capability. Furthermore, the increasing usage of computer aided engineering requests precise material parameters. Especially for sensors and actuators, an exact knowledge of mechanical as well as electrical quantities yields reliable simulations and ensures the functionality of a device. For the identification of the material parameters of piezoceramic transducers, we already introduced a so-called Inverse Method [1, 2]. Based on this method, the present contribution deals with a similar procedure to investigate the dynamic as well as thermal dependencies of mechanical material properties, namely elasticity modulus, Poisson's ratio, and a damping factor.

In most cases, the mechanical properties for a material – e.g. provided by the manufacturer – are given as static values, typically arising from tensile or bending tests. Other possible methods are the indentation of defined tips and the acoustical logging of ultrasound pulses. It is well known that material properties more or less depend on temperature and frequency. Moreover, the so-called Williams-Landel-Ferry (WLF) shift constant gives a nonlinear relation between these two dependencies [3]. Especially plastics show a significant sensitivity due to frequency and temperature. Consequently, many applications could benefit from a more precise knowledge of these relations. Some publications already dealt with this topic, mostly by exciting a specimen with a defined force and calculating material parameters with analytical descriptions (e.g., [4, 5]). Although these approaches exhibit very short computing times, they are exclusively applicable to specific sample geometries and demand for well defined experimental setups. Just as other research [6, 7], we overcome this problem by means of using finite element simulations instead of analytical relations. In principle, these methods allow for arbitrary sample geometries und various measurements quantities at the cost of computational effort. In contrast to the existing publications on this topic, our novel method mainly benefits from utilizing both, a proven and tested optimization algorithm as well as an adapted finite element method. In addition, the experimental setup covers a wider frequency and temperature range compared to previous researches. Our first approach for determining frequency dependent material parameters was published in [8, 9] applied to cylindrically shaped specimens made of silicone rubber. The present paper mainly deals with thin-walled components, namely polymer, ceramic, and aluminium plates.

The paper is organized as follows: In Sec. 2 the experimental setup to measured frequency resolved transfer function is explained. The utilized FE model and approach is briefly discussed in Sec. 3. After these fundamentals, the applied Inverse Method to determine frequency dependent material properties is described with different sub-steps (Sec. 4). Section 5 gives examples of possible investigations with the presented method and lists the results. Finally, the paper is summarized in Sec. 6, including a short outlook to our future research.

2 Experimental setup

Figure 1 shows the experimental setup to measure a vibration transfer function between two points on a test sample. The clamped specimen is harmonically excited by an electromechanical shaker (TIRA® S 5200-120) under defined environmental condition (CTS® CV-70/200). Two laser Doppler vibrometers measure the out-plane velocities v_1 as well as v_2 and the frequency resolved transfer function $\mathbf{H}_\mathrm{M}$ is given by relating the amplitudes $\hat{v}_1$ and $\hat{v}_2$:

$$|\mathbf{H}_\mathrm{M}(f)| = \frac{\hat{v}_2}{\hat{v}_1}\bigg|_f \tag{1}$$

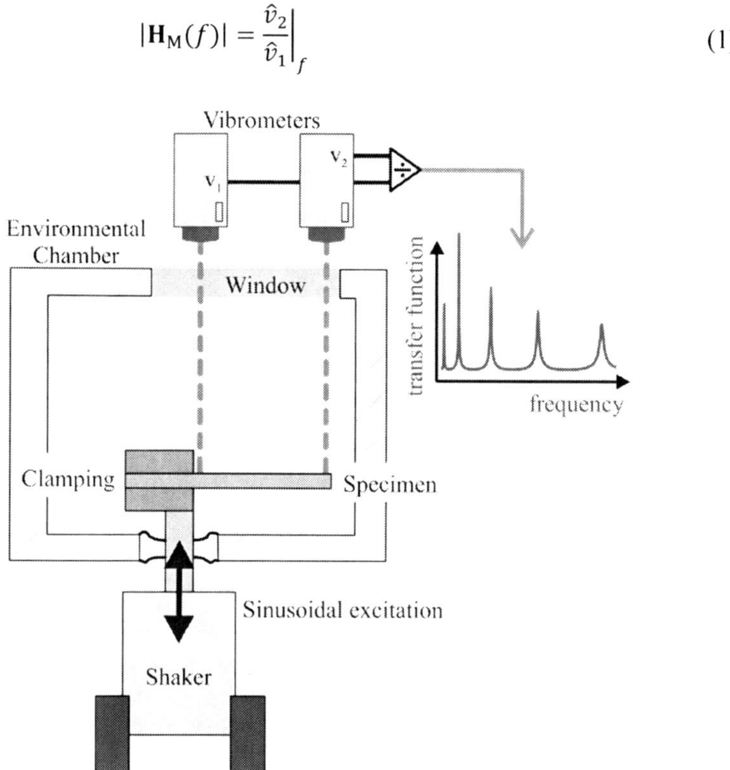

Figure 1: Experimental setup.

Although the measurement points can be chosen arbitrarily, v_1 and v_2 are acquired near the clamping and the specimens' free end in order to maximize the resonances magnification of $\mathbf{H}_\mathrm{M}(f)$. The automation and signal processing is carried out with LabVIEW® and a NI PXIe-1071 data acquisition system (National Instruments Germany GmbH).

Since thin walled components play a decisive role for constructing devices as lightweight as possible, the method is concentrating on plate shaped specimens. In principle, the measurements can be performed on various sample geometries,

though it fits especially to rotationally symmetric specimens as the computational effort (Sec. 3) can be reduced in this case [8, 9]. Figure 2 displays the three different possible excitations of a clamped beam in single direction: bending around *x*-axis (a), bending around *z*-axis (b), tension in *y*-direction (c). Cylindrically shaped samples are exclusively excited in z-direction (d). Note that the definition of the coordinate system is given in Fig. 3 and is always related to the specimen. One major problem, especially for the cases (b) and (c), is the occurrence of undesired vibrations in other directions than movement of the shakers base. The mass centre of clamping and sample do not exactly coincide with the shakers axis for various specimens. This results in an increasingly swaying movement of the shakers base and especially the free end of the sample for higher frequencies. In order to exclusively measure the vibration v_2 in the direction of the excitation v_1, an in-plane laser Doppler vibrometer is utilized (Polytec® LSV-065-306F). Velocity v_1 is always detected by means of an out-of-plane vibrometer (Polytec® OFV-303) that measures the surface normal velocity. The type of vibrometer is indicated with a double-pointed arrow in Fig. 2.

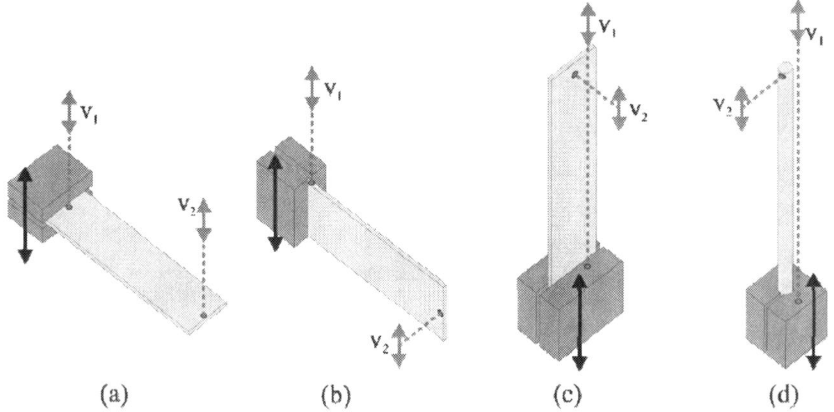

Figure 2: Different cases of excitation for the two sample geometries: bending of a plate over short edge (a) and long edge (b), tension of a plate (c), tension of a cylinder (d); v_1 and v_2 represent the two measurands.

The measurement procedure is organized as follows: At first, the desired temperature is committed to the chamber control and the actual temperature is measured by means of a PT100 near to the free end of the sample. When this sensor detects the achievement of the target temperature, a holding time of 30 minutes is applied to ensure a homogeneous temperature distribution within the specimen. Subsequent, the transfer function is acquired by measuring v_1 and v_2 for each entry in a given frequency vector. Note that these amplitudes are calculated by averaging over 100 single sinusoidal vibrations. The limitation of the frequency range depends on various quantities. First, the whole shaker setup exhibits resonance frequencies as well, depending on the masses of plunger, clamping, and specimen (manufacturer data for unloaded base: >5 kHz). Consequently, the input power to the shaker has to be increased when the first anti-resonance occurs to provide a significant movement of the clamping. Second, the vibrometers introduce some limitations, namely a lower frequency limit and a specific sensitivity that determines the possible resolution of the measurements. Typically the frequency range is limited to 10 Hz – 5 kHz, but especially for materials with a low damping factor, measurements could be carried out up to more than 10 kHz (see Sec. 5).

3 Finite element models

Additionally to the measurements, the presented Inverse Method is based on FE simulations. Therefore, a finite element model is needed that fulfils both, precise simulation results as well as minimum computing time. For the standard FE method (*h*–version) based on finite elements with second order shape functions the aspect ratio of an element is limited to avoid locking effects (see e.g. [10]). Since mainly thin walled samples are investigated, this standard method would result in a high amount of elements and as a consequence in high computing times. On account of this fact, the presented method uses FE models with hierarchic shape functions (*p*–version) of higher order ($p > 2$). Additionally, the utilized non-commercial FE tool CFS++ [11] allows for the implementation of anisotropic shape functions – here, the order in *x*- and *y*-direction is chosen higher than along the thickness. For detailed information on this topic, we refer to [12–14]. The order *p* of the shape functions is chosen by means of increasing *p* until the simulated transfer function converges towards the solution with a very fine standard FE mesh. Note that for this parameter study, the transfer function contains five resonances. An example of a FE model optimized for the *p*–version is given in Fig. 3 (left). In case of cylindrical specimens the standard *h*–version with second order shape functions is applied and the resulting model is exemplarily displayed in the right drawing of Fig. 3.

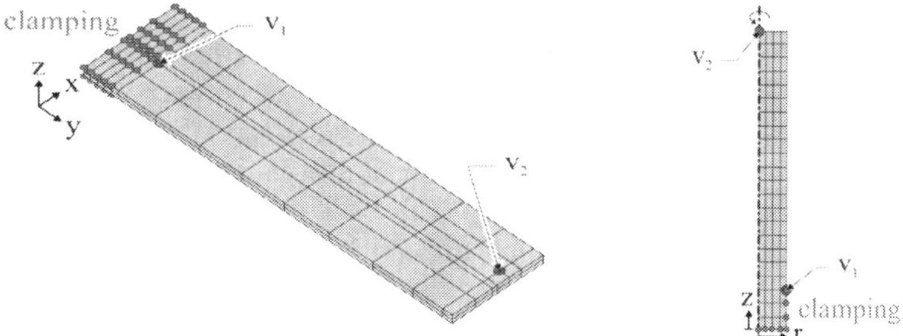

Figure 3: Finite element models for a plate shaped specimen (left) and for a cylindrical sample (right).

The simulated frequency resolved transfer function $\mathbf{H}_S(f)$ is calculated equally to the measured one (Eq. (1)). For a plate, the nodes at the bottom and the top side – with respect to z-direction – between the clamping are harmonically excited in one direction and fixed in the others. The excitation for cylindrical models is defined corresponding to the applied clamping as well (compare Figs. 2 and 3).

The material of the FE models is assumed to be globally homogeneous and isotropic. The variable material parameters are the real part of the elasticity modulus E, the damping factor ξ, and the Poisson's ratio ν. Typically, the complex Young's modulus is given by

$$\underline{E} = E' + \mathrm{j}E'' \tag{2}$$

and the damping factor or loss factor $\tan(\delta)$ is defined by the quotient between imaginary E'' and real part E':

$$2\xi = \tan(\delta) = \frac{E''}{E'} \tag{3}$$

Note that due to readability, the real part of the elasticity modulus is denoted as $E = E'$. Considering the three different cases of excitation (Fig. 2), E corresponds to the bending/flexural modulus or a tensile modulus. Only in case of isotropic material samples, E represents the Young's modulus.

4 Inverse Method

The presented measurements und FE models are necessary input quantities for the so-called Inverse Method. This procedure is implemented in MATLAB® (The MathWorks, Inc.) and can be divided into the three sub-steps described in the following. Additionally, the result of each step is given in Fig. 4. Within every step, one or more material parameters are adapted in order fit simulation results towards the measurements.

4.1 Eigenfrequency optimization

At first, the elasticity modulus E is adapted by means of two eigenfrequency simulations and the measured resonance frequencies that can be obtained from the measured transfer function. Based on the difference between two simulated eigenfrequencies $f_{\mathrm{S,n}} - f_{\mathrm{S,n}}^{*}$ with respect to a given variation of the moduli $E_{\mathrm{n}} - E_{\mathrm{n}}^{*}$, a proper initial guess $E_{\mathrm{init,n}}$ for the n-th eigenfrequency $f_{\mathrm{M,n}}$ is calculated by assuming proportionality. This first step provides the following optimization with a proper initial guess of E for every resonance (see Fig. 4) that was found in the measured transfer function. The major benefit of this preliminary step is that the whole procedure can start with a completely unknown set of material parameters.

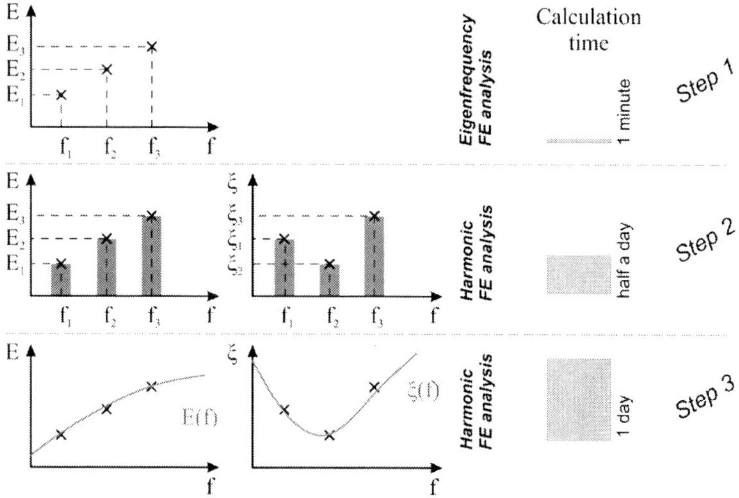

Figure 4: Different steps of the parameter identification procedure: results, types of analysis, and exemplarily calculation times.

4.2 Optimization of the transfer function

Within the second step, the transfer function around the resonances is considered for the optimization based on an iteratively regularized Gauss-Newton method (see e.g., [2, 9, 15]). This algorithm minimizes the difference between the measured and the simulated transfer function in the least square sense. For every iteration (i) the correction vector $\mathbf{s}^{(i)}$ has to be calculated that is given by [16] as

$$\mathbf{s}^{(i)} = -\left[\mathbf{F}'(\mathbf{p}^{(i)})^t\mathbf{F}'(\mathbf{p}^{(i)}) + \alpha^{(i)}\mathrm{I}\right]^{-1} \cdot \left[\mathbf{F}'(\mathbf{p}^{(i)})^t\mathbf{F}(\mathbf{p}^{(i)}) + \alpha^{(i)}\left(\mathbf{p}^{(i)} - \mathbf{p}^{(0)}\right)\right] \tag{4}$$

The vector $\mathbf{p}^{(i)}$ represents the parameters for the i-th iteration and $\mathbf{I}$ stands for the identity matrix. $\alpha^{(i)}$ is the so-called Lagrange parameter that tends towards zero for an increasing (i). The vector $\mathbf{F}(\mathbf{p}^{(i)})$ denotes the difference between the simulated and the measured transfer function:

$$\mathbf{F}(\mathbf{p}^{(i)}) = \mathbf{H}_S(\mathbf{p}^{(i)}) - \mathbf{H}_M. \tag{5}$$

$\mathbf{F}'(\mathbf{p}^{(i)})$ stands for the Jacoby-matrix with respect to the parameter vector $\mathbf{p}$ and is given by:

$$\mathbf{F}'(\mathbf{p}^{(i)}) = \left.\frac{\partial \mathbf{F}(\mathbf{p})}{\partial \mathbf{p}}\right|_{\mathbf{p}=\mathbf{p}^{(i)}} = \left.\frac{\partial \mathbf{H}_S(\mathbf{p})}{\partial \mathbf{p}}\right|_{\mathbf{p}=\mathbf{p}^{(i)}} \tag{6}$$

For step 2, the parameter vector $\mathbf{p}$ contains only the elasticity modulus and the damping factor. At the end of this step, an optimum parameter pair of E_n and ξ_n is given for every n-th resonance frequency $f_{M,n}$.

Figure 5 gives an example of the resulting transfer function around one resonance. Two statements can be made by means of this graph: First, the eigenfrequency analysis in step one provides a good initial guess that is close to the measured resonance. Second, the Inverse Method performs an exact curve fitting even in the case of noisy measurement data. Note that measured transfer functions are typically less noisy and this example represents the worst case.

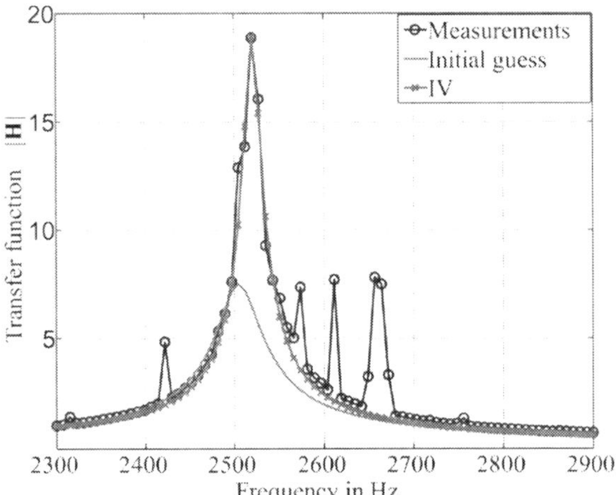

Figure 5: Result for a single resonance frequency after the second step including the initial guess obtained from step one.

4.3 Determination of functional relations

In order to use the determined material parameters for forward simulations, functional relations with respect to frequency are estimated in the least squares sense. Since plastics often show a logarithmic behaviour, we chose the following arbitrary functions:

$$E(f) = a_1 + a_2 \cdot f + a_3 \cdot \log(f + 10) \tag{7}$$

$$\xi(f) = b_1 + b_2 \cdot f + b_3 \cdot \log(f + 10) \tag{8}$$

$$\nu \neq \nu(f) \tag{9}$$

Finally, these functions can again be optimized with the above mentioned Gauss-Newton algorithm (see Sec. 4.2) considering the entire measured frequency range. Within this last step, the six coefficients $a_{1,2,3}$ and $b_{1,2,3}$ are the sought-after and optimized material parameters $\mathbf{p}$. This last step is not essential as it gives just a refinement to the results of step 2.

Note that the coefficient a_1 represents the static behaviour and corresponds to the value obtained form a static tensile or bending test. Therefore, if a_1 is well known, it can be restricted to this measured quantity to reduce the amount of unknown parameters $\mathbf{p}$ within the Inverse Method [9].

5 Results

This section gives examples of measurements that can be carried out with the presented experimental setup and the Inverse Method. First, Fig. 6 shows the temperature dependent bending modules of a high-pressure die-casted aluminium plate identified with excitation case (a) (see Fig. 2). Here, the results are obtained only from the first step of the Inverse Method. Although, the accuracy of the determined elasticity modulus depends on the frequency resolution, a good indication of the temperature dependence is given.

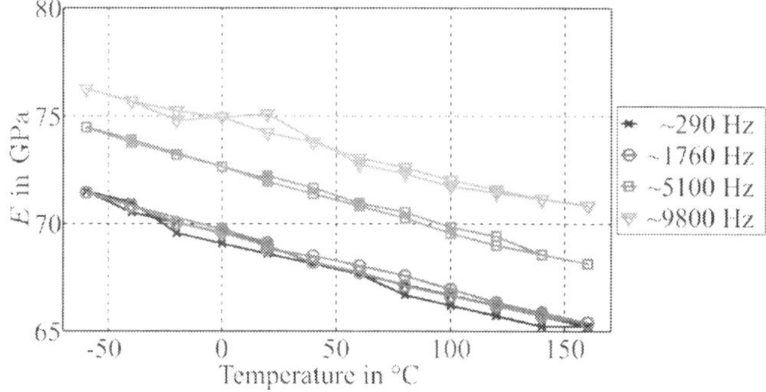

Figure 6: Temperature dependent bending modulus for the first four eigenfrequencies (Sample: high pressure die casted aluminium, 150mm x 10mm x 6mm).

The second example is given in Fig. 7. The tensile modulus as well as the damping factor of a polyetherimide (PEI) rod is given over temperature. The modulus shows a nearly linear behaviour with a marginal hysteresis. The linearity matches to the fact that the glass transition temperature is much higher (manufacturer: ~215°C). The damping exhibits a minimum around zero degrees.

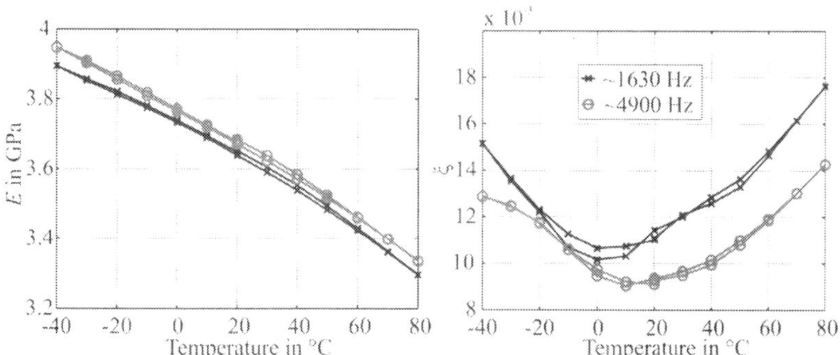

Figure 7: Temperature dependent tensile modulus (excitation case d, first two resonances; Sample: PEI rod 300mm x Ø12mm).

The last example is given in Fig. 8, investigating a glass fibre reinforced polypropylene (GFR-PP) plate. Here, the frequency dependent bending modulus is given for different temperatures. One can see that with increasing temperature, the modulus is decreasing much more with frequency. The glass transition temperature of PP is around 0°C, which can be observed in the wide distance to the -20°C and +20°C curves.

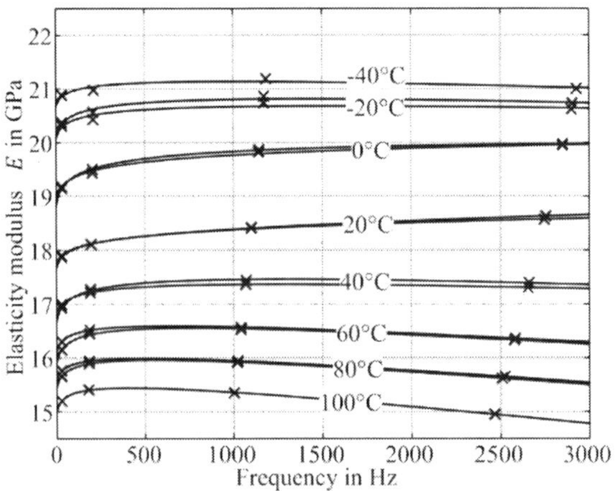

Figure 8: Frequency dependent bending modulus (excitation case a) for different temperatures (Sample: glass fibre reinforced polypropylene, TEPEX® dynalite104 2mm).

Finally, the frequency dependence of the material properties is quantitatively expressed according to Section 4.3. The determined coefficients of functional relations for various investigated materials are summarized in Tab. 1. The value f_{max} indicates the upper frequency limit of the measurements and consequently gives the range in which the functions are valid.

Table 1: Coefficients of the functional relations (Eqs (4) and (5)) for various investigated specimens, representing the bending modulus (case a) at room temperature (20°C).

	a_1 (Pa)	a_2 (Pa·s)	a_3 (Pa)	b_1	b_2 (s)	b_3
PEI	$3.08\cdot10^{9}$	$-4.85\cdot10^{4}$	$1.21\cdot10^{8}$	$7.11\cdot10^{-2}$	$1.31\cdot10^{-5}$	$-2.51\cdot10^{-2}$
GFR-PP	$1.74\cdot10^{10}$	$1.44\cdot10^{4}$	$4.09\cdot10^{8}$	$3.20\cdot10^{-2}$	$-2.37\cdot10^{-6}$	$-6.20\cdot10^{-3}$
Aluminium	$6.83\cdot10^{10}$	$6.08\cdot10^{5}$	$2.05\cdot10^{8}$	$2.90\cdot10^{-3}$	$6.90\cdot10^{-8}$	$-7.21\cdot10^{-4}$
	$\sim f_{max}$	length	width	thickness		
PEI	2.5 kHz	280 mm	40 mm	5 mm		
GFR-PP	3.0 kHz	200 mm	40 mm	2 mm		
Aluminium	10.0 kHz	150 mm	10 mm	6 mm		

6 Conclusion and outlook

We presented a novel method to determine the frequency as well as temperature dependence of the elasticity modulus and a damping factor. The experimental setup allows for dynamic bending and tensile loading of plate shaped specimens. In addition, the necessary finite element model is capable of producing reliable simulation results for thin-walled samples by applying the *p*–version of the FE method. Both simulations and measurements are the input quantities for the utilized Inverse Method, wherein simulation results are adapted appropriately to match the experimental data.

The determined material properties are quantified as functional relations versus frequency at room temperature for different specimens. The thermal dependencies of the elasticity modulus and the damping factor are graphically presented in different ways considering a single frequency or the whole frequency range. The investigated materials exhibit a more or less distinct frequency and temperature dependence. Especially for polymers – as one could expect – the dynamic and thermal properties have to be considered with a view to an accurate design of devices incorporating these materials. However, also aluminium shows a significant frequency and particularly temperature dependence. Well knowledge of both relations yields reliable simulation results and can help to reduce the weight of a device in a very early stage of development.

The focus of our future work lies on including Poisson's ratio in the Inverse Method and, furthermore, concentrating on transversally isotropic materials.

Acknowledgements

This research is supported by the Deutsche Forschungsgemeinschaft (DFG) in context of the Collaborative Research Centre/Transregio 39 PT-PIESA, subproject C6.

References

[1] Rupitsch, S. J. and Lerch, R.: Inverse Method to estimate material parameters for piezoceramic disc actuators. *Applied Physics A*, **97(4)**, pp. 735–740, 2009.

[2] Rupitsch, S.J., Ilg, J., Lerch, R., Enhancement of the Inverse Method Enabling the Material Parameter Identification for Piezoceramics. *Proceedings of the IEEE Ultrasonic Symposium*, Orlando, pp. 357–360, 2011.

[3] Ferry, J., *Viscoelastic Properties of Polymers*. Wiley-Interscience: Chichester, England, 1980.

[4] Madigosky, W.M. and Lee, G.F., Improved resonance technique for materials characterization. *The Journal of the Acoustical Society of America*, **73(4),** pp. 1374–1377, 1983.

[5] Gibson, R.F., Modal vibration response measurements for characterization of composite materials and structures. *Composites Science and Technology*, **60(15)**, pp. 2769–2780, 2000.

[6] Willis, R.L., Lei Wu, L. and Berthelot, Y.H., Determination of the complex Young and shear dynamic moduli of viscoelastic materials. *The Journal of the Acoustical Society of America*, **109(2)**, pp. 611–621, 2001.

[7] Kim, S.-Y. and Lee, D.-H., Identification of fractional-derivative-model parameters of viscoelastic materials from measured FRFs. *Journal of Sound and Vibration*. 2009, **324(3-5)**, pp. 570–586.

[8] Ilg, J., Rupitsch, S.J., Sutor, A. and Lerch, R., Determination of Dynamic Material Properties of Silicone Rubber Using One-Point Measurements and Finite Element Simulations. *IEEE Transactions on Instrumentation and Measurement*, **61(11)**, pp. 3031–3038, 2012.

[9] Rupitsch, S.J., Ilg, J., Sutor, A., Lerch, R. and Döllinger, M., Simulation based estimation of dynamic mechanical properties for viscoelastic materials used for vocal fold models. *Journal of Sound and Vibration*, **330**, pp. 4447–4459, 2011.

[10] Bathe, K.-J., *Finite Element Procedures*. Prentice Hall: New Jersey, 1996.

[11] Kaltenbacher, M., Advanced Simulation Tool for the Design of Sensors and Actuators. *Proceedings Eurosensors XXIV*, **5**, pp. 597–600, 2010.

[12] Duester, A., Broeker, H. and Rank, E., The p-version of the finite element method for three-dimensional curved thin walled structures. *Int. Journal for Numerical Methods in Engineering*, **52**, pp. 673–703, 2001.

[13] Šolín, P., Segeth, K. and Doležel, I., *Higher-Order Finite Element Methods*. Chapman and Hall: New York, 2004.

[14] Hauck, A., Kaltenbacher, M. and Lerch, R., Simulation of Thin Piezoelectric Structures Using Anisotropic Hierarchic Finite Elements. *Proceedings of the IEEE Ultrasonics Symposium*, Vancouver, pp. 476–479, 2006.

[15] Kaltenbacher, B., Neubauer, A. and Scherzer, O., *Iterative Regularization Methods for Nonlinear Ill-Posed Problems*. de Gruyter: Berlin New York, 2008.

[16] Rupitsch, S.J., Wolf, F., Sutor, A. and Lerch, R., Reliable modeling of piezoceramic materials utilized in sensors and actuators. *Acta Mechanica*, **223(8)**, pp. 1809–1821, 2012.

Vibration tests and metamodelling for composite material characterisation

S. Syngellakis[1] & R. Setiawan[2]
[1]*Wessex Institute of Technology, UK*
[2]*Institut Teknologi Bandug, Indonesia*

Abstract

The mechanical characterisation of laminates is achieved through a methodology based on vibration test measurements. The relation between effective mechanical properties and vibrational characteristics is initially generated by multiple finite element simulations and then approximated through the development of a surrogate model or meta-model. The inverse problem of finding the properties from given experimental data is solved through an iterative optimisation scheme within which the frequencies for a particular set of material parameters are predicted by the meta-model. This results in a versatile and time-saving characterisation process, which can be applied to a large number of specimens with a complicated geometry.
Keywords: composites, characterisation, vibration, experiments, meta-modelling.

1 Introduction

Advanced composites have become realistic alternatives to conventional materials for the design of safe and durable structures since they can meet the requirements for high strength- or stiffness-to-weight ratios and other structural integrity objectives. A reliable dynamic performance is another important requirement on composites in many applications. For an accurate analytical assessment of composite behaviour under loads, their mechanical properties need to be known with sufficient confidence and precision. Such properties can be determined by static tests, according to established procedures [1], which involve several specimens cut from the same material.

WIT Transactions on Engineering Sciences, Vol 77, © 2013 WIT Press
www.witpress.com, ISSN 1743-3533 (on-line)
doi:10.2495/MC130101

Early work on composite characterisation through dynamic tests was based on a relatively simple experimental technique according to which the composite properties are determined iteratively from experimentally obtained modal parameters; this approach was initially developed for and applied to orthotropic plates [2, 3]. The mode shapes associated with the measured natural frequencies may be included in the characterisation process. A direct approach [4, 5] required a precise mode shape measurement but another [6] only qualitative examination of the mode shapes associated with the measured natural frequencies. Various optimisation algorithms can be used for the solution of the inverse problem of determining properties from experimental vibration output [7–9].

The development of components for various applications demands composites with increasingly complicated geometry, such as laminates with stiffening, sandwich structures, plates with variable thickness. Moreover, mass production of composite components requires fast, inexpensive characterisation, preferably carried out in the laboratory, subsequent to the dynamic tests [10]. Existing strategies are computationally expensive when applied to samples with complex geometries.

This paper presents an alternative strategy for the solution to this problem by making use of meta-modelling to approximate the relation between natural frequencies and material properties. Such an approximate model is versatile with respect to geometric and material parameter selection and efficiently applicable to the required solution of an inverse problem, leading to fast, in-situ composite characterisation.

2 Methodology

The material property parameters and their ranges are initially selected. This process is either guided by or guides the preparation of the specimens. Dynamic tests are performed resulting in sets of measured natural frequencies. An FE program relating material to vibration parameters is built and validated by comparing its predictions to the experimental data. A number of material property sets are generated by a scheme that guarantees a good spread of data over their range. Multiple FE analyses are then carried out with these property sets as inputs and natural frequencies sets as outputs. Thus, data sets comprising both independent and dependent variables are generated and subsequently used to train meta-models. The latter are simple approximate surrogates of the more expensive and time consuming FE analysis. Finally, an optimisation algorithm is used for the solution of the inverse problem, namely the determination of material properties from available vibration data.

The adopted material parameter set consisted of the in-plane elastic constants E_x, E_y, G_{xy} and ν_{xy} of laminated, orthotropic plate specimens. A large number of such sets were generated according to a random sampling method, subject to the constraints that the material properties lie within specified ranges and the resulting stiffness matrix of the material is positive-definitive. For the latter condition to be satisfied, the four in-plane constants obtained from the sampling were complemented by assumed values for the through-thickness properties.

Each material set was input to the FE vibration analysis yielding a set of frequencies f_i. The rectangular plate specimens were modelled by a 12×12 mesh of quadrilateral, 8-node shell elements. The adopted element type is based on a first order shear deformation theory (FSDT), which is expected to give accurate natural frequency predictions for plates with moderate thickness to width ratio. It requires input of effective laminate properties, including the four in-plane elastic constants.

The independent variable set actually employed in the characterisation process comprised the values of the frequency parameter, defined by

$$\bar{f}_i = f_i h \sqrt{\rho} \tag{1}$$

where h is the thickness of the specimen and ρ the average density of the material.

The developed meta-models provided an approximation to the relation between material and vibration parameters generated by FE modelling. They were based on the Gaussian radial basis function (RBF), which takes the form,

$$y(\mathbf{x}) = \sum_{j=1}^{m} w_j \exp\left(-\frac{\left(\mathbf{x} - \mathbf{c}_j\right)^2}{r^2} \right) \tag{2}$$

where $y(\mathbf{x})$ is the approximated function, m is the number of basis functions; w_j are the weights; $\mathbf{x}$ is the array of input variables; $\mathbf{c}_j$ the array of the centres, and r is the width parameter, here taken as constant for all centres.

Eqn (2) was evaluated for each input-output pair of the training data with the centres being defined as coincident with the training data. The weights, w_j, can be found by minimising a cost function defined by the sum of squared errors (SSE) of model responses (here the natural frequencies) between the training data and those predicted by the meta-model. There are various techniques for improving the accuracy and efficiency of meta-models, including optimising the choice of training data to be used [11].

For the characterisation, the meta-models need to be based on specific specimen geometry, leaving only the material properties as the variables to be identified. For plate specimens, the length-to-thickness and aspect ratios can be considered constant, so that the same meta-model can be applied to specimens having such ratios with the same values. The thickness and density are incorporated in the meta-model output, that is, the frequency parameters as defined by eqn (1).

An accuracy test method, based on the R^2 accuracy measure, may be employed. This measure is given by

$$R^2 = 1 - \frac{\sum_{i=1}^{N} \left(y_i - \hat{y}_i\right)^2}{\sum_{i=1}^{N} \left(y_i - \bar{y}_i\right)^2} \tag{3}$$

where $y_i, \hat{y}_i, \bar{y}$ are the RBF predictions, true and the mean of the true values, respectively, observed at p validation points.

The optimisation problem involved the minimisation of the error function

$$\phi = \sum_{i=1}^{N} \frac{\left(\bar{f}_i - \tilde{f}_i\right)^2}{\tilde{f}_i^2} \tag{4}$$

where N is the number of frequencies being considered and $\tilde{f}_i$ the experimentally obtained frequency parameters. The search for the optimum solution to the characterisation problem was subject to the same constraints on material properties as those imposed on the selection of training data. These constraints were implemented within the optimisation algorithm using the external penalty function method, which tackles solutions that violate them by adding a penalty and forcing the solution in the direction of the feasible region. By introducing these external penalty terms and thus formulating a revised objective function, an unconstrained optimisation algorithm can be used to solve a constrained optimisation problem. An algorithm called BFGS (Broyden, Fletcher, Goldfarb, Shanno) [12], was used; this algorithm falls into the non-linear unconstrained optimisation category, based on the steepest descent method with global convergence capability.

3 Experiments

The vibration experiments were designed to simulate free edge conditions for rectangular plate specimens. Such conditions lead to accurate measurement of natural frequencies; for this reason, they have been widely adopted in the past despite the absence of an explicit solution to the corresponding theoretical problem.

Setting up a versatile experimental arrangement yielding accurate estimates of the lower natural frequencies required the design of supports approximating free-edge conditions and the selection of excitation apparatus, natural frequency measurement methods as well as appropriate sensing devices. These features should have a negligible constraint or inertia effect on the measured frequencies; if this is not possible, the effect should be compensated or accounted for in the modelling. The information on the mode shapes associated with the measured natural frequencies also contributes to the accuracy of the characterisation. Therefore, an experimental arrangement was achieved yielding this information.

One possible, simple method of replicating ideal free-edge conditions is the soft-pad support [3]. This reduces the dominance of the rigid body vibration mode but the imposed lateral constraint can be considerable even if a spiked-shape sponge is used to minimise contact between support and specimen. For this reason, the more widely used method of specimen suspension [2, 7, 13, 14], as shown in fig. 2, was also adopted here.

The experiments were carried out using a mechanical shaker, which imposed continuous pseudo-random excitation on the specimen and, as sensor, a laser vibrometer, which allows non-contacting sequential measurement at several points thus reducing the possibility of overlooking any vibration mode. The natural frequencies for each specimen were extracted from the average of all measurements taken from the entire specimen domain, not just from a single point. From the response spectra and phases of all points of measurement, the vibration mode shape at each frequency can be systematically deduced and viewed on the computer screen through post-processing analysis.

The laser vibrometer equipment consisted of a laser-emitter/digital-camera unit, signal processor/controller unit and a PC as a display, post processing and storage device as schematically shown in fig. 3. The plate was suspended from wires and the excitation was imposed at approximately the centre point of the lower left quadrant of the specimen. The shaker was screwed to the specimen to provide a sufficiently fixed attachment.

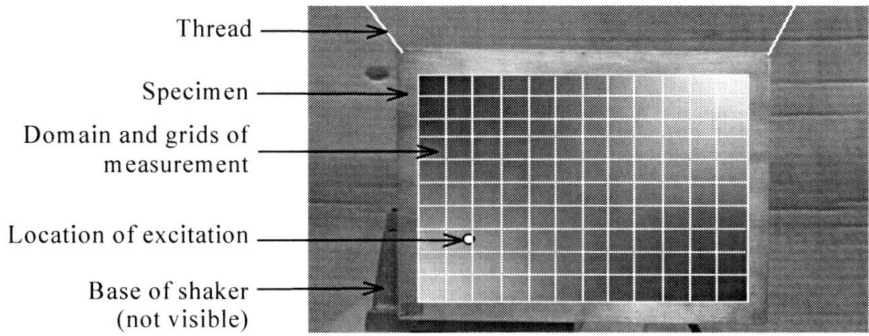

Figure 1: Dynamic response of suspended specimen as shown in the equipment display.

First, a pseudo-random excitation signal was defined within a certain frequency range. The image of the specimen was captured through the digital camera and displayed on the PC screen. Then, a piece of software within the PC unit (5) defined a mesh over the valid measurement area on the specimen as shown in fig. 2. The commands were then realised by the controller (4), to generate the excitation signal that was sent to the signal amplifier (6) and finally the shaker (2). The commands related to data acquisition were sent to the laser/camera unit (3).

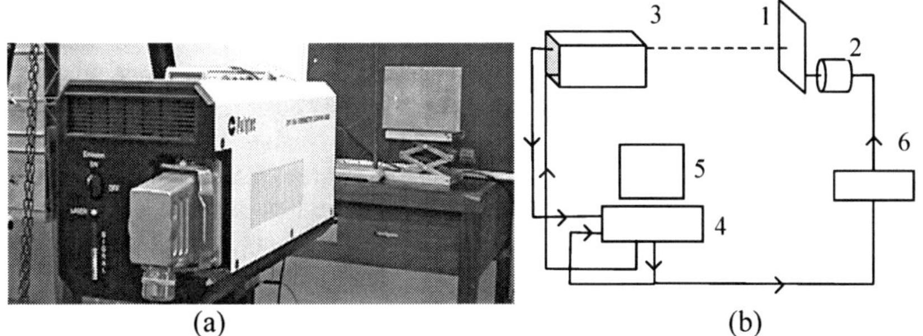

Figure 2: (a) Laser vibrometer and (b) schematic diagram of the corresponding set up: specimen (1), mechanical shaker (2), laser/camera unit (3), controller (4), PC unit (5), signal amplifier (6).

The effectiveness of the test arrangements to generate accurate frequency measurement was tested by using initially a 155×200×1.15 mm^3 aluminium plate as the specimen. The modulus of elasticity, shear modulus and Poisson's ratio were taken equal to 72.4 GPa, 28 GPa and 0.33, respectively. The natural frequencies were extracted from the average spectrum shown in fig. 4. As can be seen in this figure, the averaging process results in a high signal-to-noise ratio.

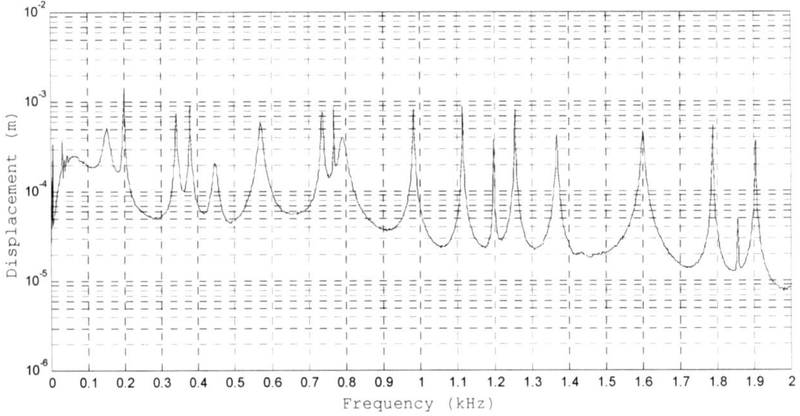

Figure 3: Average frequency response function for the aluminium specimen.

The frequency peaks are clearly identified and accepted as natural frequencies. Frequency peaks below 100 Hz were probably associated with either rigid body vibration modes or electromagnetic interference. The first significant frequency peak is at approximately 150 Hz. The mode shapes at that frequency and below indicated that the one at 150 Hz corresponded to the twisting mode, thus interpreted as the fundamental frequency; frequencies below 100 Hz corresponded to rigid body vibration modes.

The main issue with a mechanical shaker, attached to the specimen by a connector, is the possibility of the excitation having a significant effect on the plate vibration characteristics. For this reason, the experimental results were compared with predictions from two FE models, namely one including the modelling of the shaker-connector arrangement (FE 1) and another without it (FE 2). The frequencies extracted from fig. 4 are listed in table 1 together with the corresponding analytical results with their percentage differences given in square brackets. The mode shapes are shown in grey shade, the experimental ones from white to dark and the analytical ones from dark to white and to dark.

The results of table 1 clearly indicate that FE 1, that is, the FE model that includes the shaker-connector arrangement generally leads to predictions in better agreement with the experimental natural frequencies and associated vibration modes. With this model, the biggest discrepancy between theory and experiment occurs in the sixth mode (9.34%) whereas, neglecting the connector effect, the most significant prediction shift compared with the respective experimental result occurs in the ninth frequency, where the difference is 15.63%.

Table 1: Experimental and analytical natural frequencies and associated mode shapes of a suspended aluminium plate.

Mode	Natural frequency (Hz)			Mode shape		
	Experiment	FE 1	FE 2	Experiment	FE 1	FE 2
1	153.13	158.01 [3.19]	162.38 [-5.70]	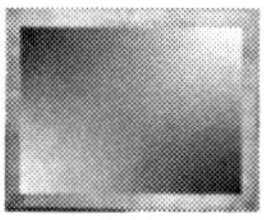	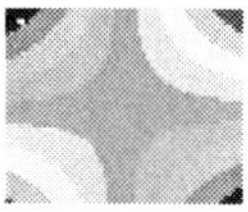	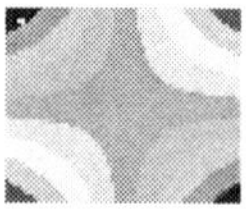
2	200.00	202.21 [1.11]	204.56 [-2.23]	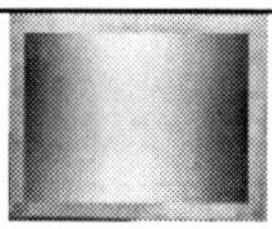	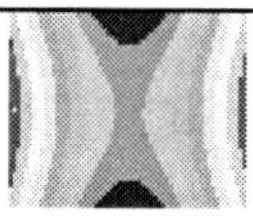	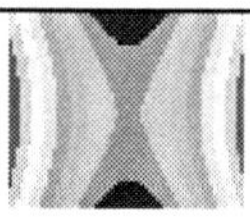
3	342.19	327.62 [-4.26]	345.82 [-1.05]	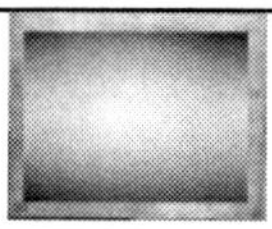	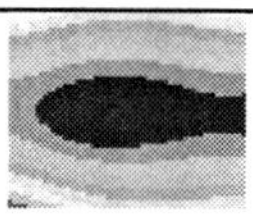	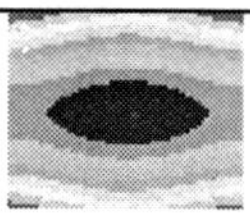
4	381.25	387.74 [1.70]	392.34 [-2.83]	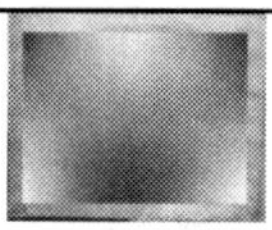	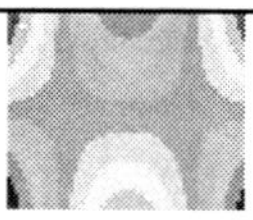	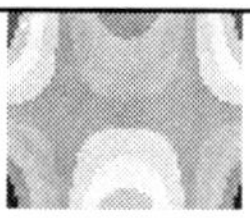
5	448.44	450.18 [0.39]	462.89 [-3.12]	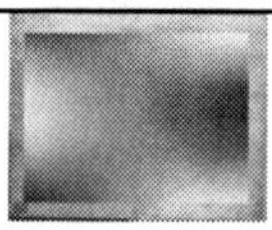	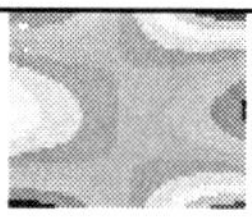	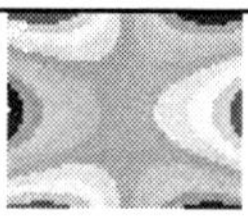
6	571.88	518.45 [-9.34]	598.93 [-4.52]	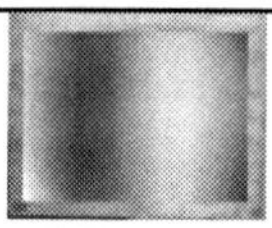	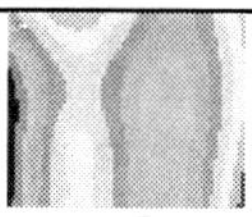	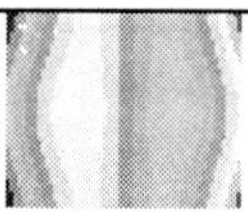
7	737.50	728.00 [-1.29]	761.52 [-3.15]	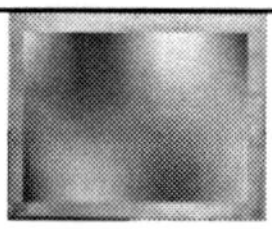	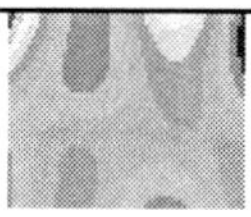	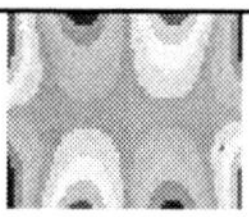
8	768.75	773.27 [0.59]	786.88 [-2.30]	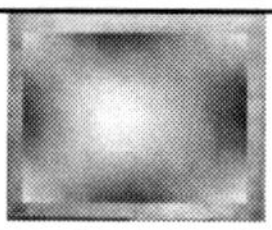	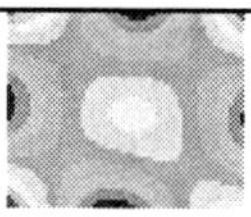	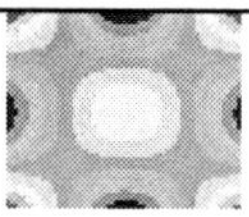
9	792.19	812.73 [2.59]	938.94 [-15.63]	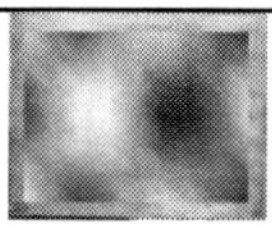	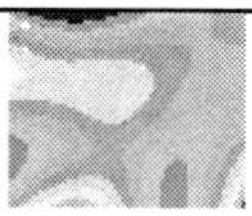	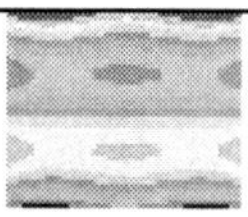
10	984.38	984.09 [-0.03]	1070.59 [-8.05]	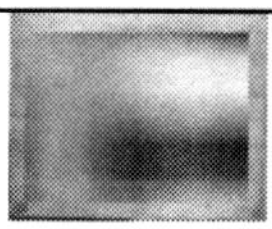	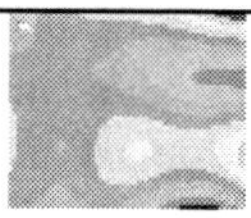	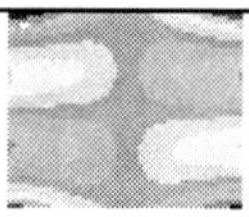

4 Characterisation results

The proposed characterisation strategy was applied, as a test case, to a generic glass-fibre/epoxy-resin based composite plate; this was consistent with the characteristics of the available specimens. The adopted ranges of material properties were

$$\begin{aligned} 2 \le E_x &\le 45 \qquad \text{(GPa)} \\ 2 \le E_y &\le 30 \qquad \text{(GPa)} \\ 1 \le G_{xy} &\le 20 \qquad \text{(GPa)} \\ 0.05 \le v_{xy} &\le 0.8 \end{aligned} \tag{5}$$

The bounds in the ranges (5) were specified according to available design data for this particular type of material so that they were as narrow as possible; this is expected to improve the accuracy of metamodelling.

The specimens were laminates with various ply arrangements and thicknesses; their properties were within the bounds (5). The characteristics of the four specimens tested are presented in table 2. The length-to-thickness ratio of the specimens was intended equal to either 55 or 110 while the aspect ratio was intended equal to 1.4. For this reason, two meta-models were built, based on training data generated using these values.

Table 2: Details of the four test laminate specimens.

Laminate	V_f	Description
A	0.46	3-layer 600 g/m^2 woven roving (WR)
B	0.19	3-layer 300 g/m^2 chopped strand mat (CSM)
C	0.33	6-layer 225 g/m^2 CSM / 600 g/m^2 WR (±45)
D	0.18	3-layer 600 g/m^2 chopped strand mat (CSM)

Measurement Data:

Laminate	h (mm)	a (mm)	b (mm)	Mass (g)	Density (kg/m^3)
A	1.68	204	145.5	82.5	1653.0
B	1.85	195	141	70.5	1387.1
C	2.92	161.5	117	85.3	1542.1
D	3.89	206.3	149.3	166.1	1387.0

As with the aluminium specimen, the natural frequencies of each composite specimen were extracted from the average of all measurements taken over the entire specimen domain, not just from a single point. The natural frequencies extracted from these average spectra are listed in table 3 for up to twelve modes, excluding the low frequencies at around 50 Hz. The frequency parameters were calculated from eqn (1) using the measured thickness and density values listed in table 2.

Table 3: Measured natural frequencies and respective frequency parameters.

Mode	Laminate A		Laminate B		Laminate C		Laminate D	
	f(Hz)	$\bar{f}$	f(Hz)	$\bar{f}$	f(Hz)	$\bar{f}$	f(Hz)	$\bar{f}$
1	92.50	6.324	107.50	7.401	267.50	30.700	200.00	28.949
2	137.50	9.400	131.25	9.036	298.75	34.286	239.06	34.603
3	233.75	15.980	241.25	16.609	328.75[b]	37.729	478.13	69.206
4	262.50	17.946	260.00	17.900	352.50[b]	40.455	504.69	73.051
5	315.00	21.535	326.25	22.461	603.75	69.290	604.69	87.525
6	372.50	25.466	368.75	25.387	673.75	77.323	720.31	104.261
7	470.00	32.131	490.00	33.734	771.25	88.513	918.75	132.984
8	492.50	33.669	497.50	34.251	926.25	106.301	1006.25	145.649
9	583.75	39.908	545.00	37.521	1227.50	140.874	1143.75	165.551
10	743.75	50.846	716.25	49.311	1326.25	152.208	1354.69	196.083
11	756.25	51.700	780.00	53.700	1422.50	163.254	1407.81	203.772
12	797.50	54.520	836.25	57.572	1673.75	192.089	1568.75	227.067

[b] Ambiguous results.

The experiments were not properly prepared to provide the output required for the application of the characterisation scheme accounting for mode shape information according to the modal assurance criteria (MAC) approach [11]. Thus, the characterisation of the tested laminates was carried out without rigorous consideration of experimentally obtained mode shapes.

The pairs of laminates (A, B) and (C, D) were characterised using the metamodels built with $a/h = 110$ and $a/h = 55$, respectively. Since however the a/h values for these specimens were not exactly equal to either 110 or 55, correction factors were introduced to compensate for such discrepancies. For this purpose, the variation of a/h with the frequency parameter was obtained using a Ritz vibration analysis based on classical laminate plate theory (CLPT); this meant that higher mode natural frequencies could have been over-estimated. The value of aspect ratio, a/b, of the tested specimens was very close to 1.4, that is, the value adopted in the metamodels, therefore the effect of its variation on the characterisation results was ignored. It is worth noting that a metamodel could have been built for each specimen adopting the exact values of the corresponding a/h and a/b ratios.

The optimum data for metamodels for both cases of a/h ratio can be found in table 4 where m is the number of optimum training data chosen from the 721 admissible training data, σ^2_{CP} is the error variance of the confirmation points prediction, $\bar{f}_{CP}^{\min}$ and $\bar{f}_{CP}^{\max}$ represent, respectively, the minimum and maximum values of the predicted frequency parameter from, e_m is the maximum error across the confirmation points, and R^2 is an accuracy measure defined by eqn (3).

Table 4: Number of optimum training data for metamodelling and selected measures of accuracy, for each mode.

Optimum metamodelling information for case I ($a/h = 110$)									
Mode no.	1	2	3	4	5	6	7	8	9
m	173	179	176	203	183	208	101	199	140
σ^2_{CP}	0.0195	0.0311	0.1069	0.1511	0.1438	0.3903	1.308	0.7499	1.3948
$\bar{f}^{min}_{CP}$	3.9273	5.6105	9.0927	12.844	14.222	17.04	21.951	24.559	30.134
$\bar{f}^{max}_{CP}$	16.642	17.85	35.943	38.509	43.336	64.16	71.447	81.731	98.545
e_m (%)	6.94	10.66	7.36	9.17	13.92	10.10	9.56	8.69	18.84
R^2	0.9979	0.9949	0.9962	0.9949	0.9953	0.9936	0.9864	0.9931	0.991
Optimum metamodelling information for case II ($a/h = 55$)									
m	192	207	200	220	158	115	168	175	173
σ^2_{CP}	0.2957	0.4618	1.5787	2.6703	2.8020	11.455	14.770	12.886	22.368
$\bar{f}^{min}_{CP}$	14.961	15.198	29.798	33.593	41.627	42.503	61.069	68.462	81.991
$\bar{f}^{max}_{CP}$	67.425	71.975	148.98	157.86	174.39	255.01	294.4	324.83	388.3
e_m (%)	7.08	8.81	9.30	10.33	13.17	14.62	11.74	9.01	18.18
R^2	0.998	0.9955	0.9966	0.9945	0.9943	0.9886	0.9905	0.9927	0.9911

Table 5 shows the characterisation results for the four tested specimens. Each result is the best among those obtained from the five initial guesses. Since there was some uncertainty concerning some frequency measurements, combinations of the most probable frequency parameters among those listed in table 3 were considered in the optimisation process. In the case of laminate C, for instance, the analysis was performed with and without the ambiguous results included in table 3. In such cases, the selection of the most plausible solution can be based on knowledge of the specimen's architecture and composition. The values for E_x and E_y obtained from static bending tests are included in table 5 for comparison; they should not however be considered as benchmarks for selecting the true solution. The characterisation result meets the expectations arising from the consideration of the tested laminate compositions and architecture.

Table 5: Characterisation results (static test values in brackets).

	E_x (GPa)		E_y (GPa)		G_{xy} (GPa)	ν_{xy}
Laminate A	17.67	(14.79)	12.96	(14.71)	3.98	0.05
Laminate B	8.19	(7.08)	6.01	(6.43)	4.57	0.09
Laminate C[a]	3.51 <u>8.99</u>	(7.27)	2.37 <u>10.77</u>	(6.87)	6.91 <u>5.66</u>	0.26 <u>0.15</u>
Laminate D	6.39	(7.46)	9.77	(7.66)	4.80	0.15

[a]The underlined predictions for laminate C were obtained with the two ambiguous measurements of 328.75 and 352.50 Hz excluded.

5 Conclusions

Each of the various experimental and analytical procedures constituting the proposed characterisation scheme may contribute errors to the final outcome. The mass of the moving part of the mechanical shaker and the stiffness of the connector affect the vibration characteristics of the specimen and need to be taken into account in the development of a reliable metamodel. Modelling the essential features of the shaker-connector arrangement for the case of aluminium specimen improved the agreement between experimental and analytical results. Thus, this modelling can be incorporated in the FE analysis for the training data generation. As an alternative, non-contact excitation, for instance, through a magnetic shaker or a speaker generating acoustic waves, would obviate the need for complex modelling and lead to experimental output very close to that predicted for vibrating specimens with free edges.

The use of a laser vibrometer makes possible the application of a scheme that uses experimental mode shape information; such a scheme has been successfully tested with simulated experimental data [11] and should be applied with vibration output from real tests. With regard to the signal processing, a higher frequency resolution would enhance frequency peaks and thus remove ambiguities from natural frequency readings from the frequency response functions. Specimen dimensions and density should be accurately measured and consistent with the respective values used in the metamodel.

The natural frequencies of plates with all-edges free have low sensitivity to Poisson's ratio variation [8]; this inherently leads to a poor result for this material parameter. Such a problem could be addressed by adopting a multi-step characterisation strategy, based on the assumption that natural frequencies have different sensitivity to material properties variation under different support conditions. An early attempt was made at experimental replication of simple-support boundary conditions for the dynamically tested specimens. However, it was not possible to achieve an experimental set-up yielding reliable frequency output corresponding to simple-support conditions. The multi-step strategy could still be feasible if a more sophisticated experimental apparatus replicating various combinations of support conditions could be devised. For the characterisation of through-thickness mechanical properties, the plate specimen should be sufficiently thick, and the dynamic analysis should take into account the thickness effects accurately.

References

[1] ASTM, ASTM *Standards and Literature References for Composite Materials,* ASTM: 1990.

[2] De Wilde, W.P. and Sol, H., Anisotropic material identification using measured resonant frequencies of rectangular composite plates. *Composite Structures 4*, ed. I.H. Marshal, Paisley College of Technology: Scotland, pp. 317-324, 1987.

[3] Deobald, L.R. and Gibson, R.F., Determination of elastic constants of orthotropic plates by a modal analysis/Rayleigh-Ritz technique. *Journal of Sound and Vibration*, **124(2)**, pp. 269-283, 1988.

[4] Grediac, M. and Paris, P.A., Direct identification of elastic constants of anisotropic plates by modal analysis: Theoretical and numerical aspects. *Journal of Sound and Vibration*, **195(3)**, pp. 401-415, 1996.

[5] Grediac, M., Fournier, N., Paris, P.A. and Surrel, Y., Direct identification of elastic constants of anisotropic plates by modal analysis: Experimental results. *Journal of Sound and Vibration*, **210(5)**, pp. 643-659, 1998.

[6] Gaul, L., Willner, K. and Hurlebaus, S., Determination of material properties of plates from modal ESPI measurements. *Proceedings of the 17th International Modal Analysis Conference*, Society for Experimental Mechanics: Bethel, pp. 1756-1752, 1999.

[7] Pedersen, P. and Frederiksen, P.S., Identification of orthotropic material moduli by a combined experimental/numerical method. *Measurement*, **10(3)**, pp. 113-118, 1992.

[8] Frederiksen, P.S., Numerical studies for the identification of orthotropic elastic constants of thick plates. *European Journal of Mechanics A-Solids*, **16(1)**, pp. 117-140, 1997.

[9] Sol, H., Hua, H., De Visscher, J., Vantomme, J. and De Wilde, W.P., A mixed numerical/experimental technique for the non-destructive identification of stiffness properties of fibre reinforced composite materials. *NDT and E International*, **30(2)**, pp. 85-91, 1997.

[10] Gibson, R.F., Modal vibration response measurements for characterization of composite materials and structures. *Composites Sciences and Technology*, **60**, pp. 2769 - 2780, 2000.

[11] Setiawan, R., Syngellakis, S. and Hill, M., A metamodeling approach to mechanical characterization of anisotropic plates. *Journal of Composite Materials*, **43(21)**, pp. 2333-2349, 2009.

[12] Kelley, C.T., *Iterative Methods for Optimization*, SIAM: Philadelphia, 1999.

[13] Araújo, A.L., Mota Soares, C.M. and Moreira de Freitas, M.M., Characterization of material parameters of composite plate specimens using optimization and experimental vibrational data. *Composites: Part B*, **27B**, pp. 185-191, 1996.

[14] De Visscher, J., Sol, H., De Wilde, W.P. and Vantomme, J., Identification of damping properties of orthotropic composite materials using a mixed numerical experimental method. *Applied Composite Materials*, **4**, pp. 13-33, 1997.

Evaluation of energy absorbing materials under blast loading

H. Bornstein & K. Ackland
Defence Science and Technology Organisation, Melbourne, Australia

Abstract

Traditional blast protection systems present a hard metallic surface, which acts to deflect and attenuate the blast loading. However, more recently, lightweight energy absorbing materials, including cellular materials, are being considered. These materials reduce the transmitted force by deforming elastically or plastically under the blast load. In this study, four commercially available energy absorbing materials were considered for application in blast protection. The materials considered were aluminium foam, Nomex® honeycomb, Skydex® and expanded polystyrene. The impulse transmitted by the rear face of experimental test panels under blast loading was assessed by positioning a mass on the rear face of each of the four panels, the velocity of which was measured using high speed imaging. Measurements of the permanent deformation of the rear face of the panels were also taken. The aluminium foam recorded both the lowest permanent deformation and impulse transfer to the mass of the materials tested, making it the most suitable material of those tested for use as an appliqué panel in this particular scenario. Analysis of the panels suggested that for the particular experimental setup used, minimal energy was absorbed by plastic deformation or crushing of the energy absorbing materials. Instead, the permanent deformation of the panels may be related to the mechanical properties of the materials in the plane perpendicular to the blast load.
Keywords: energy absorbing materials, blast, blast testing.

1 Introduction

Energy absorbing materials have been investigated for a variety of different applications when subjected to blast and impact loading. Metallic and non-metallic foams and honeycombs are designed to crush under loading, thus

WIT Transactions on Engineering Sciences, Vol 77, © 2013 WIT Press
www.witpress.com, ISSN 1743-3533 (on-line)
doi:10.2495/MC130111

absorbing energy and reducing the load transfer to the primary structure behind them. These materials or structures are generally characterised by their compressive stress-strain curves, which typically show an initial elastic region, followed by a plateau stress representing plastic yielding (Figure 1). Densification occurs when the material has been significantly compressed and is identified by a sharp rise in stress.

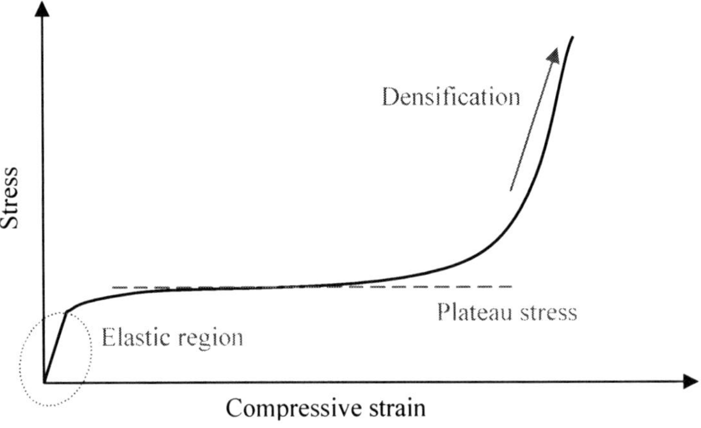

Figure 1: Typical shape of a stress-strain curve of a crushable energy absorbing material or structure.

Karagiozova *et al.* [1] investigated the properties of aluminium honeycomb and polystyrene sandwich panels subjected to near-field blast loading using both experimental testing and numerical modelling. They showed that as the loading increases, the densification of the foam can result in high force transmissions to the back faces of sandwich panels. This suggests that energy absorbing materials can only be optimised for a particular loading and may not provide good performance once overmatched. They also found that the force transmitted to the back face of the sandwich panels increased with density for thinner panels. This assessment is also supported by Zhu [2], who compared a range of aluminium honeycomb and aluminium foams. For thicker panels, the force transmitted was related to the core efficiency (the ratio between the energy absorbed per unit mass of the core and the density of the core). This leads to the suggestion that lighter materials with high plateau stresses might be the best core materials, when it is feasible to use a thick core.

Zhu [2] also performed an analytical study into the deformation of the panels, which relied on a simplifying assumption that divides the structural response of the panel into three distinct loading phases. These are defined as follows:

1. The blast load is transmitted to the front face, which attains a given velocity, while the rest of the structure is stationary.
2. The core is compressed, while the back face remains stationary.
3. The back face begins to deform and the structure is brought to rest by plastic bending and stretching.

www.witpress.com, ISSN 1743-3533 (on-line)

As the structure is brought to rest by bending and stretching, the permanent deformation of a sandwich panel is likely to be affected by the compressive and shear strength and moduli in the planes perpendicular to the loading direction of the core material in addition to its properties under axial compression.

Other studies investigating the use of energy absorbing materials under blast loading include Warne *et al.* [3] and Hanssen *et al.* [4]. Warne *et al.* compared the permanent deformation of the rear face of sandwich panels under blast loading using expanded polystyrene, polyethylene 30, Nomex® honeycomb and a bare steel panel. It was found that the Nomex® honeycomb provided the best performance in terms of back face deformation. Hanssen *et al.* investigated the use of aluminium foam panels attached to a pendulum and subjected to near-field blast loading. They found that the addition of the aluminium foam panel resulted in an increase in the impulse delivered to the pendulum, although the authors were not able to provide a definitive reason for this. They also conducted an analytical study, which showed that whilst the peak force was reduced by the panels, the loading duration increased, suggesting that whilst foam panels can control the peak stresses they cannot reduce the momentum transfer to a structure.

The use of energy absorbing materials is not limited in application to protecting structures from blast loading, but is also used to protect personnel. Wang *et al.* [5] investigated the use of Nomex® honeycomb and aluminium honeycomb for use in landmine resistant boots. They observed that whilst both provided a reduction in the transmitted force, the Nomex® honeycomb, which had a lower plateau stress, provided better performance. Wang *et al.* [6] compared the use of Nomex® honeycomb and polystyrene for the use in false floors in armoured vehicles to protect against lower limb injuries during landmine loading. It was observed under drop testing conditions that the Nomex® honeycomb provided slightly better performance than the polystyrene. Work by McKay [7] compared a range of aluminium honeycomb materials and aluminium foam for use in false floors. His research showed that the plateau stress of a material must be tuned to be just below the threshold for lower leg injury to minimise the risk of lower leg injury. The plateau stress of the Nomex® honeycomb tested by Wang *et al.* was very similar to the best performing aluminium honeycomb tested by McKay. It should be noted that these tests subjected the panel to axial compressive loads only.

This paper investigates the performance of four different commercially available energy absorbing materials in the role of an appliqué panel to be placed on a structure. The comparison between the materials is performed by analysing the impulse transmitted by the back face of sandwich panels under blast loading as well as the deformation of the rear structure. The study was conducted using an experimental blast test.

2 Experimental method

An experiment was designed and conducted around the use of four commercially available energy absorbing materials. The materials selected for comparison

were Nomex® honeycomb, Skydex® convoy decking, Alporas aluminium foam and expanded polystyrene (EPS). The Skydex® convoy decking was selected as it is a commercially available energy absorbing foot pad used in US military vehicles such as the MRAP [8]. The selected materials are shown in Figure 2.

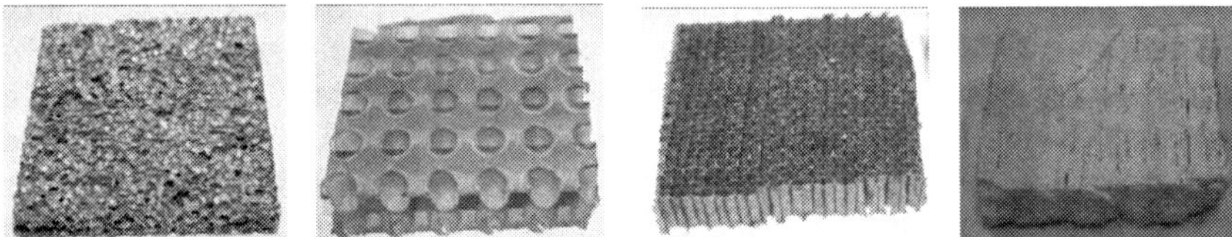

Figure 2: (Left to right) aluminium foam, Skydex®, Nomex® honeycomb and EPS.

As there were limitations surrounding the availability of identical thicknesses of the four materials, the material thicknesses varied slightly. Table 1 provides the densities and thicknesses of the materials. The Skydex® came in the form of a double layer convoy deck, which was modified to a single layer without the top rubber layer for use in a sandwich panel. The stress-strain properties under uniaxial compression of the four materials are presented in Figure 3. The stress-strain curves for the aluminium foam, EPS and Nomex® were sourced from literature [1, 6, 9], while the Skydex® curve was taken from information provided by the company [10] for the single layer convoy decking.

Table 1: Energy absorbing material densities, thicknesses and elastic moduli.

Material	Density (kg/m^3)	Thickness (mm)	Areal density (kg/m^2)	Elastic modulus (GPa)
Aluminium foam	217	14	3.04	69
Skydex	170	25	4.25	1.4
Nomex	36	20	0.72	29
EPS	28	22	0.62	<3

Sandwich panels were made for each material with a 500 mm × 500 mm × 1.7 mm steel panel as the front face, a 500 mm × 500 mm energy absorbing layer, and a 600 mm × 600 mm × 1.7 mm steel panel as the back face. Sikaflex® 291 was used to bond the materials to make the sandwich panels. The structure of the panels is shown in Figure 4, together with a schematic of the test layout.

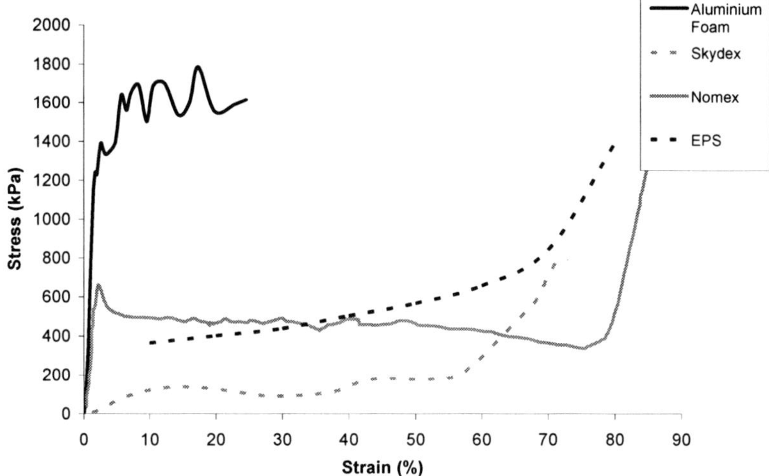

Figure 3: Stress-strain properties of materials under uniaxial compression (as obtained from the literature).

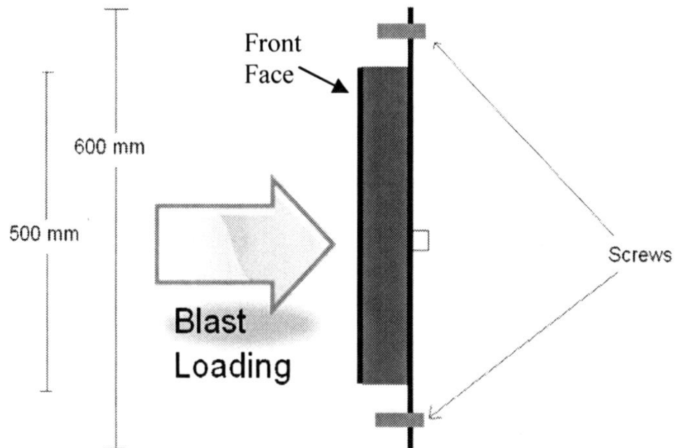

Figure 4: Schematic of sandwich panel setup for experiments.

The rear face of each sandwich panel was then attached to the target stand using 16 evenly spaced screws to represent a structure with an energy absorbing appliqué. The setup of the sandwich panels on the target stand is shown in Figure 5. A 12.8 kg cubic PE4 charge was placed on a stand with the centre of the charge 500 mm above the ground. The target stand was positioned such that the centre of the stand was 5 m from the charge centre and was perpendicular to the face of the charge. This distance was selected as it is on the edge of the fireball for this charge size and hence would allow high-speed video to record the back face of the panels.

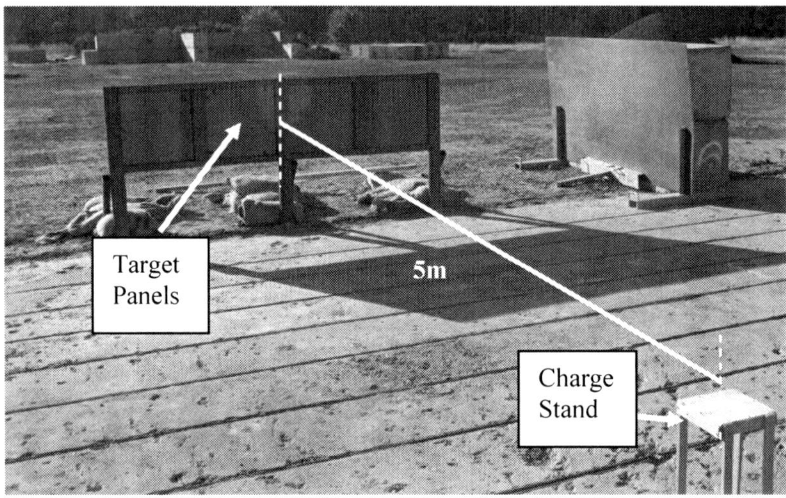

Figure 5: Setup of target panels and explosive charge.

A nail was glued in position at the centre of the rear face of each sandwich panel. This allowed an aluminium cylinder to be mounted flush with the rear surface of the plate. The aluminium cylinders were 30 mm in diameter and 20 mm in length. They were painted in bright colours for identification on the high-speed video footage. The video was taken at 6000 frames per second to record the velocity of the cylinders. Images were taken with reference markers prior to the event for each panel to calibrate distances within the high-speed video footage. Figure 6 shows the rear face of the sandwich panels with the aluminium cylinders in position.

Traditionally the performance of sandwich panels is evaluated experimentally by using the permanent rear face deformation. The aluminium cylinders were used to provide a measurement of the impulse transmitted by the rear face of a sandwich panel to an object in direct contact with it. This measurement technique was used to determine whether there was any correlation between the load transfer to the rear face of the panel and the permanent rear face deformation.

Due to the cubic shape of the explosive charge and the shape of the target stand, there was some asymmetry between the loading on the inner (Skydex® and Nomex® honeycomb) and outer (EPS and aluminium foam) sandwich panels on the target fixture. However, preliminary simulations performed in AUTODYN predicted only a small difference ($\leq 5\%$) in the cylinder velocities for a bare steel panel at the inner and outer panel locations.

The residual deformation from each sandwich panel was measured using a 3D laser scanner.

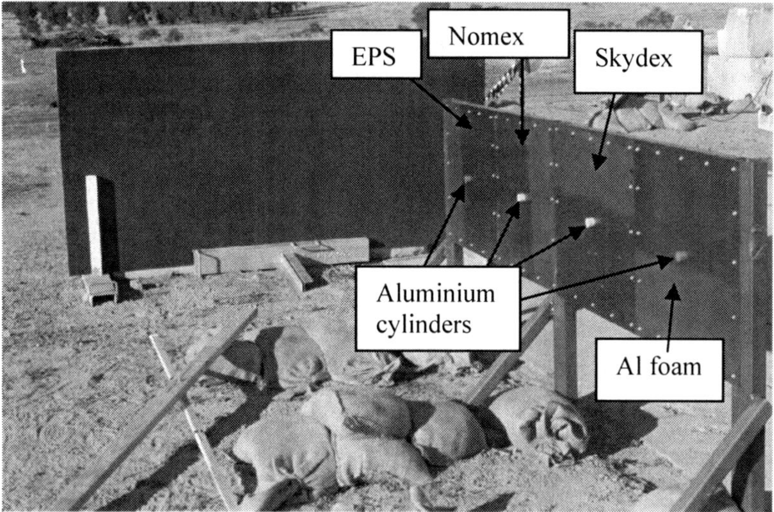

Figure 6: Rear face of sandwich panels.

3 Experimental results

The high-speed video footage was analysed to record the velocities of the aluminium cylinders for each of the panels. Unfortunately, due to limitations in the experimental design, the panels became dislodged during the event. This affected the boundary conditions of the experiment, and made it difficult to draw any firm conclusions based on the cylinder velocities. The velocities of the aluminium cylinders are presented in Figure 7. There appears to be two distinct acceleration phases on the curve, which could represent different loading mechanisms by the panels on the cylinders. The potential acceleration mechanisms are acceleration due to rear face deformation, global motion of the panel after failure of the boundary conditions, and load transmission through the panel due to the compressive stress wave propagation. Analysis of the high-speed video indicates that the first acceleration phase is due to rear face deformation. The high-speed video footage also shows that the aluminium foam panel breaks away from the target fixture at ~4 ms, which corresponds to the second acceleration phase of the aluminium cylinders (global motion of the panel after boundary failure). The average velocities of the aluminium cylinders were taken once the cylinder velocity had reached a plateau and are shown in Table 2. This plateau occurred after ~6 ms. Further analysis of the high-speed video indicates that the screws on all panels failed at a similar time. If the assumption is then made that the failure of the screws required minimal energy, then the impulse transmitted to the cylinders of all panels due to global motion of the panels should have been similar. Whilst definitive conclusions on the different materials performance based on the cylinder velocities cannot be drawn due to failure of the boundary conditions, the differences may still be indicative of their relative performance.

The aluminium foam panel clearly had the lowest cylinder velocity. This could be attributed to its stiffness allowing minimal panel deformation to accelerate the cylinder in the initial acceleration phase.

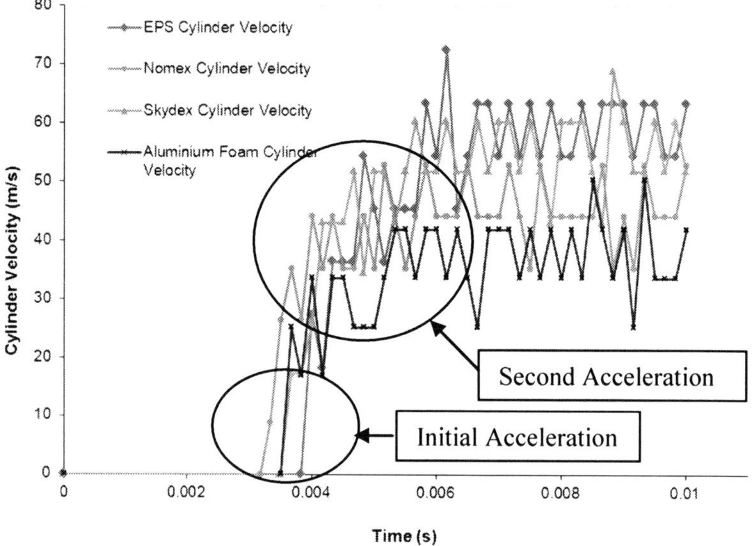

Figure 7: Aluminium cylinder velocity vs. time results.

Table 2: Aluminium cylinder velocity for energy absorbing material panels. The adjusted velocity adds an increase in velocity of 5% to allow for differences in loading due to panel location.

Material	Cylinder Velocity (m/s)	Panel Location	Adjusted velocity (m/s)
Aluminium Foam	35.2	Outer	37.0
Skydex®	50.5	Inner	50.5
Nomex® honeycomb	43.4	Inner	43.4
EPS	54.8	Outer	57.5

Figure 8 shows the rear faces of the four sandwich panels. The aluminium foam panel had almost no permanent deformation, although the high-speed video showed that there was elastic deformation of the panel. The Skydex® and EPS panels showed similar deformation profiles to those expected by a rectangular plate under blast loading. However, the Nomex® honeycomb panel is shown to have a different deformation profile to the Skydex® and EPS. Unlike the other materials, Nomex® honeycomb has different mechanical properties along each of its three axes [11, 12]. Preliminary testing by Hexcel composites [12] has shown that for the grade of Nomex® honeycomb used there is a 50% difference in the shear strength depending on the direction of the applied load. This anisotropy in the material appears to have resulted in the panel deforming preferentially in one plane. This indicated that the mechanical properties of the material in all axes are important in determining the deformation profile of the sandwich panel. Another possible contributing factor to the deformation profiles is the variation in boundary conditions between the panels due to the panels coming away from the target stand. However as all the panels were dislodged before they had finished their structural response, it is unlikely that the boundary conditions significantly affected the difference in deformation profiles seen between the panels.

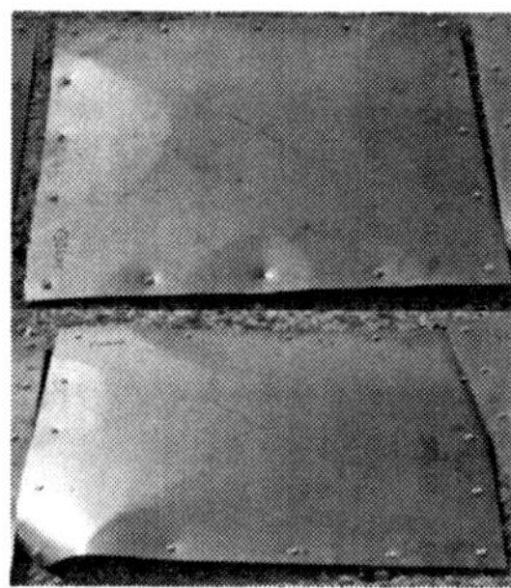

Figure 8: Rear faces of the sandwich panels; (top left) aluminium foam (top right) Skydex® (bottom left) Nomex® honeycomb (bottom right) EPS.

The deformation profiles were analysed using a 3D laser scanner with the resulting deformations presented in Figure 9. These profiles were measured across the plate centres between the midpoints of two opposite edges. The Skydex® and EPS not only had similar deformation profiles but the magnitude of their deformation profiles was found to be very similar. Skydex® and EPS have a tendency to recover post event, hence negligible plastic deformation was observed in the materials after the event. It is likely that due to their low elastic modulus, both the Skydex and EPS absorbed some energy under compression during the event.

The Nomex® honeycomb panel recorded a larger permanent deformation than any of the other panels. Post event analysis of the Nomex® honeycomb also revealed that very little plastic deformation of the material had occurred. As mentioned previously, Nomex® honeycomb is largely anisotropic. The compressive strength and modulus of honeycomb structure in the directions perpendicular to the blast can be less than 5% of the through-thickness strength [12]. These properties make the Nomex® particularly susceptible to large deformations under bending loads.

The aluminium foam panel had the lowest permanent deformation and the foam was not found to have compressed during the event. This suggests that in this particular experiment, the benefit it provided in comparison to the other materials was its higher strength and moduli in all directions. This is shown by its higher compressive strength (see Figure 3) compared to the other materials in conjunction with its isotropic properties.

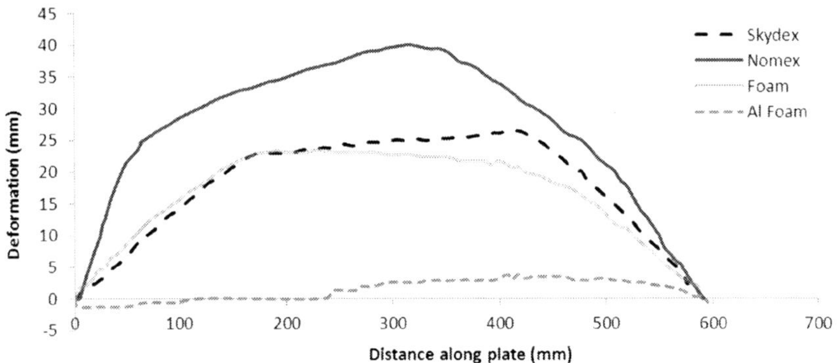

Figure 9: Deformation profiles of the rear faces of the sandwich panels. The Nomex® was assessed in the direction of maximum deformation.

None of the materials showed evidence of plastic deformation under the compressive loading. This may have been due to the rear panels deforming and coming away from the fixture before the materials were able to compress. Alternatively, as the panels were outside the fireball, the loading on the panels may not have been sufficient to cause compression of the materials.

4 Discussion on the experimental design

The experiment, even with the issues highlighted below, identified that the mechanical properties of the panels in the planes perpendicular to the blast direction are important when examining the permanent deformation of the rear face, as well as their properties in direction of the blast. As the intention of the experimental setup was to look at appliqué panels under conditions where large deformations were expected, materials selected in future work should have reasonable compressive and shear strength and moduli in all directions. Data on such materials is therefore required in both quasi-static and high strain rate regimes in multiple directions to assess their suitability for use as an appliqué. Materials such as Nomex® honeycomb may not be suitable for this type of application. It is also recognised that the rear face of the panel should be thicker than the front face of the panel to represent a more realistic appliqué armour scenario.

The intended use of the cylinder velocity was to record whether there was a correlation between the performance of different materials when comparing the load transferred through a panel and the permanent deformation of the panel. Whilst there appeared to be a difference, this observation was not conclusive due to the detachment of the panels from the supporting frame during the event. The results suggested that the cylinder velocity was generated by the initial velocity of the centre of the rear face of the panel, as well as the global velocity of the plate after detachment. The detachment of the panels at the boundaries could be rectified in future by an improved boundary design, which does not allow detachment.

A number of design improvements in the setup of the experiment would have to produce more meaningful analysis of the different core materials for sandwich panels under blast loading. The panels coming away from the target stand during the event provided a modification in boundary conditions that made meaningful analysis of the results difficult, especially the cylinder velocities. This could be rectified by mounting the rear face of the panels to the front of a heavier frame.

The experiment was conducted outside the fireball, which may have limited the blast loading to a level that did not allow compression of the energy absorbing materials. It is intended that future experiments focus on scenarios with smaller charges but in the near-field as is more frequently performed in the literature.

A number of materials were compared for their suitability as appliqué armour but there was no baseline test without an appliqué material. A baseline test would have allowed measurement of the improvement provided by each appliqué. If numerical modelling of the experiment was to be conducted it would also provide a suitable data point for validation of the applied loading conditions. Validating the experimental loading conditions in a numerical model with complex material properties such as the appliqués used here would present a much greater challenge.

5 Conclusions

The experimental results presented in this paper showed that the aluminium foam provided superior performance to the other materials tested in terms of use as an appliqué panel for these particular test conditions. It recorded both the lowest permanent deformation and impulse transfer to the cylinder of the materials tested. For the scenario evaluated, due to the minimal energy absorption via compression of the panels, it appeared that performance of the materials was governed by more than just the compressive stress-strain behaviour of the material. Evidence of the anisotropic effects of honeycomb materials was shown by the asymmetric deformation profile of the Nomex® panel. This is particularly important when looking at an appliqué panel for structures undergoing significant deformation as the core material of the sandwich panel will be subjected to significant bending forces. A number of experimental design improvements were found, most notably improvements to the panel boundary fixtures.

Acknowledgements

The authors would like to acknowledge the assistance of Steve Pattie, Frank Marian, Andrew McLean, Darren Wiese, Emily Frain and Chris Townsend and the staff at P&EE Graytown for their assistance in setting up and conducting the blast trial. The authors would also like to thank Stuart Cannon, Leo de Yong,

Craig Flockhart, Stephen Cimpoeru and Brian Dixon for their independent advice.

References

[1] Karagiozova, D., Nurick, G., Langdon, G., Chung Kim Yuen, S., Chi, Y. and Bartle, S., Response of flexible sandwich-type panels to blast loading, *Composites Science and Technology,* **69(6)**, pp. 754-763, 2009.

[2] Zhu, F., Impulsive loading off sandwich structures with cellular cores, PhD thesis, Swinburne University of Technology, 2008.

[3] Warne, S., Anderson, C. and Ackland, K., The use of energy absorbing materials within a blast scenario, *l8th International Conference on Shock and Impact Loads on Structures,* Adelaide, 2009.

[4] Hanssen, A., Enstock, L. and Langseth, M., Close-range blast loading of aluminium foam panels, *International journal of impact engineering,* **27,** pp. 593-618, 2002.

[5] Wang, J., Bird, R., Swinton, B. and Krstic, A., Experimental and computational research for conceptual design of mine-resistant boots, *Journal of battlefield technology,* **6(1)**, 2003.

[6] Wang, J., Bird, R., Swinton, B. and Krstic, A., Protection of lower limbs against floor impact in army vehicles experiencing landmine explosion, *Journal of battlefield technology,* **4(3)**, 2001.

[7] McKay, B., Development Of lower extremity injury criteria and biomechanical surrogate to evaluate military vehicle occupant injury during an explosive blast event, *Wayne state university dissertations,* **Paper 146,** 2010.

[8] Skydex® applications, <http://www.skydex.com/technology/applications>, Accessed 31 May 2012.

[9] Danneman, K. and Lankford Jr, J., High strain rate compression of closed-cell aluminium foams, *Materials science and engineering,* **293(1-2)**, pp. 157-164, 2000.

[10] Metzer, C., Personal communication, 23 February 2011, Skydex Technologies Inc.

[11] Foo, C., Chai, G. and Seah, L. Mechanical properties of Nomex material and Nomex honeycomb structure, *Composite Structures,* **80**, pp. 588-594, 2007.

[12] Hexcel composites, HexWeb™ honeycomb attributes and properties, http://www.hexcel.com/Resources/DataSheets/Brochure-Data-Sheets/Honeycomb_Attributes_and_Properties.pdf, accessed 10 July 2012.

After 150 years of research, fatigue still causes 85% of failures

J. Volak[1] & V. Mentl[2]
[1]*Research and Testing Institute Ltd., Pilsen, Czech Republic*
[2]*University of West Bohemia, Pilsen, Czech Republic*

Abstract

The fatigue phenomenon is said to be responsible for more than 80% of structural failures and operational accidents. One of the most important branches of industry, where unexpected outages can cause extremely expensive costs is the power producing industry. The steam turbine rotors represent large, expensive and, in the case of failure, potentially dangerous components. Their local properties generally differ from one forging to another, or if we compare head and bottom parts of the original ingot, or central and circumferential locations of one rotor body respectively, or if we compare the properties of separate discs, e.g. in the case of welded rotors. These differences stem from both even slight changes in the chemical composition (of separate heats or even within one ingot), heat treatment and in the differences in technology with respect to the real shape and size of the forgings in question.

At present, questions on quantitative evaluation of remaining lifetime (to avoid failure) become urgent and, because fatigue is the most frequent cause of material degradation and resulting structural components failures, the remaining lifetime assessment is based on the evaluation of mechanical properties before and after some time of service. The original material data are usually tested by means of classic test specimens, which can hardly be used in the case of components in service, where there is no possibility to withdraw a sufficient volume of representative material for the manufacturing of a classic test specimen. In such cases today, methods of semi-destructive removal of a small volume of test material are utilized. This makes it possible to produce miniature test specimens.

This paper describes the results of fatigue tests performed on miniature test specimens in comparison with the classic fatigue tests for several alloys applied

WIT Transactions on Engineering Sciences, Vol 77, © 2013 WIT Press
www.witpress.com, ISSN 1743-3533 (on-line)
doi:10.2495/MC130121

in the power producing industry. The miniature test specimens were produced by water-jet cutting from the large test specimens. With respect to the specimen shape, stress concentration had to be taken into consideration for the purposes of comparison with the classic test specimens.
Keywords: fatigue, failure risk assessment, remaining lifetime evaluation.

1 Introduction

Fatigue is still the prevailing cause of failure of metallic, plastic and ceramic components. In cases of cyclic loading of test specimens or structural components, the cumulation of cyclic plastic deformation causes failure below ultimate strength and even yield point values.

At present, reliable quantitative evaluation of remaining lifetime becomes urgent and because fatigue is the most frequent cause of material degradation and resulting structural components failures, the most important means of remaining lifetime assessment is the evaluation of fatigue properties before and after some time of service. The original material data are usually tested by means of traditional (standardized) test specimens, which can be hardly used in case of components in service, where there is no possibility to withdraw sufficient volume of representative material for the traditional test specimen manufacturing and testing. In such cases today, new untraditional methods of semi-destructive removal of a small volume of test material by grinding or an electro-discharge method are utilized. This makes it possible to produce a miniature test specimens. The direct comparison of the results must be then based on reliable correlation between traditional and miniature test specimen results.

The basic mechanical characteristics of fatigue behaviour of structural materials are:

a) Woehler (S–N) curve describing the relationship between the amplitude of loading and the resulting number of cycles to failure,

b) Manson-Coffin (ε–N) curve describing the relationship between the amplitude of deformation and the resulting number of cycles to failure,

c) Paris-Erdogan law describing the relationship between the crack growth rate and the stress intensity factor.

All the three abovementioned material characteristics are used in engineering practice. The Woehler (S–N) and Manson-Coffin curves treat the material as a mechanical continuum and include all the stages of the fatigue process, the Paris–Erdogan equation describes the macroscopic crack growth only.

It is well known that the specimen size has a negligible effect in the case of uniaxial tensile loading, and thus the yield point and ultimate strength values are not dependent on the cross section size. On the other hand, the failure process during fatigue loading depends on the specimen or component size, i.e. the increasing specimen size has a negative effect on the specimen life. This fact is caused primarily by the stress gradient in the respective cross section, e.g. at bending loading, or due to stress concentrators, or due to the simple fact that it is not possible to produce real components free of defects.

At present, questions on quantitative evaluation of remaining lifetime become urgent and the most important means of remaining lifetime assessment is the evaluation of fatigue properties before and after some time of service.

Prediction of total service life of cyclically loaded structural components or their remaining lifetime evaluation after some time of service represent important problem both in components, structures and machines design and exploitation.

Theoretically, the fatigue life can be divided into four stages:

1. Changes of mechanical properties at the beginning of fatigue loading resulting in saturation of hysteresis loops.
2. Initiation of numerous crystallographic microcracks mostly on the component surface.
3. Stage of macrocrack propagation.
4. Fracture.

In practical situations often, the loaded component contains imperfections of various kind and origin, structural and technological notches, geometrical discontinuites, etc. and total fatigue life of a component in service thus may be approximated by the third stage only, i.e. macrocrack propagation described by the Paris-Erdogan Law. And, whereas the macroscopic cracks can be observed by numerous experimental techniques, the material degradation is not clearly visible. Some NDT methods nevertheless exist, e.g. X-Ray diffraction [3–5].

There are several problems related to the material properties degradation assessment during service:

1. The material of a component is not homogenous (as a result of applied technology) as far as its chemical composition and structure is concerned. As a suitable example, large castings and forgings are this case.
2. Components are of complex geometry, contain geometry discontinuities, notches, etc., so that even at simple (e.g. uniaxial) loading and uniform temperature field, individual localities of the component can reveal different degree of material properties degradation.
3. Original (virgin) mechanical properties are not known from various reasons. The problem is complicated also by point 1. These data should be then stored during the whole service life of the component.
4. During component operation it is not possible to withdraw sufficient volume of material from the component, because the classic test specimens used acc. to existing standards for the mechanical properties measurement are mostly too large.
5. Alternative ways of mechanical properties measurement have to be used based e.g. on the use of miniature test samples or semidestructive test methods. The results of such tests must then be correlated to the classic ones and these correlations must be based on the abovementioned items.

Thus the goal of the presented research was to establish correlation(s) of fatigue test results between classic and miniature test specimens.

2 Method of material sampling of real components

For the purposes of material properties evaluation in case of components in service, a small volume of material can be withdrawn by means of special device (see Fig. 1).

Figure 1: Test material sampling from outer surface sampling machine [6].

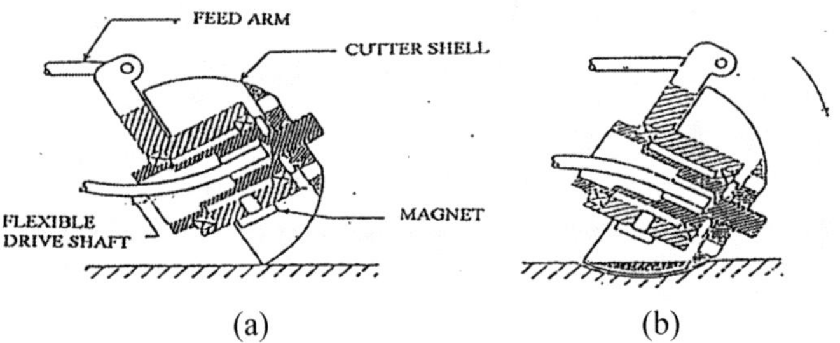

Figure 2: The principle of the sampling by SSamTM-2 sampling machine [6].

The principle is shown in Fig. 2. The resulting surface damage after the material removal is relatively small and in many cases it can be neglected causing no substantial stress concentration or wall thickness reduction. For example, shallow cracks discovered during in-service inspections in steam boiler tubing are also frequently removed by grinding. The process of material sample removal is slow and takes place at intensive cooling, so that material properties are not affected. An example of the location of material removal is then seen in Fig. 3.

www.witpress.com, ISSN 1743-3533 (on-line)

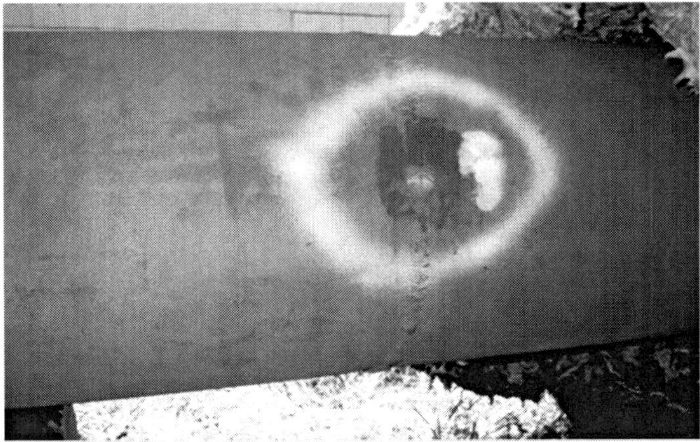

Figure 3: An example of test sampling location [7].

The piece of material removed from the component (see Fig. 4) can be used for chemical and microscopic investigation and also for the purposes of the mechanical testing, e.g. for the determination of yield point and ultimate strength values, creep properties at elevated temperatures, transition temperatures, fracture toughness, and also the fatigue properties. All the abovementioned values are nevertheless measured by means of non-standardized test specimens and thus the obtained results must be correlated with the classic ones.

Figure 4: A sample of removed material [6].

The necessary correlations between classic and miniature characteristics must be known prior to remaining lifetime evaluation and can be determined in laboratory without material removal from real components. The test specimens can be produced by mechanical manufacturing or water jet cutting (see Fig. 6).

Figure 5: Miniature specimens for small punch and fatigue tests.

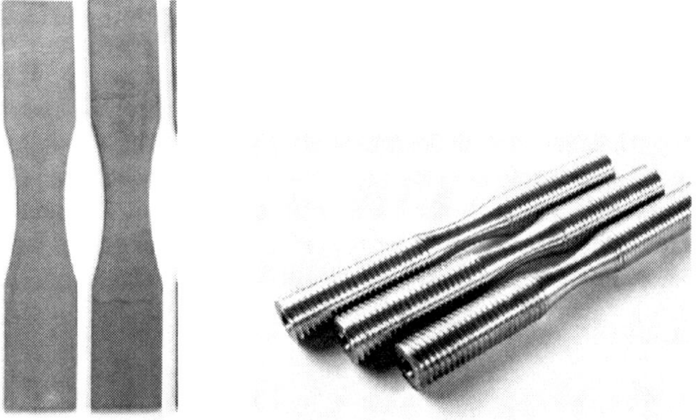

Figure 6: Classic fatigue test specimens.

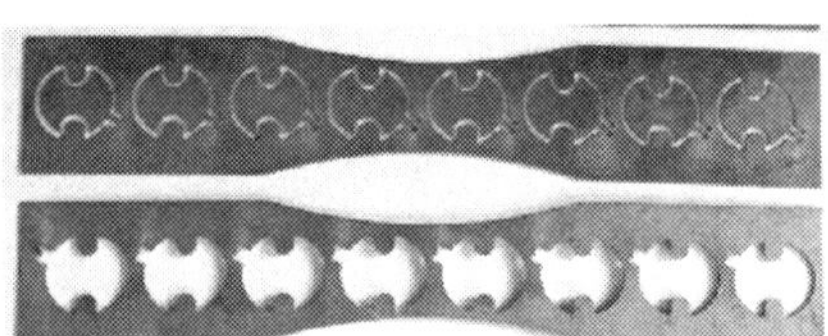

Figure 7: Miniature fatigue test specimens.

The geometry of the miniature test specimens used for the fatigue test causes stress concentration within the cross section and it must be taken into consideration when assessing and correlating the test results. The performed FEM analysis revealed that the stress concentration factor in case of the miniature test specimens used here is k = 1.33 (see Fig. 8).

3 Experimental procedure

Several steels and Al-alloys were fatigue tested at room temperature for the purposes of correlation determination between classic and miniature test specimens test results. (Because of the commercial origin, the details of the tested materials are not stated here.)

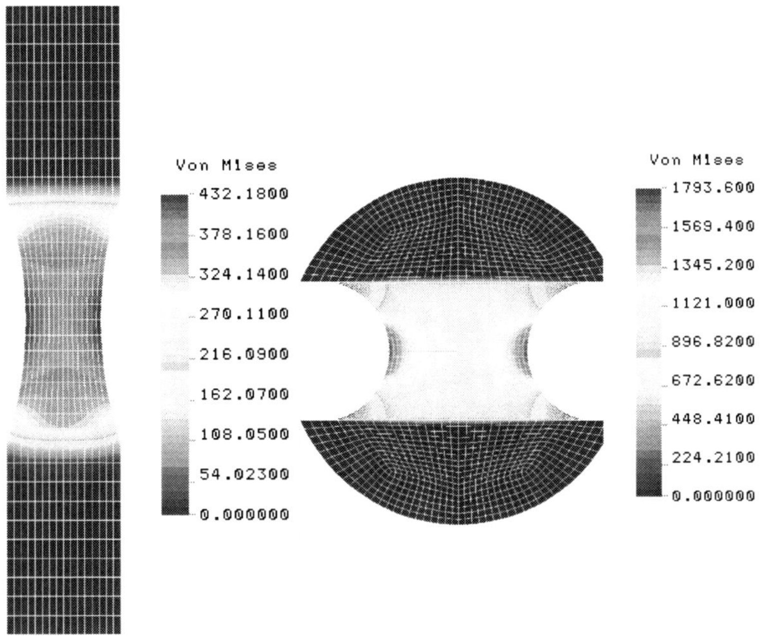

Figure 8: Stress concentration determined by FEM model.

In the diagrams in Figures 9–12, the "classic fatigue" represents the data obtained by means of traditional standardized test specimens. The "nominal stress" represents the data obtained on miniature test specimens, where no stress concentration was taken into consideration, and "stress concentration" represents the data obtained on miniature test specimens, where the stress concentration factor (determined by FEM) of k = 1.33 was taken into consideration.

4 Conclusion

It is well visible from the experimental results, that the S–N (Woehler) curves of miniature test specimens have a different (steeper) course in comparison with the traditional ones for all the tested materials. At present, no unique generally valid correlation between the classic and miniature test specimen results has been found.

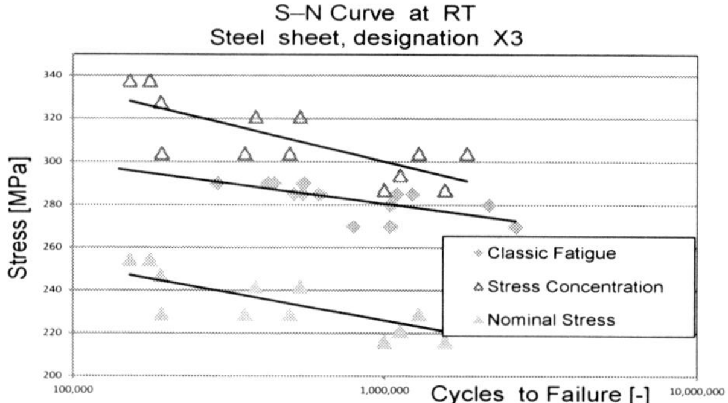

Figure 9: S–N (Woehler) curves, steel "X3".

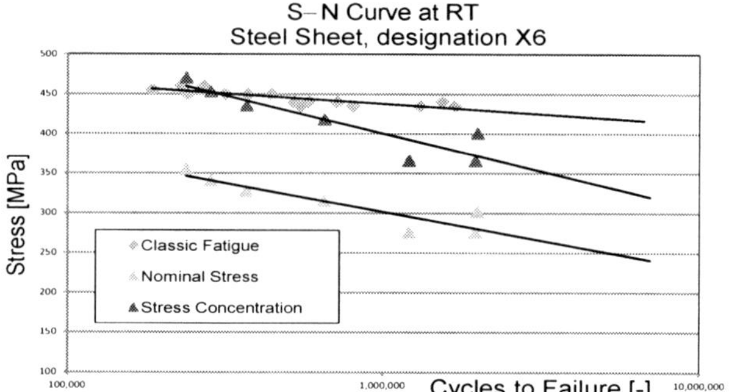

Figure 10: S–N (Woehler) curves, steel "X6".

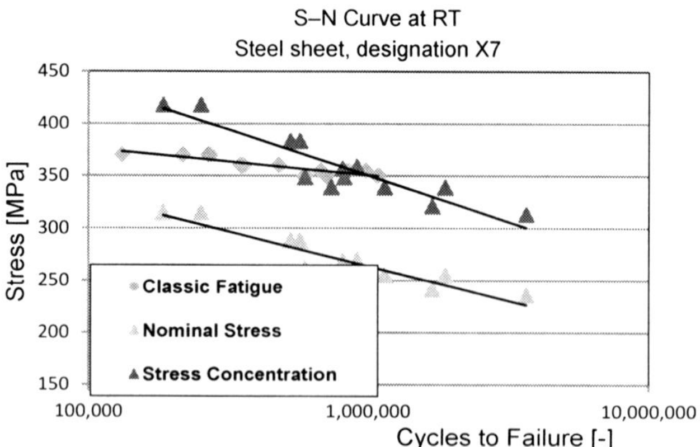

Figure 11: S–N (Woehler) curves, steel "X7".

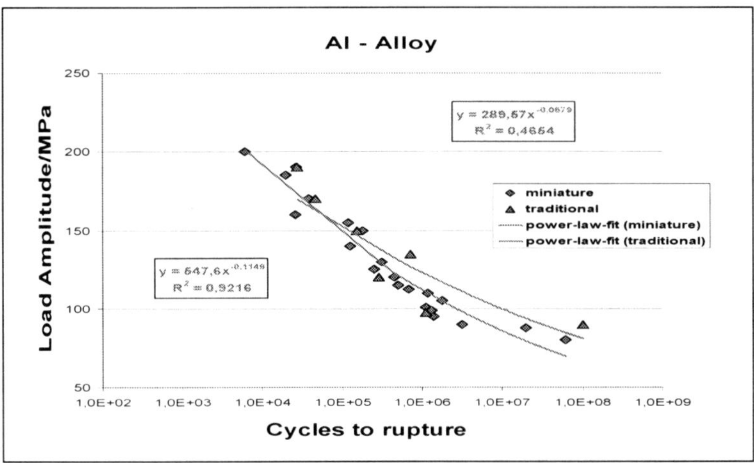

Figure 12: S–N (Woehler) curves, Al–alloy.

Thus, the recommended solution (steps) for the failure prevention and the remaining lifetime evaluation might be as follows:

1. To establish correlation between classic and miniature test results prior to service in the situation when the material is in "virgin" state and no service material properties degradation exists.

2. This correlation must, in case of large components, e.g. steam turbine rotors, be established in the locations, where the test samples will be in future removed for the purposes of material degradation assessment. The material mechanical properties can differ substantially within the forging.

3. The relevant data must be stored for the purposes of future comparison with the data measured after certain service period. In this situation, miniature test specimens can be used only and the previously established correlation should be used. (The assumption that the correlation between classic and miniature test specimens is the same both for virgin and degraded material should be proved.)

4. It is recommended to make out a "birth-certificate" of the component, where all the relevant data are summarized and, during service inspections and performed investigation, recorded and stored.

Such a procedure is based on real experimental data in opposite to the remaining lifetime assessment methods based on many (often risky) assumptions or the permanent storing of numerous variables data on pressure, temperatures, water quality etc. (e.g. in case of steam turbine boilers) for many years and their evaluation by methods based on theories, which in the time of the remaining

lifetime evaluation are already 20, 30 or even more years old and maybe surmounted.

The necessary costs for the abovementioned procedure will be balanced by lowering the risk of an unexpected failure and inevitably related economic losses and more precise evaluation of remaining component or structure lifetime, what makes it possible to assess the risk of prolonged operation, to plan and minimize structure outages and make use of the closed loop information with respect to the future production and maintenance.

Acknowledgements

The research was supported by Czech Ministry of Industry and Commerce, project No. FR-TI4/602.

References

[1] Mentl V.: Mapping of Fracture Toughness Values in Large Forgings, Proc. ECF 18, Dresden, Germany, 2010, ISBN 978-3-00-031802-3.
[2] Volak J., Mentl V.: A Comparison of Fatigue Characteristics of Structural Steels Measured by means of Classic and Miniature Test Samples, Proc. ICMFF9, Parma, Italy, 6/2010, ISBN 978-88-95940-31-1.
[3] Fiala J., Mentl V.: Degradation Assessment of Fatigue by X-Ray Diffraction Technique, Int. Conf. Fatigue Design 2007, Senlis, France, 2007.
[4] Fiala J., Mentl V.: Degradation Assessment of Welded Joints by X-Ray Diffraction Technique, Int. Conf. "WELD", Geesthacht, Germany, 2005.
[5] Kolarik V. *et al.* An Innovative Lifetime Monitoring and Improving Method for High Temperature Components: Xpection, Int. Conf. "Plant Life Extension", Cambridge, UK, 2004.
[6] Small punch test method for metallic materials, CEN Workshop Agreement, CWA 15627, 2007.
[7] Kubon, Z.: Small sampling test method, Vítkovice-Research and Development, Ltd., 2005.

Section 3
Computational models and experiments

Physisorption of molecular hydrogen in curved carbon nanomaterials: a computational study

D. J. Durbin[1,2], N. L. Allan[2] & C. Malardier-Jugroot[1]
[1]*Department of Chemistry and Chemical Engineering, Royal Military College of Canada, Canada*
[2]*School of Chemistry, University of Bristol, UK*

Abstract

Hydrogen physisorption on carbon nanomaterials is a promising method of hydrogen storage because carbon materials are cheap, abundant and light weight. However, storage is difficult because dispersion forces between C and H are weak. Curved carbon substrates are more promising than planar systems because the increased level of sp^3-hybridization enhances H_2 physisorption. The present study uses density functional theory to model large fullerenes, single-walled carbon nanotubes and graphene to investigate the interaction with H_2; decorating platinum is also considered. We conclude that H_2 can be stored in fullerenes without an energy input if the H_2 molecules are more than 3 Å from the carbon surface and more than 2 Å from each other. In addition, confinement effects are observed when hydrogen is stored in fullerenes rather than nanotubes – storage in nanotubes is more favourable for systems with small diameters.
Keywords: hydrogen storage, curvature, confinement, fullerenes, carbon nanotubes, graphene, platinum, physisorption, fuel cells.

1 Introduction

Hydrogen gas has high potential as a fuel in fuel cells vehicles because of its high gravimetric energy density; unfortunately, it also has a very low volumetric density. Compact and efficient storage techniques are required. Traditional methods of compression and liquefaction are inadequate because of their high cost and inability to achieve sufficient compression. Alternative storage methods of metal and non-metal hydrides and absorption to carbon substrates are all currently of considerable current interest [1–4].

WIT Transactions on Engineering Sciences, Vol 77, © 2013 WIT Press
www.witpress.com, ISSN 1743-3533 (on-line)
doi:10.2495/MC130131

Hydrogen physisorption on carbon nanomaterials is one promising possibility. This storage technique has the advantage of using a cheap, light weight and abundant carbon nanomaterial, but it does not involve potentially awkward chemical reactions. Unfortunately, $C\cdots H_2$ dispersion forces are weak so the materials are often activated by the introduction of sp^3-hybridized carbon or addition of metals, while ensuring that chemisorption is suppressed [5–7]. Even still, low temperatures and high pressures are required to achieve acceptable hydrogen storage capacities. Previous *ab initio* studies have found hydrogen gravimetric storage capacities as large as 19 wt% on packed carbon nanotubes (CNTs) at 77 K and 100 atm [8].

Curved rather than planar carbon materials are beneficial because a larger sp^3-hybridized content in the structure makes hydrogen physisorption more favourable [7, 9]. Many studies have investigated such curved carbon surfaces. Most of these analyses were performed using only a small carbon surface with as little as five atoms [10–12]. Due to the small surface, these studies can miss the effects of multiple $C\cdots H_2$ interactions and they are vulnerable to edge effects.

To prevent this, the present study uses Density Functional Theory (DFT) to model entire C_n fullerenes where $n = 20, 60, 180, 540, 960$. These fullerenes are compared to infinite (m,m) single-walled CNTs (SWCNTs) where $m = 3, 5, 9$ and to large graphene fragments to study the effects of confinement. Finally, the effects of exohedral and substitutional platinum and non-metal substitutional dopants (boron, nitrogen and oxygen) are considered in an effort to increase the favourability of $C\cdots H_2$ interactions. The results presented in this paper use the B3LYP functional and LANL2MB basis set. More detailed results incorporating dispersion effects with more extensive basis sets will be presented in a future paper by the authors.

2 Computational method

Calculations were performed using DFT as implemented in *Gaussian 09* [13]. The hybrid B3LYP functional was used because of its reliability for modelling transition metals and it is not very computationally expensive [14, 15]. The LANL2MB basis set was also employed because its use of previously-determined effective core potentials [16–18] greatly decreases computation time. The B3LYP functional and LANL2MB basis set have been shown to produce reliable results for systems containing similar carbon substrates [19, 20]. Still, they do not consider the important dispersion effects. To account for this, the B97D functional was used in more detailed studies; these results will be presented in a future work. Systems studied with B3LYP and with B97D show similar trends; B3LYP is, therefore, acceptable for the purposes of this paper.

Calculations were performed on carbon systems containing one H_2 molecule, one exohedral Pt atom and one substitutional dopant atom where applicable. The carbon systems include entire C_n fullerenes ($n = 20, 60, 180, 540, 960$), (m,m) SWCNTs ($m = 3, 5, 9$) and graphene fragments. The nanotubes were represented as 8.5 Å segments with 1D periodic boundary conditions; the graphene sheet measured 14.5 x 14.8 Å and was hydrogen-terminated. This area is large enough that edge effects should be unimportant. Similar systems containing a large carbon surface with single dopants have been used successfully in previous modelling studies [19, 20].

Coordinates for the structures of the smaller fullerenes ($20 \leq n \leq 540$) were obtained from the *Computational Chemistry List, Ltd.* on-line fullerene database compiled by Cramer at the University of Minnesota. Coordinates for the C_{960} structure were provided by Henrard at the University of Namur. Coordinates for the SWCNTs were generated by TubeGen On-line (version 3.4) created by Frey and Doren at the University of Delaware. All systems were optimized without constraints.

3 Results and discussion

This study investigated the $C\cdots H_2$ interaction in a variety of C_n fullerenes (n = 20, 60, 180, 540, 960) (fig. 1) by analysing $C\cdots H_2$ and $H_2\cdots H_2$ bond distances and the fullerene–H_2 interaction energy, ΔE. This is given by eqn. (1) for systems containing only one H_2 molecule and by eqn. (2) for systems with multiple H_2.

$$\Delta E = E_{fullerene:H_2} - \left[E_{fullerene} + E_{H_2}\right] \quad (1)$$

$$\Delta E = E_{fullerene:nH_2} - \left[E_{fullerene:(n-1)H_2} + E_{H_2}\right] \quad (2)$$

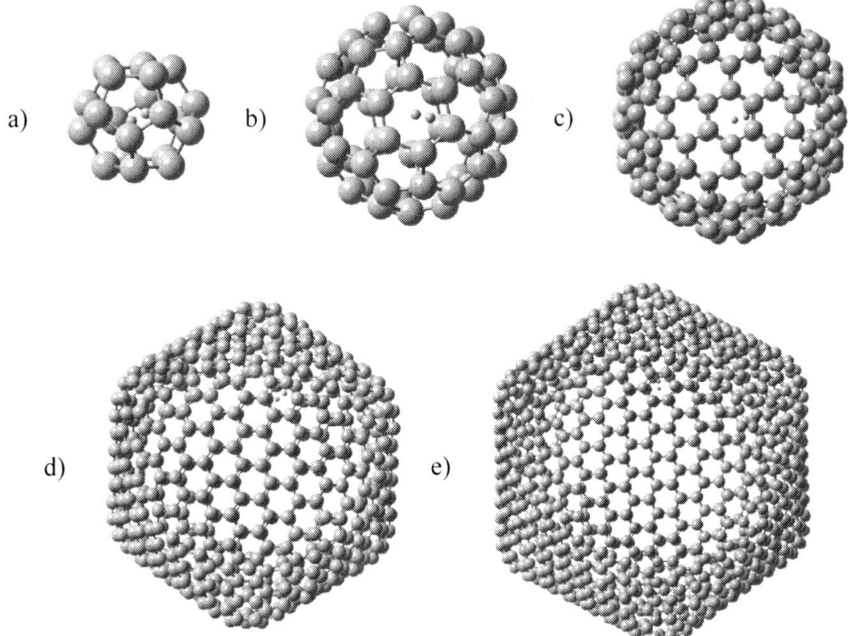

Figure 1: C_n fullerenes optimized with one internal H_2 molecule (n = 20, 60, 180, 540 and 960 for (a) – (e) respectively) (C = grey, H = blue).

Systems containing exohedral or substitution platinum (fig. 2) or no Pt were analysed with H2 stored inside the fullerene to determine the optimal C···H2 distance. C60 and C180 were further analysed to determine the possible effects of boron, nitrogen and oxygen substitutional dopants as well as multiple (2 – 6) H2 molecules (section 3.1).

Although platinum was included in the systems, Pt–H bonds are not formed because the initial position of H_2 inside the fullerene presents a large geometric and energetic barriers to chemisorption.

To understand the systems better, the smaller fullerenes (n = 20, 60, 180) were compared with (m,m) SWCNTs (m = 3, 5, 9) of similar diameter to determine the effect of confinement on hydrogen storage. The larger fullerenes, C_{540} and C_{960}, were compared with graphene to determine if these fullerenes are so large that confinement becomes negligible to the point that graphene itself is a good model for the system; in addition, large fullerenes contain flat surfaces that can be compared directly to flat graphene (section 3.2).

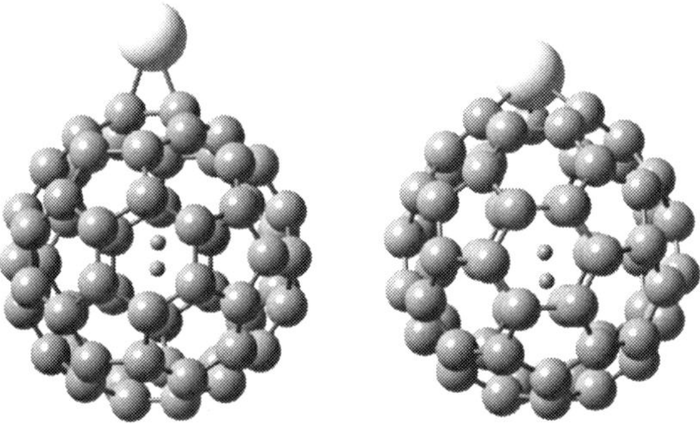

Figure 2: C_{60} fullerenes optimized with one internal H_2 molecule and exohedral (left) or substitutional (right) platinum (Pt = yellow, C = grey, H = blue).

3.1 Optimal C···H_2 distance

The C···H_2 distance is critical to determining the stability of a hydrogen storage system. This is observed, for example, with C_{20} fullerenes where confining even one H_2 molecule is energetically unfavourable by 525 kJ/mol with respect to the separate systems. This might be expected because the maximum possible C···H_2 distance is 1.9 Å. This is as much as 1 Å less than the combined van der Waals radii of carbon and hydrogen (1.7 and 1.2 Å respectively) and so leads to unfavourable Pauli repulsion.

When H_2 is placed inside larger fullerenes, the C···H_2 distance can be greater than 2.9 Å and so there are not significant repulsive interactions between carbon and hydrogen. The addition of further hydrogen to relatively small fullerenes such as C_{60}, however, becomes unfavourable because of the H_2···H_2 interactions. When there is only one H_2 molecule in C_{60}, the shortest C···H_2 distance is 2.9 Å. When additional H_2 molecules are added, the C···H_2 distances decrease to as

www.witpress.com, ISSN 1743-3533 (on-line)

little as 2.3 Å for 2 – 4 H_2 molecules and 2.2 Å for 5 – 6 H_2. Correspondingly, there is a repulsive interaction of 105 – 131 kJ/mol per H_2 for 2 – 4 H_2 molecules and 184 – 236 kJ/mol for 5 – 6 H_2 (fig. 3).

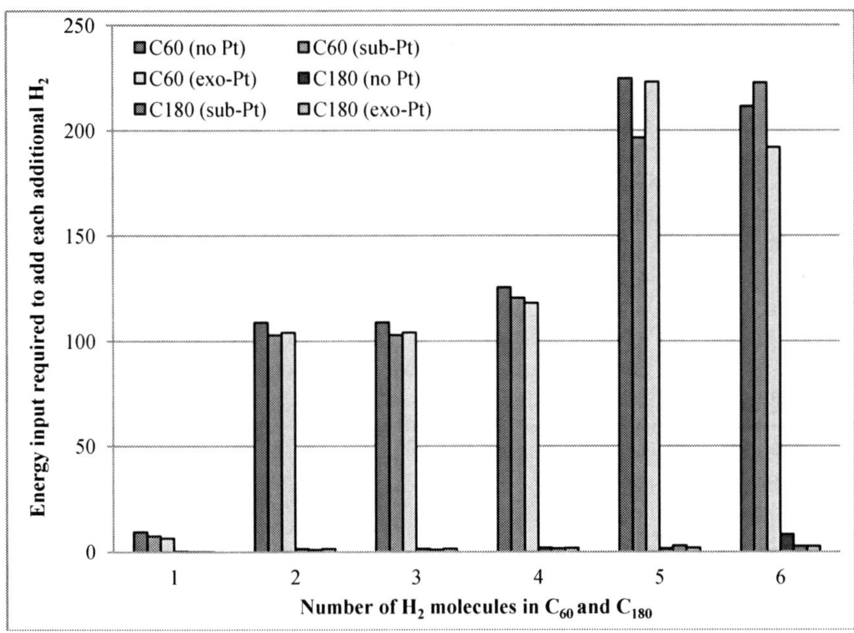

Figure 3: The energy input required to add additional H_2 molecules to C_{60} containing no platinum or substitutional or exohedral platinum dopants for 1 – 6 H_2 molecules.

In all cases the H_2 molecules orient in the centre of the fullerene to minimize their contact with the carbon surface and with each other (fig. 4). The molecules

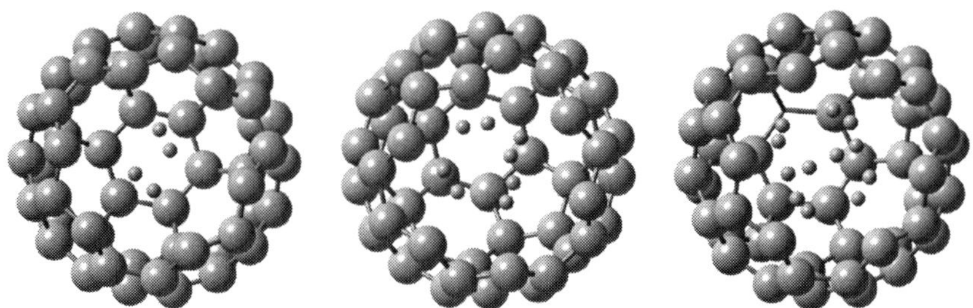

Figure 4: C_{60} fullerenes optimized with two (left), four (middle) and six (right) internal H_2 molecules. The topmost six carbon atoms have been excluded so that the H_2 molecules can be observed more clearly (C = grey, H = blue).

cluster in the centre in an attempt to keep the $H_2 \cdots H_2$ distances greater than 2 Å and the $C \cdots H_2$ distances greater than 2.9 Å. When the distances become smaller than these optimal values, repulsive interactions dominate.

Similar central clustering of the H_2 molecules is observed in C_{180} systems. However, up to six H_2 molecules (inclusive) can be placed within the fullerene without causing significant unfavourable $C \cdots H_2$ or $H_2 \cdots H_2$ interactions (fig. 3); this is consistent with the $C \cdots H_2$ distance remaining above 3 Å. This suggests that hydrogen can be stored in fullerenes if the system is large enough to allow the H_2 molecules to remain at least 3 Å from the carbon surface.

The additions of platinum and non-metal (boron, nitrogen, oxygen) dopants have a negligible effect on the $C \cdots H_2$ interactions. In previous studies of H_2 and CO adsorbed on a platinum/carbon substrate, alterations to the substrate had a notable effect on the substrate–gas interaction [19, 20]. The same effect is likely not observed here because in the previous study, the H_2 molecules were able to easily interact with the platinum; in contrast, there are significant physical boundaries in the present study. In addition, the weak H_2 interaction with the substrate is not influenced by the metal–carbon or dopant–carbon interactions as was observed in previous studies [21].

3.2 Confinement effects

The effects of 2D vs. 3D confinement on hydrogen storage within carbon nanomaterials were studied by comparing fullerenes with (*m,m*) single-walled carbon nanotubes of similar diameters (fig. 5). C_{20}, C_{60} and C_{180} were compared with (3,3), (5,5) and (9,9) SWCNTs respectively.

In addition, the larger fullerenes, C_{540} and C_{960}, were compared with graphene. Because the shape of fullerenes changes from spherical to polyhedral (fig. 1) and the volume within the fullerene increases as the number of carbon atoms increases, flat graphene is often taken as a good approximation of these systems. If this is correct then curvature and confinement effects should be negligible in large fullerenes.

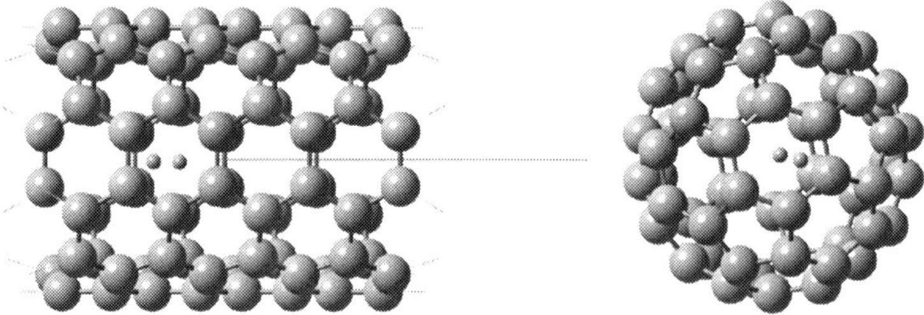

Figure 5: SWCNT (5,5) (left) and C_{60} (right) optimized with one internal H_2 molecule showing the comparable diameters of the systems (C = grey, H = blue, vector used for periodic boundary calculations = red).

www.witpress.com, ISSN 1743-3533 (on-line)

Analysis of H_2 in C_{20} and SWCNT (3,3) shows that it is unfavourable to store H_2 within either small-diameter system; however, the interaction energy is lower in the nanotube. The energy required to put a H_2 molecule into C_{20} (3D confinement) is over twice that required to place a H_2 into SWCNT (3,3) (2D confinement) (499 kJ/mol for C_{20} vs. 236 kJ/mol for SWCNT (3,3)). This is likely to result from the larger number of $C \cdots H_2$ interactions in the fullerene.

As expected, larger systems that have negligible $C \cdots H_2$ interactions in the highly confined structure (C_{60}, C_{180}) also show minimal interactions in their comparable nanotubes (SWCNT (5,5), SWCNT (9,9) respectively). When the larger polyhedron fullerenes, C_{540} and C_{960}, are compared with a graphene segment, negligible $C \cdots H_2$ interactions are observed in all systems. Therefore, it appears that confinement and curvature are insignificant in such large fullerenes and so flat graphene is an acceptable approximation.

4 Conclusion

The present study analysed H_2 stored within fullerenes and single-walled carbon nanotubes to understand $C \cdots H_2$ interactions in curved carbon nanomaterials. *Ab initio* calculations were used to model C_n fullerenes (n = 20, 60, 180, 540, 960) to determine the optimal $C \cdots H_2$ and $H_2 \cdots H_2$ distances. Hydrogen storage in fullerenes was also compared to that in SWCNTs (m,m) with similar diameters to determine the effects of confinement. Finally, the impact of exohedral and substitutional platinum as well as non-metal (boron, nitrogen, oxygen) substitutional dopants was also considered.

$C \cdots H_2$ interactions become negligible outside the combined van der Waals radii of carbon and hydrogen (1.7 and 1.2 Å respectively); $H_2 \cdots H_2$ interactions are unfavourable below 2 Å. Therefore, H_2 can be stored in fullerenes without an energy input when these distances are respected. The addition of platinum (exohedral, substitutional) or non-metal (substitutional) dopants does not affect this. Endohedral platinum was not considered because H_2 chemisorbed on Pt creates strong Pt–H bonds; these would require a large energy input to break to release H_2 and so are also undesirable in a storage material. The optimal Pt–H interaction would be an intermediate between chemisorption and the physisorption analysed in this study.

Confinement effects are observed when spherical fullerenes are compared to SWCNTs. This is most prominent in smaller structures as can be observed with H_2 stored in C_{20} versus SWCNT (3,3); storage in the nanotube requires less than half of the energy input per H_2 added. Confinement (and curvature) appears to be negligible in large fullerenes, such as C_{540} and C_{960}. It is possible that the addition of metals dopants has a larger effect in systems with lower confinement. These results are still under investigation, as are studies with more extensive basis sets and explicit account taken of dispersion.

www.witpress.com, ISSN 1743-3533 (on-line)

References

[1] Demirel, Y., *Energy production, conversion, storage, conservation and coupling:* Springer, 2012.

[2] Khan, M.T.I., M. Monde, and T. Setoguchi, *Hydrogen gas filling into an actual tank at high pressure and optimization of its thermal characteristics. Journal* of Thermal Science, 2009. **18**(3): p. 235-240.

[3] Jorgensen, S.W., *Hydrogen storage tanks for vehicles: Recent progress and current status.* Current Opinion in Solid State and Materials Science, 2011. **15**(2): p. 39-43.

[4] Tozzini, V. and V. Pellegrini, *Prospects for hydrogen storage in graphene.* arXiv preprint:1207.5703, 2012.

[5] Yürüm, Y., A. Taralp, and T.N. Veziroglu, *Storage of hydrogen in nanostructured carbon materials.* International Journal of Hydrogen Energy, 2009. **34**(9): p. 3784-3798.

[6] Xu, W.C., et al., *Investigation of hydrogen storage capacity of various carbon materials.* International Journal of Hydrogen Energy, 2007. **32**(13): p. 2504-2512.

[7] Tibbetts, G.G., G.P. Meisner, and C.H. Olk, *Hydrogen storage capacity of carbon nanotubes, filaments, and vapor-grown fibers.* Carbon, 2001. **39**(15): p. 2291-2301.

[8] Assfour, B., et al., *Packings of carbon nanotubes – new materials for hydrogen storage.* Advanced Materials, 2011. **23**(10): p. 1237-1241.

[9] Sofo, J.O., A.S. Chaudhari, and G.D. Barber, *Graphane: A two-dimensional hydrocarbon.* Physical Review B, 2007. **75**(15): p. 153401.

[10] Elias, D.C., et al., *Control of graphene's properties by reversible hydrogenation: evidence for graphane.* Science, 2009. **323**(5914): p. 610-613.

[11] Tozzini, V. and V. Pellegrini, *Reversible hydrogen storage by controlled buckling of graphene layers.* Journal of Physical Chemistry C, 2011. **115**(51): p. 25523-25528.

[12] Wang, L. and R.T. Yang, *Hydrogen storage properties of carbons doped with ruthenium, platinum, and nickel nanoparticles.* Journal of Physical Chemistry C, 2008. **112**(32): p. 12486-12494.

[13] Frisch, M., et al., *Gaussian 09*, 2009, Gaussian Inc.: Wallingford, CT.

[14] Lee, C., W. Yang, and R.G. Parr, *Development of the Colle-Salvetti correlation-energy formula into a functional of the electron density.* Physical Review B, 1988. **37**(2): p. 785.

[15] Chen, Z. and J. Yang, *The B3LYP hybrid density functional study on solids.* Frontiers of Physics in China, 2006. **1**(3): p. 339-343.

[16] Hay, P.J. and W.R. Wadt, *Ab initio effective core potentials for molecular calculations. Potentials for the transition metal atoms Sc to Hg.* Journal of Chemical Physics, 1985. **82**(1).

[17] Hay, P.J. and W.R. Wadt, *Ab initio effective core potentials for molecular calculations. Potentials for K to Au including the outermost core orbitals.* Journal of Chemical Physics, 1985. **82**: p. 299.

[18] Wadt, W.R. and P.J. Hay, *Ab initio effective core potentials for molecular calculations. Potentials for main group elements Na to Bi.* Journal of Chemical Physics, 1985. **82**: p. 284.

[19] Durbin, D.J. and C. Malardier-Jugroot, *Density functional theory analysis of metal/graphene systems as a filter membrane to prevent CO poisoning in hydrogen fuel cells.* Journal of Physical Chemistry C, 2010. **115**(3): p. 808-815.

[20] Durbin, D.J. and C. Malardier-Jugroot, *Theoretical investigation of the use of doped graphene as a membrane support for effective CO removal in hydrogen fuel cells.* Molecular Simulation, 2012. **38**(13): p. 1061–1071.

[21] Groves, M.N., C. Malardier-Jugroot, and M. Jugroot, *Improving platinum catalyst durability with a doped graphene support.* Journal of Physical Chemistry C, 2012. **116**(19): p. 10548-10556.

Development of two-dimensional magneto-hydrodynamic simulation code in cylindrical geometry using the discontinuous Galerkin finite element method

K. Lee[1], K.-J. Chung[2], D.-K. Kim[3], S.-G. Lee[1] & Y. S. Hwang[1,2]
[1]*Department of Nuclear Engineering, Seoul National University, Korea*
[2]*Center for Advance Research in Fusion Reactor Engineering, Seoul National University, Korea*
[3]*Agency for Defense Development, Korea*

Abstract

Discontinuous Galerkin finite element method (DG-FEM) is applied for modelling magneto-hydrodynamics of electrically discharging plasma channel (DPC). For the exact application of a material model for DPC and surrounding media, the Lagrangian forms of equations are formulated in cylindrical geometry. This paper presents procedures of formulation and development of two-dimensional simulation code which includes prescribed features with details. To simplify algorithms, element-matrix relations are obtained analytically using linear triangular element and linear shape function. Numerical schemes and sequence of calculation are also discussed with a numerical result to demonstrate the feasibility of proposed method.
Keywords: discontinuous Galerkin finite element formulation, Lagrangian simulation, magneto-hydrodynamics, electrical discharge.

1 Introduction

In the early stage of electrical discharge process in water, exploding material quickly turns into the plasma state of high pressure, which is called non-ideal plasma. Analysis of non-ideal plasmas requires advanced modelling of material properties and accurate numerical simulation due to its unsteady behaviour, as one can observe in underwater electrical wire explosion or water gap arc

WIT Transactions on Engineering Sciences, Vol 77, © 2013 WIT Press
www.witpress.com, ISSN 1743-3533 (on-line)
doi:10.2495/MC130141

discharge [1]. Also, equation of state (EOS) and transport parameters such as thermal and electrical conductivities of non-ideal plasma play an important role in the numerical simulation [2]. Therefore, research groups on this subject had developed their own magneto-hydrodynamic (MHD) simulation code [3, 4] independently to accommodate their material model for non-ideal plasma. Arbitrary Eulerian-Lagrangian (ALE) or Lagrangian formulation was adopted in those codes for the effective description of unsteady non-ideal plasmas with surrounding media.

In terms of numerical method, techniques solving MHD equations had been extensively discussed in various fields of science and engineering. The finite difference method (FDM) is most classical approach in the field of computational fluid dynamics (CFD) due to its ease of implementation. Also, the finite volume method (FVM) became popular recently, since it inherently satisfies the conservation laws. However, both methods contain essential limitations in the treatment of complex geometries and inhomogeneous material with higher order accuracy. On the other hand, finite element method (FEM) is accepted as a more powerful tool instructural and electromagnetic (EM) analysis because it is easy to handle complicated systems with unstructured mesh [5]. In virtue of its generality came from Galerkin approximation, it is even applicable to MHD problems. However, the size of an assembled global matrix increases rapidly with the size of system. Solving a large size of matrix equation in each time step is too expensive.

In recent years, the Discontinuous Galerkin (DG) method has been developed and quite impressive progress has been made both fields of CFD [6, 7] and EM [8]. DG method introduces numerical fluxes instead of assembling process, allowing us to perform element-wise calculation. In addition to all the merits of standard FEM, it is free to choose basis functions and shape of elements, and easy to parallelize the algorithms [6]. Thus the combination of DG and FEM, so called DG-FEM, is a truly attractive method to solve unsteady MHD problem.

As a part of code development, this paper presents Lagrangian formulation of resistive MHD equations in axi-symmetric cylindrical geometry using DG-FEM. The detailed discretization process for the MHD equation sets with DG-FEM using simple shape of element and shape function is described in Sec. 2. In Sec. 3, procedure of code development and numerical scheme are discussed with some remarks on time advancement techniques. Validation of the code for simple shock tube problem is also described in Sec. 3.

2 Formulation

2.1 Governing equations

We consider inviscid, compressible unsteady flow for discharging plasma channel. If we also assume the plasma is a single fluid, the single fluid MHD equations can be organized. By rearranging and substitutions of momentum conservation, Maxwell equations, Ohm's law and energy conservation, following

www.witpress.com, ISSN 1743-3533 (on-line)

set of MHD equations is derived in Lagrangian form. The meaning of notations and their dimensions are given in table 1.

Equation of motion: $$\vec{u} = \frac{\partial \vec{r}}{\partial t} , \quad (1)$$

Momentum equation: $$\rho \frac{D\vec{u}}{Dt} = -\vec{\nabla} p + \vec{J} \times \vec{B} , \quad (2)$$

Induction equation: $$\frac{D\vec{B}}{Dt} = \eta \nabla^2 \vec{B} + (\vec{u} \cdot \vec{\nabla}) \vec{B} , \quad (3)$$

Energy equation: $$\rho \frac{D\varepsilon}{Dt} = - p \vec{\nabla} \cdot \vec{u} + \vec{\nabla} \cdot (\kappa \vec{\nabla} T) + \frac{J^2}{\sigma} - Q_{rad} , \quad (4)$$

Ampere's law: $$\vec{\nabla} \times \vec{B} = \mu \vec{J} , \quad (5)$$

Equation of state: $$p = p(\rho, T) , \ \varepsilon = \varepsilon(\rho, T) , \quad (6)$$

Transport parameters: $$\kappa = \kappa(\rho, T) , \ \sigma = \sigma(\rho, T) . \quad (7)$$

Table 1: List of symbols with dimension.

Notation	Meaning	Dimension
$\vec{r}$	Position	[m]
$\vec{u}$	Velocity	[m/s]
ρ	Mass density	[kg/m^3]
v	Specific volume	[m^3/kg]
p	Pressure	[Pa]
$\vec{J}$	Electric current density	[A/m^2]
$\vec{B}$	Magnetic flux density	[T]
σ	Electrical conductivity	[S/m]
κ	Thermal conductivity	[W/m·K]
η	Magnetic diffusivity	[m^2/s]
ε	Internal energy density	[J/kg]
Q_{rad}	Radiative power loss per unit volume	[W/m^3]

2.2 Discontinuous Galerkin finite element formulation

Azimuthal-symmetry $(\partial / \partial \theta = 0)$ in cylindrical geometry is necessary to reduce those general equations to two dimensional equations. According to Ampere's law, there is only azimuthal component for magnetic flux density. Splitting second-order equations into two first-order differential equations allows us make a system of first-order partial differential equations. We introduced new notations $\varepsilon_T = \partial \varepsilon / \partial T$, $\varepsilon_v = \partial \varepsilon / \partial v$ to express energy equation in terms of temperature by chain rule, since most EOS models and material data are organized with respect to mass density (ρ) and temperature (T) [1].

Derivation of weak forms of eqns (2)–(4) with Galerkin FEM are given as eqns (8)–(15). Note that the eqns (10), (11), (13), (14) show the definition of the newly introduced variables q_T and q_B. Also, Ψ stands for a set of weighting functions which belongs to the same function space (V_h) with the shape functions when Galerkin method is applied. Thus the subscript h means approximation with basis of V_h.

Momentum equations:

$$\int_{\Omega_e}\left(\rho\dot{u}_{r,h}+\frac{\partial p_h}{\partial r}+\frac{B_\theta}{\mu}\frac{\partial B_{\theta,h}}{\partial r}+\frac{B_\theta}{\mu r}B_{\theta,h}\right)\Psi\, rdr\, dz=0\ , \tag{8}$$

$$\int_{\Omega_e}\left(\rho\dot{u}_{z,h}+\frac{\partial p_h}{\partial z}+\frac{B_\theta}{\mu}\frac{\partial B_{\theta,h}}{\partial z}\right)\Psi\, rdr\, dz=0\ . \tag{9}$$

Induction equations:

$$\int_{\Omega_e}\left(q_{Br,h}-\frac{\partial B_{\theta,h}}{\partial r}\right)\Psi\, rdr\, dz=0\ , \tag{10}$$

$$\int_{\Omega_e}\left(q_{Bz,h}-\frac{\partial B_{\theta,h}}{\partial z}\right)\Psi\, rdr\, dz=0\ , \tag{11}$$

$$\int_{\Omega_e}\Big(\dot{B}_{\theta,h}-\eta\left[\frac{\partial q_{Br,h}}{\partial r}+\frac{\partial q_{Bz,h}}{\partial z}+\frac{q_{Br,h}}{r}-\frac{B_{\theta,h}}{r^2}\right] -\left[u_r q_{Br,h}+u_z q_{Bz,h}\right]\Big)\Psi\, rdr\, dz=0\ . \tag{12}$$

Energy equations:

$$\int_{\Omega_e}\left(q_{Tr,h}-\kappa\frac{\partial T_h}{\partial r}\right)\Psi\, rdr\, dz=0\ , \tag{13}$$

$$\int_{\Omega_e}\left(q_{Tz,h}-\kappa\frac{\partial T_h}{\partial z}\right)\Psi\, rdr\, dz=0\ , \tag{14}$$

$$\int_{\Omega_e}\left(\varepsilon_T\dot{T}_h+(p+\varepsilon_v)\dot{v}\ -v\left[\vec{\nabla}\cdot(\vec{q}_T)+\frac{J^2}{\sigma}-Q_{rad}\right]\right)\Psi\, rdr\, dz=0\ . \tag{15}$$

In order to make clear and simple formulation using DG-FEM, we chose linear triangular element and linear shape function. Feature of arbitrary element is shown in fig. 1. DG numerical fluxes [9] were adopted to treat boundary line integration of element. Notations for jump [] and mean value {} are also defined to specify the numerical fluxes. Variables ϕ and $\vec{q}$ stand for general variable and its gradient, respectively.

Jumps and Mean values:

$$\begin{aligned}&\{\phi\}=\frac{1}{2}(\phi+\phi_{NB})\ ,\quad \{\vec{q}\}=\frac{1}{2}(\vec{q}+\vec{q}_{NB})\ ,\\ &[\phi]=(\phi-\phi_{NB})\cdot\hat{n}\ ,\quad [\vec{q}]=(\vec{q}-\vec{q}_{NB})\cdot\hat{n}\ ,\end{aligned} \tag{16}$$

Numerical fluxes (DG model):

$$\hat{\phi}_h = \{\phi_h\} + \mathbf{C}_{12}[\phi_h] - C_{22}[\vec{q}_h],$$
$$\hat{\vec{q}}_h = \{\vec{q}_h\} - C_{11}[\phi_h] - \mathbf{C}_{12}[\vec{q}_h]. \quad (17)$$

Now, above equations can be discretized in space by applying the Gauss-divergence theorem to treat the area integral over the element, as we do in usual finite element formulation [11]. In order to simplify the notations, all the terms are categorized into seven types. Please refer to the Appendix for the definition of notations. The subscript l denotes the coordinates, i.e. $l=r$, z.

Type I:

$$\frac{\partial \phi_h}{\partial r} \rightarrow \sum_{s=1}^{3} \mathbf{D}_{r,s}\underline{\phi} + \sum_{s=1}^{3} \mathbf{A}_{r,s}\underline{\phi}_{NB,s} - \sum_{s=1}^{3} \mathbf{J}_{rr,s}\underline{q}_r + \sum_{s=1}^{3} \mathbf{J}_{rz,s}\underline{q}_z - \mathbf{H}_r\underline{\phi}, \quad (18)$$

Type II:

$$\frac{\partial \phi_h}{\partial z} \rightarrow \sum_{s=1}^{3} \mathbf{D}_{z,s}\underline{\phi} + \sum_{s=1}^{3} \mathbf{A}_{z,s}\underline{\phi}_{NB,s} - \sum_{s=1}^{3} \mathbf{J}_{rz,s}\underline{q}_r + \sum_{s=1}^{3} \mathbf{J}_{zz,s}\underline{q}_z - \mathbf{H}_z\underline{\phi}, \quad (19)$$

Type III:

$$\frac{\partial q_{hl}}{\partial r} \rightarrow \sum_{s=1}^{3} \mathbf{A}_{l,s}\underline{q}_l + \sum_{s=1}^{3} \mathbf{D}_{l,s}\underline{q}_{l,NB,s} - \sum_{s=1}^{3} \mathbf{G}_{ll,s}\underline{\phi} + \sum_{s=1}^{3} \mathbf{G}_{ll,s}\underline{\phi}_{NB,s} - \mathbf{H}_r\underline{q}_l, \quad (20)$$

Type IV:

$$\frac{\partial q_{hl}}{\partial z} \rightarrow \sum_{s=1}^{3} \mathbf{A}_{l,s}\underline{q}_l + \sum_{s=1}^{3} \mathbf{D}_{l,s}\underline{q}_{l,NB,s} - \sum_{s=1}^{3} \mathbf{G}_{ll,s}\underline{\phi} + \sum_{s=1}^{3} \mathbf{G}_{ll,s}\underline{\phi}_{NB,s} - \mathbf{H}_z\underline{q}_l, \quad (21)$$

Type V:

$$\vec{\nabla} \cdot \vec{q} \rightarrow \sum_{s=1}^{3} \mathbf{A}_{r,s}\underline{q}_r + \sum_{s=1}^{3} \mathbf{A}_{z,s}\underline{q}_z + \sum_{s=1}^{3} \mathbf{D}_{r,s}\underline{q}_{r,NB,s} + \sum_{s=1}^{3} \mathbf{D}_{z,s}\underline{q}_{z,NB,s}$$
$$- \sum_{s=1}^{3} \mathbf{Q}_s\underline{\phi} + \sum_{s=1}^{3} \mathbf{Q}_s\underline{\phi}_{NB,s} - \mathbf{H}_r\underline{q}_r - \mathbf{H}_z\underline{q}_z, \quad (22)$$

Type VI:

$$\phi_h,\ q_{lh} \rightarrow r_m\mathbf{E}\underline{\phi} \ \ and \ \ r_m\mathbf{E}\underline{q}_l, \quad (23)$$

Type VII:

$$\text{c (const.)} \rightarrow c r_m \underline{Y}. \quad (24)$$

Substituting expressions (18~24) into each terms appear in eqns (8)–(15) leads us to obtain the element-matrix relations for each equation. Note that the non-linear terms related to magnetic flux density have been linearized. Also, mean radius approximation is taken in the integrands of eqns (8)–(15) to avoid singularity [12] at the axis. Finally, the resultant element-matrix relations are obtained in eqns. (25)–(32).

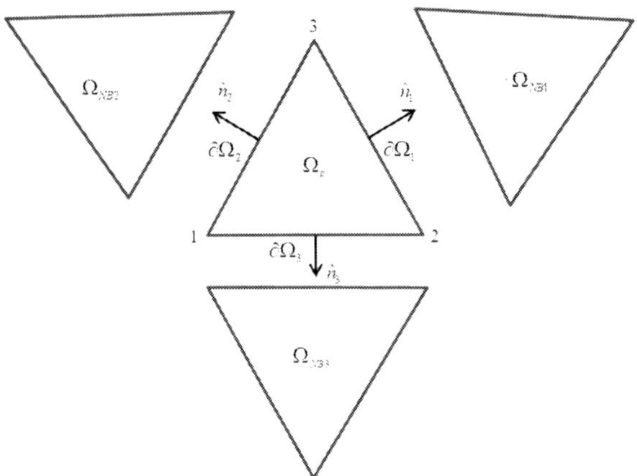

Figure 1: Sketch of an element and its neighbors.

Momentum equations:

$$\left(\mathbf{E}\underline{\dot{u}}_r\right) = -\frac{\nu}{r_m}\left(\sum_{s=1}^{3}\mathbf{N}_{r,s}\underline{p} + \sum_{s=1}^{3}\mathbf{N}_{r,s}\underline{p}_{NB,s} - \mathbf{H}_r\underline{p}\right)$$
$$-\frac{\nu B_m}{\mu r_m}\left(\sum_{s=1}^{3}\mathbf{D}_{r,s}\underline{B} + \sum_{s=1}^{3}\mathbf{A}_{r,s}\underline{B}_{NB,s} - \sum_{s=1}^{3}\mathbf{J}_{rr,s}\underline{q}_{Br} + \sum_{s=1}^{3}\mathbf{J}_{rz,s}\,\underline{q}_{Bz} - \mathbf{H}_r\underline{B}\right), \tag{25}$$

$$\left(\mathbf{E}\underline{\dot{u}}_z\right) = -\frac{\nu}{r_m}\left(\sum_{s=1}^{3}\mathbf{N}_{z,s}\underline{p} + \sum_{s=1}^{3}\mathbf{N}_{z,s}\underline{p}_{NB,s} - \mathbf{H}_z\underline{p}\right)$$
$$-\frac{\nu B_m}{\mu r_m}\left(\sum_{s=1}^{3}\mathbf{D}_{z,s}\underline{B} + \sum_{s=1}^{3}\mathbf{A}_{z,s}\underline{B}_{NB,s} - \sum_{s=1}^{3}\mathbf{J}_{rz,s}\underline{q}_{Br} + \sum_{s=1}^{3}\mathbf{J}_{zz,s}\,\underline{q}_{Bz} - \mathbf{H}_z\underline{B}\right), \tag{26}$$

Induction equations:

$$\left(\mathbf{E}\underline{q}_{Br}\right) = \frac{1}{r_m}\left(\sum_{s=1}^{3}\mathbf{D}_{r,s}\underline{B} + \sum_{s=1}^{3}\mathbf{A}_{r,s}\underline{B}_{NB,s} - \sum_{s=1}^{3}\mathbf{J}_{rr,s}\underline{q}_{Br} + \sum_{s=1}^{3}\mathbf{J}_{rz,s}\,\underline{q}_{Bz} - \mathbf{H}_r\underline{B}\right), \tag{27}$$

$$\left(\mathbf{E}\underline{q}_{Bz}\right) = \frac{1}{r_m}\left(\sum_{s=1}^{3}\mathbf{D}_{z,s}\underline{B} + \sum_{s=1}^{3}\mathbf{A}_{z,s}\underline{B}_{NB,s} - \sum_{s=1}^{3}\mathbf{J}_{rz,s}\underline{q}_{Br} + \sum_{s=1}^{3}\mathbf{J}_{zz,s}\,\underline{q}_{Bz} - \mathbf{H}_z\underline{B}\right), \tag{28}$$

$$\left(\mathbf{E}\underline{\dot{B}}\right) = \frac{\eta}{r_m}\left(\sum_{s=1}^{3}\mathbf{A}_{r,s}\underline{q}_{Br} + \sum_{s=1}^{3}\mathbf{D}_{r,s}\underline{q}_{Br,NB,s} - \sum_{s=1}^{3}\mathbf{G}_{rr,s}\underline{B} + \sum_{s=1}^{3}\mathbf{G}_{rr,s}\underline{B}_{NB,s} - \mathbf{H}_r\underline{q}_{Br}\right.$$
$$\left.+\sum_{s=1}^{3}\mathbf{A}_{z,s}\underline{q}_{Bz} + \sum_{s=1}^{3}\mathbf{D}_{z,s}\underline{q}_{Bz,NB,s} - \sum_{s=1}^{3}\mathbf{G}_{zz,s}\underline{B} + \sum_{s=1}^{3}\mathbf{G}_{zz,s}\underline{B}_{NB,s} - \mathbf{H}_z\underline{q}_{Bz}\right) \tag{29}$$
$$+\mathbf{E}\left(\frac{\eta}{r_m}\underline{q}_{Br} - \frac{\eta}{r_m^{\,2}}\underline{B} + u_{rm}\underline{q}_{Br} + u_{zm}\underline{q}_{Bz}\right),$$

Energy equations:

$$\left(\mathbf{E} + \frac{1}{r_m}\sum_{s=1}^{3}\mathbf{J}_{rr,s}\right)\underline{q}_{Tr} = \frac{1}{r_m}\left(\sum_{s=1}^{3}\mathbf{D}_{r,s}\kappa\underline{T} + \sum_{s=1}^{3}\mathbf{A}_{r,s}\kappa\underline{T}_{NB,s} + \sum_{s=1}^{3}\mathbf{J}_{rz,s}\,\underline{q}_{Tz} - \mathbf{H}_r\kappa\underline{T}\right), \tag{30}$$

$$\left(\mathbf{E}-\frac{1}{r_m}\sum_{s=1}^{3}\mathbf{J}_{zz,s}\right)\underline{q}_{Tz}=\frac{1}{r_m}\left(\sum_{s=1}^{3}\mathbf{D}_{z,s}\kappa\underline{T}+\sum_{s=1}^{3}\mathbf{A}_{z,s}\kappa\underline{T}_{NB,s}-\sum_{s=1}^{3}\mathbf{J}_{rz,s}\underline{q}_{Tr}-\mathbf{H}_z\kappa\underline{T}\right), \tag{31}$$

$$\begin{aligned}\varepsilon_T\left(\mathbf{E}\dot{\underline{T}}\right)&=-\dot{v}\left(\mathbf{E}\underline{p}\right)-\left(\dot{v}\varepsilon_v-v\frac{J^2}{\sigma}+vQ_{rad}\right)\left(\underline{Y}\right)+\frac{v}{r_m}\left(\sum_{s=1}^{3}\mathbf{A}_{r,s}\underline{q}_r+\sum_{s=1}^{3}\mathbf{A}_{z,s}\underline{q}_z\right.\\&\left.+\sum_{s=1}^{3}\mathbf{D}_{r,s}\underline{q}_{r,NB,s}+\sum_{s=1}^{3}\mathbf{D}_{z,s}\underline{q}_{z,NB,s}-\sum_{s=1}^{3}\mathbf{Q}_s\kappa\underline{T}+\sum_{s=1}^{3}\mathbf{Q}_s\kappa_{NB}\underline{T}_{NB,s}-\mathbf{H}_r\underline{q}_r-\mathbf{H}_z\underline{q}_z\right).\end{aligned} \tag{32}$$

3 Numerical schemes and code development

3.1 Numerical schemes

Eqns (26), (29) and (32) require time integration for a numerical solution of transient problem. Explicit schemes are usually preferred such as second- or higher-order Runge-Kutta methods (RK2, RK3,...), since they enhance the advantages of DG-FEM formulation. As a beginning phase of development, we adopted Explicit Euler (EE) method for the velocity and temperature, and RK2 for the position. Temporal accuracy can be improved later when the feasibility and suitability of formulation are acquired.

The induction equation is basically classified into magnetic diffusion equation. Unfortunately, magnetic diffusion time is seriously short for some cases such as electrical wire explosion in laboratory (~10^{-16} sec). This time scale is much smaller than that of momentum and energy equations. Hence, using common time step satisfying CFL condition for induction equation is too weary to solve entire governing equations simultaneously. Therefore, we chose to solve equations one by one at a time.

In fact, various techniques to solve induction equation based on FEM had been reported. Thus, we can try one of those methods very flexibly, depending on the physical features of the problem. For example, magnetic diffusion in Z-pinch physics is handled with standard FEM formulation with implicit time stepping scheme [3] due to its large electric current (~MA) and complicated system. On the other hand, explicit scheme "cross" with FVM formulation has been used to solve all the MHD equations in the same time step for the one-dimensional UEWE analysis [4]. Also, the Crank-Nicolson time stepping method could be used with DG-FEM formulation [13] by dividing sub-regions.

From now on, we will drop the further discussions related to solving induction equations and will treat them in a future work. Note that even if we dropped the induction equations from the set of governing equations, we can still handle the problems such as simple thermodynamic problem such as cylindrical shock tube problem or MHD phenomena with mild current (~kA) such as UEWE and water gap arc discharges. Now the models for EOS and transport coefficients became much more important than the calculation of magnetic fields, for the problems just mentioned.

3.2 Computational mesh

As a part of pre-process, linear triangular area mesh is created using *Gmsh 2.6.1* [14]. Although this program is freeware, it provides four modules: geometry, mesh, solver and post-processing. But we utilized it only for generating the mesh file.

The drawback of *Gmsh* is that we cannot control the exact number of elements. Only the length of sides of edge element can be specified instead. However, as of our current status, this does not cause any serious problem.

3.3 Code and validation

Code has been written in C language. As already mentioned, all equations are solved one by one, based on the state of previous time step (see fig. 2.). First, the code reads mesh file and finds out the total number of elements and nodes. Then, required memory can be allocated to node and element data structures. After that, the code initialises variables. Geometrical data are also calculated such as surface outward normal vectors, length of sides, coordinates of center of mass, and the area of element from the data of mesh file. Physical data for the time, subzones, nodes and elements also can be initialized depending on the problem and material. Since we are using the linear triangular element, Jacobian and all the derivatives can be calculated analytically. Hence there is no need to use Gaussian quadrature for the numerical integration. This fact makes the code much nice and fast. After initialization, solver begins element-wise calculations for momentum equations and energy equations by direct inversion of element matrix relations. Now, thermodynamic properties can be obtained from EOS model. Thermal and

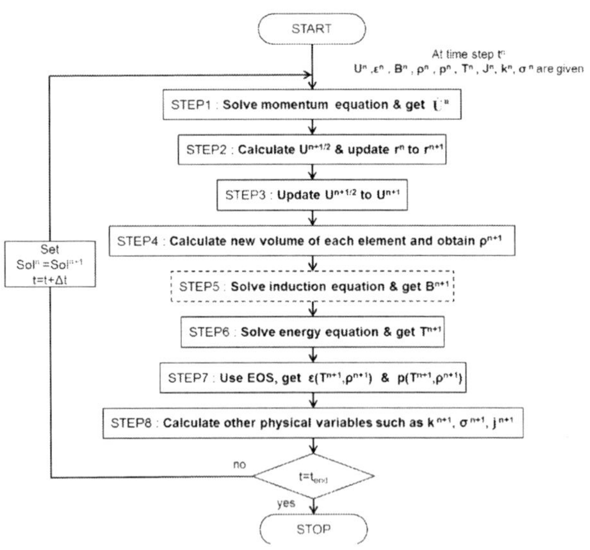

Figure 2: Flowchart of calculation.

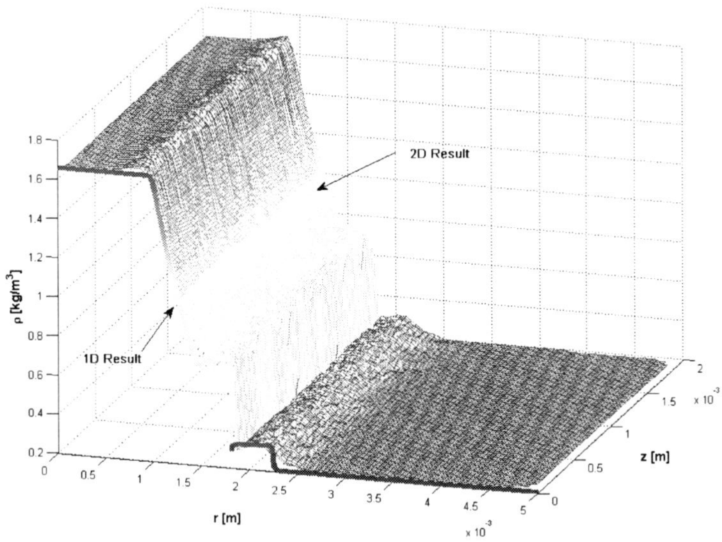

Figure 3: Numerical result of cylindrical shock tube problem (t=1.6 μs).

electrical conductivities are also calculated from their physics models. The code will repeat the time advancement based on the data of current time step.

As a part of validation, crude numerical results for the cylindrical shock tube problem are presented in Fig. 3. The test gas is Argon and ideal EOS is applied. Both sides are initially in thermal equilibrium and the diaphragm was located at r=1.5mm. As we can see, results from 1D and 2D code agree to each other very well. However, there exists a bump behind the rarefaction wave, which is originated from numerical error. It should be eliminated by optimisation of code.

More strict validation can be made by comparing with the experimental results such as UEWE or water gap arc discharges with various EOS and transport parameter models [1, 15]. It should be the next work after this one.

4 Conclusion

The DG-FEM formulation of the resistive MHD equations and the development of simulation code have been described. The noticeable features of our code are: (i) to solve Lagrangian form of equations in cylindrical geometry, (ii) to handle inhomogeneous materials, and (iii) to accommodate independent EOS and transport parameter models. Although this is not the first trial to solving MHD equations with DG-FEM [16–18], the distinct procedure for the numerical implementation with those unique features has been established in this work. Further studies such as rigorous validation of the code with experimental results and advanced techniques solving induction equation will be handled in the future work.

Acknowledgements

This work was supported by the Defense Research Laboratory Program of the Defense Acquisition Program Administration and the Agency for Defense Development of Republic of Korea.

Appendix

$$\underline{Y} = \int_{\Omega_e} \Psi \, dr\, dz = \frac{S_e}{3}\begin{bmatrix}1 & 1 & 1\end{bmatrix}^T , \quad \mathbf{E} = \int_{\Omega_e} \Psi\Psi^T \, dr\, dz = \frac{S_e}{12}\begin{bmatrix}2 & 1 & 1\\ 1 & 2 & 1\\ 1 & 1 & 2\end{bmatrix},$$

where $S_e = \frac{1}{2}\left| r_{31}\cdot z_{21} - r_{21}\cdot z_{31}\right|$, $(z_{ij} = z_i - z_j,\ r_{ij} = r_i - r_j)$

$$\mathbf{H}_r = \int_{\Omega_e} r_m \frac{\partial \Psi}{\partial r}\Psi^T \, dr\, dz = \frac{r_m}{6}\begin{bmatrix} z_{23} & z_{23} & z_{23}\\ z_{31} & z_{31} & z_{31}\\ z_{12} & z_{12} & z_{12}\end{bmatrix}, \quad \mathbf{H}_z = \int_{\Omega_e} r_m \frac{\partial \Psi}{\partial z}\Psi^T \, dr\, dz = \frac{r_m}{6}\begin{bmatrix} r_{32} & r_{32} & r_{32}\\ r_{13} & r_{13} & r_{13}\\ r_{21} & r_{21} & r_{21}\end{bmatrix},$$

$$\mathbf{S}_s = \int_{\partial\Omega_s} \Psi\Psi^T ds\ , \ (s = 1,2,3\ ; \text{"side index"});$$

$$\mathbf{S}_1 = \frac{L_1}{6}\begin{bmatrix}0 & 0 & 0\\ 0 & 2 & 1\\ 0 & 1 & 2\end{bmatrix}, \ \mathbf{S}_2 = \frac{L_2}{6}\begin{bmatrix}2 & 0 & 1\\ 0 & 0 & 0\\ 1 & 0 & 2\end{bmatrix}, \ \mathbf{S}_3 = \frac{L_3}{6}\begin{bmatrix}2 & 1 & 0\\ 1 & 2 & 0\\ 0 & 0 & 0\end{bmatrix},$$

$$\mathbf{N}_{r,s} = (\frac{1}{2}) r_m n_r \int_{\partial\Omega_s} \Psi\Psi^T ds = (\frac{1}{2}) r_m n_{r,s} \mathbf{S}_s \ ,$$

$$\mathbf{N}_{z,s} = (\frac{1}{2}) r_m n_z \int_{\partial\Omega_s} \Psi\Psi^T ds = (\frac{1}{2}) r_m n_{z,s} \mathbf{S}_s \ ,$$

$$\mathbf{D}_{r,s} = (\frac{1}{2} + C_{12}) r_m n_r \int_{\partial\Omega_s} \Psi\Psi^T ds = (\frac{1}{2} + C_{12}) r_m n_{r,s} \mathbf{S}_s \ ,$$

$$\mathbf{D}_{z,s} = (\frac{1}{2} + C_{12}) r_m n_z \int_{\partial\Omega_s} \Psi\Psi^T ds = (\frac{1}{2} + C_{12}) r_m n_{z,s} \mathbf{S}_s \ ,$$

$$\mathbf{A}_{r,s} = (\frac{1}{2} - C_{12}) r_m n_r \int_{\partial\Omega_s} \Psi\Psi^T ds = (\frac{1}{2} - C_{12}) r_m n_{r,s} \mathbf{S}_s \ ,$$

$$\mathbf{A}_{z,s} = (\frac{1}{2} - C_{12}) r_m n_z \int_{\partial\Omega_s} \Psi\Psi^T ds = (\frac{1}{2} - C_{12}) r_m n_{z,s} \mathbf{S}_s \ ,$$

$$\mathbf{J}_{rr,s} = \left(C_{22} r_m n_r\, n_r\right)\int_{\partial\Omega_s} \Psi\Psi^T \, ds = \left(C_{22} r_m n_{r,s}\, n_{r,s}\right)\mathbf{S}_s \ ,$$

$$\mathbf{J}_{rz,s} = \left(C_{22} r_m n_r\, n_z\right)\int_{\partial\Omega_s} \Psi\Psi^T \, ds = \left(C_{22} r_m n_{r,s}\, n_{z,s}\right)\mathbf{S}_s \ ,$$

$$\mathbf{J}_{zz,s} = \left(C_{22} r_m n_z\, n_z\right)\int_{\partial\Omega_s} \Psi\Psi^T \, ds = \left(C_{22} r_m n_{z,s}\, n_{z,s}\right)\mathbf{S}_s \ ,$$

$$\mathbf{G}_{rr,s} = \left(C_{11} r_m n_r\, n_r\right)\int_{\partial\Omega_s} \Psi\Psi^T \, ds = \left(C_{11} r_m n_{r,s}\, n_{r,s}\right)\mathbf{S}_s \ ,$$

$$\mathbf{G}_{zz,s} = \left(C_{11} r_m n_z\, n_z\right) \int_{\partial\Omega_s} \Psi\Psi^T \, ds = \left(C_{11} r_m n_z\, n_z\right) \mathbf{S}_s \ ,$$

$$\mathbf{Q}_s = \left(C_{11} r_m\right) \int_{\partial\Omega_s} \Psi\Psi^T \, ds = \left(C_{11} r_m\right) \mathbf{S}_s \ .$$

References

[1] Chung, K. J. and Hwang,Y. S., Thermodynamic Properties and Electrical Conductivity of Water Plasma, *Contrib. Plasma Phys. (to be published),* vol. 53, 2013.

[2] Sheftman,D. and Krasik,Y. E., Evaluation of electrical conductivity and equations of state of non-ideal plasma through microsecond timescale underwater electrical wire explosion, *Physics of Plasmas,* vol. 18, pp. 092704-8, 2011.

[3] Garasi, C. J., Bliss, D. E., Mehlhorn, T. A., Oliver, B. V., Robinson, A. C., and Sarkisov, G. S., Multi-dimensional high energy density physics modeling and simulation of wire array Z-pinch physics, *Physics of Plasmas,* vol. 11, pp. 2729-2737, 2004.

[4] Grinenko, A., Gurovich,V. T., Saypin, A., Efimov, S., Krasik, Y. E., and Oreshkin, V. I., Strongly coupled copper plasma generated by underwater electrical wire explosion, *Physical Review E,* vol. 72, p. 066401, 2005.

[5] Jin, J. M., *The finite element method in electromagnetics*, 2nd ed., Wiley: New York, 2002.

[6] Li, B. Q., *Discontinuous Finite Elements in Fluid Dynamics and Heat Transfer*, Springer-Verlag: Germany, 2006.

[7] Cockburn, B., Discontinuous Galerkin methods, *ZAMM - Journal of Applied Mathematics and Mechanics,* vol. 83, pp. 731-754, 2003.

[8] Gedney, S., Chong, D. L., Guernsey, B., Roden, J. A., Crawford, R., and Miller,J. A., The Discontinuous Galerkin Finite Element Time Domain Method (DGFETD): A High Order, Globally-Explicit Method for Parallel Computation, in *EMC 2007. IEEE International Symposium on Electromagnetic Compatibility, 2007.* , 2007, pp. 1-3.

[9] Kirby, R. and Karniadakis, G., Selecting the Numerical Flux in Discontinuous Galerkin Methods for Diffusion Problems, *Journal of Scientific Computing,* vol. 22-23, pp. 385-411, 2005.

[10] Castillo, P., Cockburn, B., Perugia, I., and Schötzau, D., An A Priori Error Analysis of the Local Discontinuous Galerkin Method for Elliptic Problems, *SIAM Journal on Numerical Analysis,* vol. 38, pp. 1676-1706, 2000.

[11] Lyon, P. E., Axisymmetric finite element modeling for the design and analysis of cylindrical adhesive joints based on dimensional stability, Master of Science, Mechanical and Aerospace Engineering, UTAH STATE UNIVERSITY, Logan, Utah, 2011.

[12] Clayton, J. D. and Rencis, J. J., Numerical integration in the axisymmetric finite element formulation, *Adv. Eng. Softw.,* vol. 31, pp. 137-141, 2000.

[13] Smolentsev, S., Morley, N., and Abdou, M., Code development for analysis of MHD pressure drop reduction in a liquid metal blanket using insulation technique based on a fully developed flow model, *Fusion Engineering and Design,* vol. 73, pp. 83-93, 2005.

[14] Geuzaine,C. and Remacle, J.-F., http://geuz.org/gmsh/

[15] Kim, I., Kim, D.-K., Baek, S.-H., and Song, S.-Y., Magnetohydrodynamic Behaviour of Warm Dense Plasmas Created by Underwater Wire Explosion in *33rd EPS Conference on Plasma Phys.*, Rome, 19 - 23 June 2006, 2006, p. P5.064.

[16] Warburton, T. C. and Karniadakis, G. E., A Discontinuous Galerkin Method for the Viscous MHD Equations, *Journal of Computational Physics,* vol. 152, pp. 608-641, 7/1/ 1999.

[17] Halashi, B. K., Luo, H., Spicer, D., and MacNiece, P., A Discontinuous Galerkin Method for the Magnetohydrodynamics on Arbitrary Grids, in *50th AIAA Aerospace Sciences Meeting including the New Horizons Forum and Aerospace Exposition, 09 - 12 January 2012*, Nashville, Tennessee, 2012.

[18] Houston, P., Schoetzau, D., and Wei, X., A Mixed Discontinuous GalerkinMethod for Incompressible Magnetohydrodynamics, *Journal of Scientific Computing,* 2008.

Fracture evaluation of metallic materials at intermediate strain rates

S. J. Lim[1], K. H. Ahn[1], H. Huh[1], S. B. Kim[2] & H. W. Kim[2]
[1]*School of Mechanical, Aerospace and System Engineering, Republic of Korea*
[2]*Agency for Defense Development, Republic of Korea*

Abstract

This paper evaluates existing dynamic hardening models and fracture strains for metallic materials of typical crystalline structures: 4130 steel (BCC); OFHC copper (FCC); and Ti6Al4V (HCP). Uniaxial tensile and fracture tests of the representative materials are carried out at intermediate strain rate. In order to describe the dynamic hardening characteristics, several models are compared for each material and the most applicable model for a specific material has been recommended. The fracture tests at strain rates ranging from $0.001s^{-1}$ to $100s^{-1}$ are conducted with several types of the pure shear, dog bone and grooved specimens for the strain paths of the pure shear, uniaxial tension and plain strain conditions. The fracture strain and the strain path are measured on the surface of the specimen and the Lou-Huh ductile fracture criterion is implemented to describe the fracture loci with respect to the strain rate.
Keywords: ductile fracture criterion, strain rate, 4130 steel, OFHC copper, Titanium alloy.

1 Introduction

Military weapons such as a missile explode under high impact and pressure conditions. Because weapons are designed using numerical simulations in order to save time and cost, the accurate material properties need to be secured under the extreme condition. In this paper, the uniaxial tensile and fracture tests of 4130 steel (BCC), OFHC copper (FCC) and Ti6Al4V (HCP) at strain rates ranging from $0.001s^{-1}$ to $100s^{-1}$ are executed as the first work of measuring the dynamic material properties of representative materials for a warhead.

WIT Transactions on Engineering Sciences, Vol 77, © 2013 WIT Press
www.witpress.com, ISSN 1743-3533 (on-line)
doi:10.2495/MC130151

Many researchers tried to describe the dynamic hardening characteristics of metallic materials with hardening models such as the Johnson-Cook model, Zerilli-Armstrong model and Preston-Tonks-Wallace [1] model. However, no model is accurately applicable for all materials, because each model approximates the hardening curve with the different characteristics as shown in table 1. Therefore, the applicable models have to be recommended for 4130 steel, OFHC copper and Ti6Al4V in order to be implemented in numerical simulation.

Table 1: The characteristics of typical dynamic hardening models.

		Approximation characteristics w.r.t the strain rate	
		Yield stress	**Flow stress**
Johnson-Cook model		Linear function	Proportional
Zerilli-Armstrong model	**BCC model**	Exponential function	Independent
	FCC model	Independent constant	Proportional
Preston-Tonks-Wallace model		Error function	Flexible
Modified Johnson-Cook model		Exponential function	Proportional
Lim-Huh model		Exponential function	Flexible
Modified Khan-Huang model		Exponential function	Flexible

Fracture characteristics is one of the important material properties to simulate the warhead penetration. The warhead undergoes a wide range of stress state from compressive upsetting to the balanced biaxial tension during the penetration. Because the fracture strain changes with respect to the loading path and the strain rate, fracture tests also should be conducted with respect to the loading path and strain condition. Many ductile fracture criteria have been proposed in the last decade, but those are not accurately applicable for all materials. In order to expand the generality of a ductile fracture criterion, the Mohr-Coulomb criterion and the Lou-Huh ductile fracture criterion are suggested in the space of stress triaxiality, the Lode parameter and equivalent fracture strain adopting the calibration method. The critical difference between two criteria is that a cut-off value for the stress triaxiality is extremely small for the Mohr-Coulomb criterion [2, 3] while the Lou-Huh criterion [4] proposes a constant cut-off value of $-1/3$ which is reasonable for the ductile material such as warhead.

2 Experimental conditions

2.1 Materials

For the tests, 4130 steel (Calstrip Steel Corporation), OFHC copper (Aurubis) and Ti6Al4V (TIMET) sheets with 1.2 mm thickness are used to measure the dynamic hardening properties and the fracture strains. The sheets were prepared according to AMS 6350 and the chemical composition of each material is shown in table 2.

www.witpress.com, ISSN 1743-3533 (on-line)

Table 2: Chemical composition of test materials.

4130 steel

Element	Mn	C	Cr	Mo	Si
wt%	0.5	0.32	0.95	0.19	0.19

OFHC copper

Element	Ag	S	Ni	Fe	As
wt%	0.0025	0.0015	0.001	0.001	0.0006

Ti6Al4V

Element	Al	V	Fe	O	C
wt%	6.21	4.02	0.17	0.11	0.009

2.2 Specimen shapes and tensile speeds

Many researchers have proved that the fracture strain depends on the stress triaxiality and the strain path. As excluding the hysteresis effect, the shape of a tensile specimen has to induce a proportional loading path. In order to realize three typical strain paths, the pure shear, dog bone and grooved specimens are used for the strain paths of the pure shear, uniaxial tension and plain strain conditions as shown in Figs 1 and 2. The result with the notched specimen is utilized not to calibrate the Lou-Huh model but to evaluate the validity of the model. In order to measure the fracture strain on the specimen surface, black spackle patterns are sprayed on the white background. Whole deformation during the test is recorded by a high speed camera and the fracture strain before fracture is calculated using a digital image correlation (DIC) method with a commercial

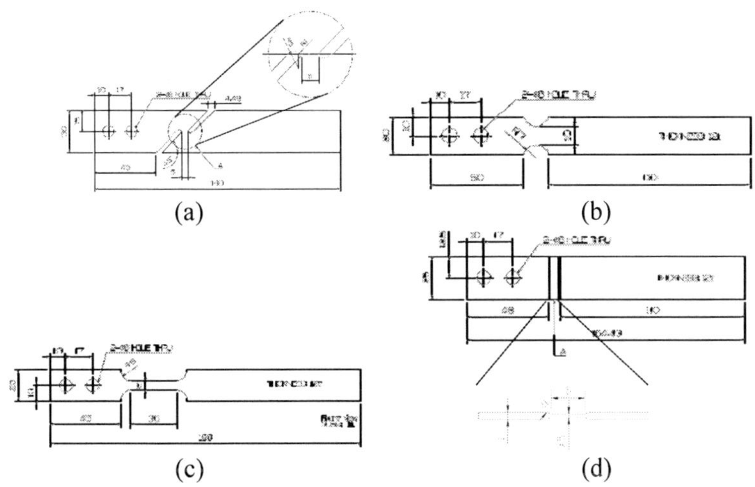

Figure 1: Specimen dimensions for the fracture test; (a) pure shear specimen; (b) notched specimen; (c) dog bone specimen; (d) grooved specimen.

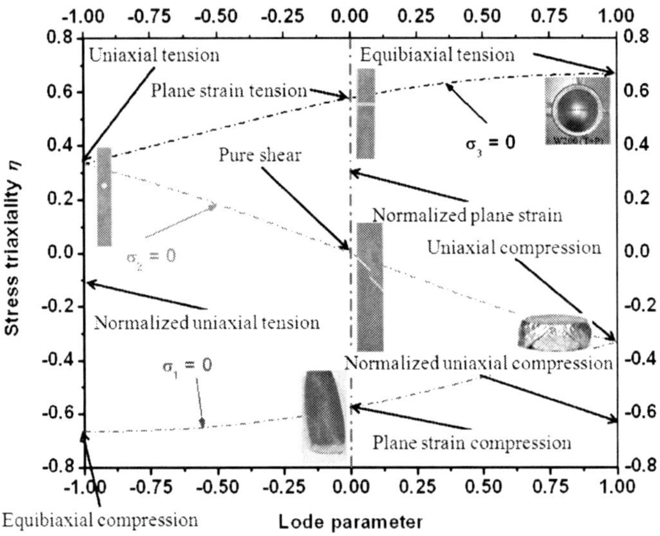

Figure 2: Stress states in the space of stress triaxiality and Lode parameter (η, L).

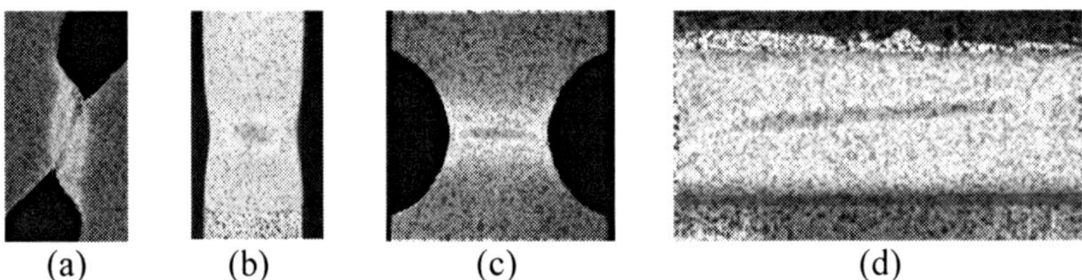

Figure 3: Calculated equivalent strain before the fracture (Ti6Al4V); (a) pure shear; (b) dog bone; (c) notched; (d) plane strain specimen.

software ARAMIS. Fig. 3 shows an example of the patterned specimen surface and calculated strain distribution before fracture.

The strain rate effect can be considered by changing the tensile speed of the test machine. The numerical simulation with each specimen is repeatedly carried out to find the tensile speed in order to obtain the equivalent strain rates as 0.001, 0.1, 1, 10, 100s^{-1} at the fractured point. Table 3 shows the calculated tensile speed with respect to the strain rate condition.

Table 3: Tensile speed with respect to the strain rate [mm/s].

Strain rate	**0.001s^{-1}**	**0.1s^{-1}**	**1s^{-1}**	**10s^{-1}**	**100s^{-1}**
Pure shear	0.004	0.4	4	40	400
Dog bone	0.03	3	30	300	3000
Notched	0.0038	0.38	3.8	38	380
Plane strain	0.0025	0.25	2.5	25	250

2.3 Lou-Huh ductile fracture criterion

The Lou-Huh ductile fracture criterion is proposed to describe ductile fracture behaviour of metals for nucleation, growth and shear coalescence of voids which has a form of:

$$\int_0^{\bar{\varepsilon}_f} \left(\frac{2\tau_{\max}}{\bar{\sigma}} \right)^{C_1} \left(\frac{\langle 1+3\eta \rangle}{2} \right)^{C_2} d\bar{\varepsilon} = C_3 \qquad \langle x \rangle = \begin{cases} x & \text{when } x \geq 0 \\ 0 & \text{when } x < 0 \end{cases} \tag{1}$$

The nucleation of voids is suggested as a function of equivalent plastic strain $\bar{\varepsilon}$, the growth of voids is calculated by the stress triaxiality as $1+3\eta$ and the coalescence of voids is determined by the normalized maximum shear stress $\tau_{\max}/\bar{\sigma}$. The fracture strain predicted by the Lou-Huh ductile fracture criterion is negative dependent on the stress triaxiality. Eqn (1) also can be transformed into the space of as below:

$$\int_0^{\bar{\varepsilon}_f} \left(\frac{2}{\sqrt{L^2+3}} \right)^{C_1} \left(\frac{\langle 1+3\eta \rangle}{2} \right)^{C_2} d\bar{\varepsilon} = C_3 \qquad \langle x \rangle = \begin{cases} x & \text{when } x \geq 0 \\ 0 & \text{when } x < 0 \end{cases} \tag{2}$$

The fracture strain, strain path and stress triaxiality of each specimen are calculated by the commercial software ARAMIS with the recorded pattern on the surface of specimen and the three coefficients are determined to reduce the difference between experimental data and the Lou-Huh criterion.

3 Experiment result

3.1 Stress-strain curves and recommended dynamic hardening models

Figs 4–6 show the stress-strain curves and variations of yield stress 4130 steel, OFHC copper and Ti6Al4V at the strain rates ranging from 0.001 to $100 s^{-1}$. The yield stress and the flow stress increase as the strain rate increases, but the amount and the tendency of the increment for the three materials are different from each other. When the strain rate increases, the yield stress of 4130 steel and OFHC copper also increase exponentially but that of Ti6Al4V increases linearly. The strain rate sensitivity of 4130 steel is quite different from that of OFHC copper. Due to these characteristics, the stress-strain curves can be accurately fitted by the Lim-Huh model (4130 steel), the Preston-Tonks-Wallace model (OFHC copper) and the modified Khan-Huang model (Ti6Al4V) as below:

The Lim-Huh model [5]

$$\sigma = \sigma_r \frac{1+q(\varepsilon)\dot{\varepsilon}^{m(\varepsilon)}}{1+q(\varepsilon)\dot{\varepsilon}_r^{m(\varepsilon)}} \qquad \text{where} \qquad q = C^{-\frac{1}{p}},\; m = \frac{1}{p} \tag{3}$$

The Preston-Tonks-Wallace (PTW) model

$$\hat{\tau} = \hat{\tau}_s + \frac{1}{p}\left(s_0 - \hat{\tau}_y\right)\ln\left[1 - \left[1 - \exp\left(-p\frac{\hat{\tau}_s - \hat{\tau}_y}{s_0 - \hat{\tau}_y}\right)\right] \times \exp\left\{-\frac{p\theta\psi}{\left(s_0 - \hat{\tau}_y\right)\left[\exp\left(-p\frac{\hat{\tau}_s - \hat{\tau}_y}{s_0 - \hat{\tau}_y}\right) - 1\right]}\right\}\right] \tag{4}$$

$$\text{where} \begin{cases} \hat{\tau}_s = s_0 - \left(s_0 - s_\infty\right) erf\left[\kappa \hat{T} \ln\left(\gamma\dot{\xi}/\dot{\psi}\right)\right] \\ \hat{\tau}_y = y_0 - \left(y_0 - y_\infty\right) erf\left[\kappa \hat{T} \ln\left(\gamma\dot{\xi}/\dot{\psi}\right)\right] \end{cases}$$

The modified Khan-Huang model [6]

$$\sigma = \left[a + B\left(1 - \frac{\ln\dot{\varepsilon}}{\ln D_0^p}\right)^{n_1} \varepsilon^{n_0}\right]\left[1 + C\left(\ln\frac{\dot{\varepsilon}}{\dot{\varepsilon}_0}\right)^p\right]\left[1 - \left(\frac{T - T_r}{T - T_m}\right)^m\right] \tag{5}$$

where $D_0^p = 10^9 / s$

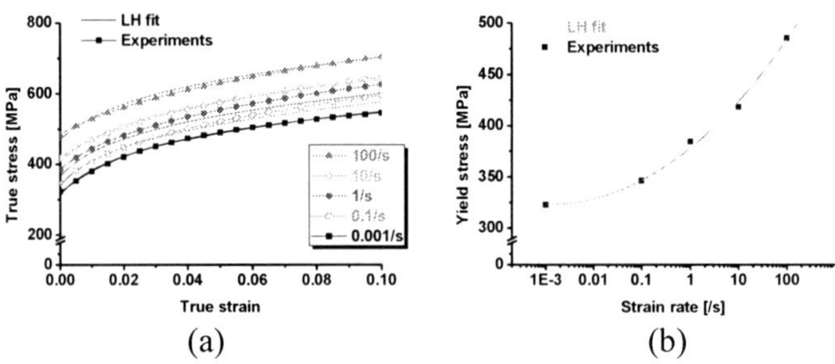

Figure 4: (a) the stress-strain curve and (b) the variation of the yield stress of 4130 steel with respect to the strain rate. The results are fitted by Lim-Huh model.

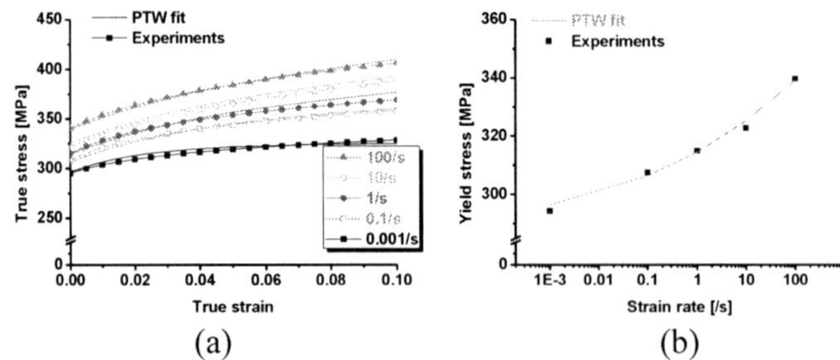

Figure 5: (a) the stress-strain curve and (b) the variation of the yield stress of OFHC copper with respect to the strain rate. The results are fitted by PTW model.

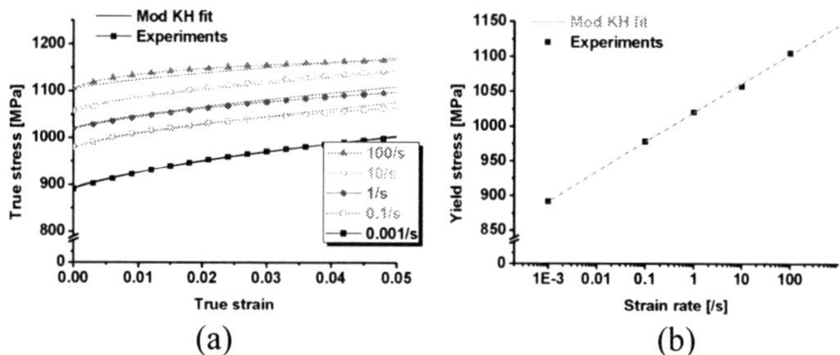

Figure 6: (a) the stress-strain curve and (b) the variation of the yield stress of Ti6Al4V with respect to the strain rate. The results are fitted by modified Khan-Huang model.

3.2 Fracture strain and fracture loci with respect to the strain rate

No ductile fracture model considering both the strain rate and the loading path has been reported yet. However, the fracture strain can be calibrated by the Lou-Huh model without considering the strain rate and then the dependence on the strain rate can be separately considered to construct a new model which incorporates both the fracture strain and the strain rate. Fig. 7 indicates where the fracture strains are measured at the four points of different stress triaxiality conditions and fitted by the Lou-Huh model. And then the result can be transformed into the forming limit diagram. Figs 8–10 show the fracture loci with respect to the strain rate in $(\bar{\varepsilon}_f, \eta)$ and $(\varepsilon_1, \varepsilon_2)$ space. As the strain rate increases from 0.001 to 100s^{-1}, the fracture strain under the uniaxial tension condition has the minimum strain at the strain rate of 1s^{-1} and the maximum strain at the 100s^{-1}. This overlap is observed in all materials. However, the fracture strain under the pure shear condition shows quite different tendency with respect to the strain rate.

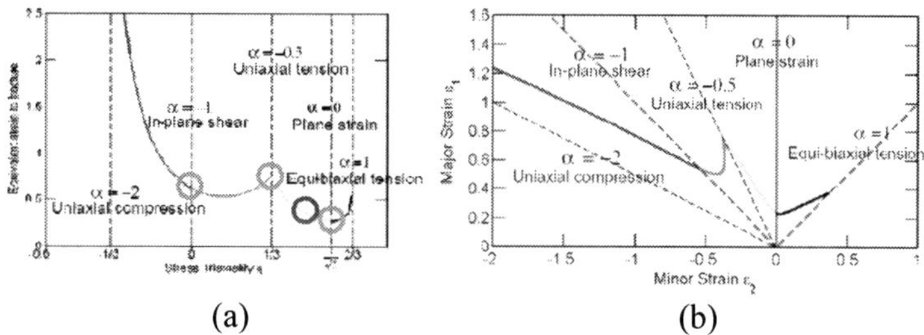

Figure 7: (a) fracture locus under plane stress condition (b) transformed fracture forming limit diagram.

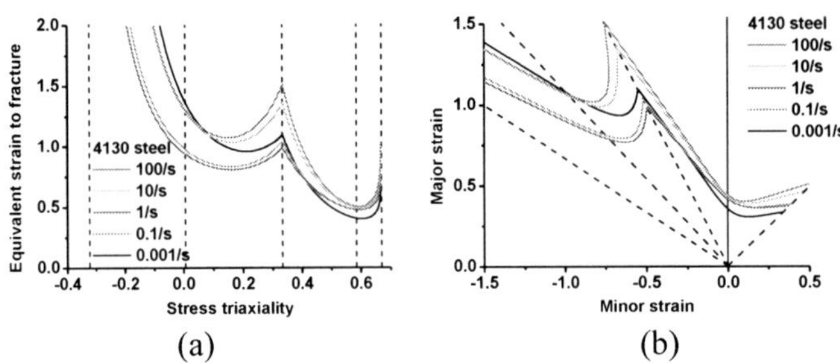

Figure 8: (a) the fracture loci, (b) fracture forming limit diagram of 4130 steel according to the strain rate.

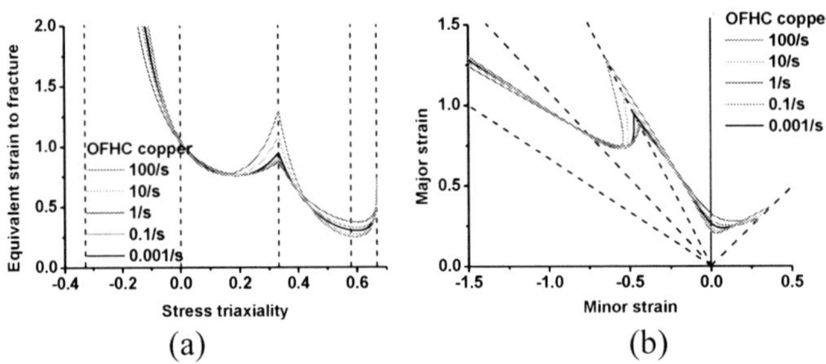

Figure 9: (a) the fracture loci, (b) fracture forming limit diagram of OFHC copper according to the strain rate.

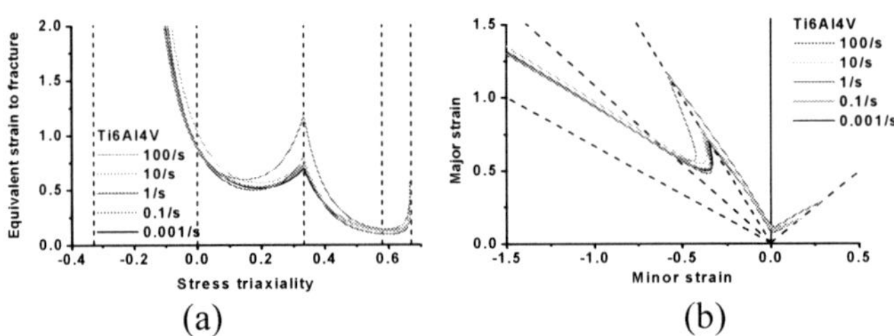

Figure 10: (a) the fracture loci, (b) fracture forming limit diagram of Ti6Al4V according to the strain rate.

4 Conclusion

The dynamic properties of the three metallic materials are investigated with respect to the strain rate. The three metallic materials show different flow stress tendencies. The Lim-Huh model, the PTW model and the modified Khan-Huang model are most applicable for 4130 steel, OFHC copper and Ti6Al4V respectively. The fracture strains are measured on the surface of a specimen with

respect to the strain rate and loading path. The fracture strain under the uniaxial tension condition has the minimum value at the strain rates ranging from 0.1 to $1s^{-1}$ and the maximum value at $100s^{-1}$. However the fracture strain under the pure shear condition shows quite different tendency with respect to the strain.

References

[1] D. L. Preston, D. L. Tonks, D. C. Wallace, Model of plastic deformation for extreme loading conditions, Journal of Applied Physics, **93**, pp. 211-220, 2003.

[2] Y. L. Bai and T. Wierzbicki, Forming severity concept for predicting sheet necking under complex loading histories, Int. J. Mech. Sci., **50**, pp. 1012-1022, 2008.

[3] Y. B. Bao and T. Wierzbicki, A comparative study on various ductile crack formation criteria, J. Eng. Mater. Tech. ASME, **126**, pp. 314-324, 2004.

[4] Y. Lou, H. Huh, S. Lim, K. Pack, New ductile fracture criterion for prediction of fracture forming limit diagrams of sheet metals, International journal of solid and structure, **49**, pp. 3605-3615, 2012.

[5] J. H. Lim, Study on dynamic tensile tests of auto-body steel sheets at the intermediate strain rate for material constitutive equations, Ph.D Dissertation, KAIST, 2005.

[6] H. J. Lee, J. H. Song, H. Huh, Dynamic tensile tests of auto-body steel sheets with the variation of temperature, Solid state phenomena, **114-117**, pp. 259-262, 2006.

The optimization of snowboarding motion and the design of a new snowboard course

F. Yang, H. Li & A. Chen
Sichuan University, China

Abstract

According to the competition rules, the criterion of the score is the judgment of an athlete's action in the air as well as the height in the air. However, the action of an athlete and the height in the air is contradictory because one cannot reach the highest height and finish the maximum twist at the same time.

To find out the best method for athletes to get a high score, we designed a new shape of a snowboard course (half-pipe) in order to maximize the time of "flying" in the air as well as optimizing the twist of an athlete in the air by mathematical computation. What's more, through mathematical analysis, we designed the optimized time of an athlete soaring through the air from the wall of the half-pipe. The analysis of the motion and model of the snowboard course are also demonstrated by computer to make the conclusion more visual.
Keywords: snowboarding, half-pipe, parabolic motion, differential equation.

1 Introduction

Snowboard movement is a stimulated event which has gradually gained in popularity in recent years. The scoring criteria of snowboard movement in the U-shaped pipe always consider two important facts-the height gain the in sky and the excellent action behave in the sky.

Actually, these two criteria require the athlete to have the ability to control himself and jump higher by taking advantage of the initial velocity and the jumping angle. On the other hand, for the audience, people wish the athlete to do great job for their entertainment.

So how to help the athlete gain the advantage is not only for athlete but also for the stadium builders. Now we will make a model to help them do better work and give them some advice.

WIT Transactions on Engineering Sciences, Vol 77, © 2013 WIT Press
www.witpress.com, ISSN 1743-3533 (on-line)
doi:10.2495/MC130161

2 Problem analysis

For consideration of such a problem, we first must know well about the whole process of how the athlete moves up and does the action in the sky. And in other aspects, we all must get the criteria for the interesting movement like the score standard and some index about the half-pipe. So, after that we have already known something about the movement. It can be divided into three parts. The athletes move up by using the initial velocity. The snowboard gets attached to the pipe. So we must consider the friction. In this process, the kinetic energy turns to be as potential energy and heat caused by the friction.

The athletes move up to a certain point which is chosen by himself. Then he will jump up with such action that the back foot punch strong on the back side of the snowboard. The reaction force will give him another velocity in a certain direction. With the inertia and this velocity, the athlete will do projectile motion right now.

The athlete will fly in the sky to make some fantastic action like twisting or rotation. The whole time is depending on the height he achieves. And for athlete himself, he must embrace like a ball to keep balance. After making action he will turn down to be ready for getting touch with the pipe again.

So after learning well about these, we find that our model will be divided into two parts, the first is about getting up along with the pipe and the second is about the projectile motion.

3 Model development

3.1 Symbols

Symbols	Meaning
F_c	Centripetal Force
f	Friction
N	Normal Force
θ	The angle corresponding to height
β	The angle corresponding to the direction of Projectile
R	Radius
$v(\theta)$	The velocity corresponding to height
v_0	Initial velocity
v_1	The additional velocity came from the push on the back of the snowboard
μ	Coefficient of friction
G	Gravitational acceleration

M	The mass of the snowboard and the athlete
E_{max}	The maximum of energy given by the athlete when he jumps up
J_{min}	The minimum of Moment of Inertia
T	The time interval of the athlete jumping in the sky
H	The highest height gained by the athlete
σ	The whole angle the athlete can use

3.2 Hypothesis

1. In the whole process, the model does not consider the existence of air resistance and the wind power.
2. When the athlete jumps up, the model does not consider the complete process about how he step the back of the snowboard.
3. We assume that every athlete have the same initial velocity v_0, jumped additional velocity v_1. And it is obvious that E_{max} and J_{min} is different for everyone but we assume it is a constant number in the model.
4. We assume the half-pipe is a perfect semicircle instead of an oval, especially for the transition path when the athlete moves up.

3.3 The deduction of the model

To build a model of the solution, at first, we apply the Newton second law to express circular motion of the athlete in the half-pipe:

$$F_c = \text{N} - mg\cos\theta = \frac{mv^2}{R} \tag{1}$$

Then, we list the expression of the friction force as follows:

$$f = \text{N}\mu \tag{2}$$

Applying the theorem of kinetic energy, we obtain the expression of the decrease in kinetic energy when the athlete is moving up from the bottom of the half-pipe:

$$\frac{1}{2}mv^2 - \frac{1}{2}mv_0^2 = \int_0^\theta f\ Rd\theta + (\text{R} - \text{R}\cos\theta)\text{mg} \tag{3}$$

Regarding θ as the independent variable, we differentiate the both sides of equation (3), and then we obtain the equation (4) as follows:

$$mvv'=f(\theta)R+mgR\sin\theta \tag{4}$$

Plugging equations (1) and (2) into the equation (4), we obtain:

$$vv'=g\mu R\cos\theta+\mu v^2+gR\sin\theta \tag{5}$$

Obviously, the format of equation (5) is accord with the format of Bernoulli equation, so we set $s=v^2$, and then equation (5) will become:

$$\frac{ds}{d\theta}=2vv'=2(g\mu R\cos\theta+\mu s+gR\sin\theta) \tag{6}$$

Throwing away the constant term of equation (5), we obtain:

$$\frac{ds}{d\theta}=2\mu s \tag{7}$$

Then we integrate the equation (7) and regard the constant term C as a function with the independent variable θ, we obtain the particular solution of equation (4) by using the Separation of variables formula:

$$v^2=s=\frac{-2gR}{1+4\mu^2}(\mu\sin\theta+(2\mu^2+1)(\cos\theta-e^{2\mu\theta}))+ce^{2\mu\theta} \tag{8}$$

Plugging the velocity at the bottom of the half-pipe into the equation (8), we can obtain the velocity equation:

$$v^2=s=\frac{-2gR}{1+4\mu^2}(\mu\sin\theta+(2\mu^2+1)(\cos\theta-e^{2\mu\theta}))+v_0^2e^{2\mu\theta} \tag{9}$$

Now we assume that an athlete will acquire perpendicular velocity v_1 when he step on the snowboard, and the tangent velocity will be damnified due to the instantaneous friction force, so we suppose the loss rate of the speed is λ, then the initial velocity of the projectile motion is:

$$v_2=\sqrt{v_1^2+(\lambda v)^2} \tag{10}$$

So we can obtain the greatest height of the athlete in the air:

$$h=\frac{(v_1^2+(\lambda v)^2)\sin^2(\pi-\theta-\arctan(\frac{v_1}{\lambda v}))}{2g}-R\cos\theta \tag{11}$$

And the time during "flying" in the air is:

$$T=\frac{2\sqrt{v_1^2+(\lambda v)^2}\sin(\pi-\theta-\arctan(\frac{v_1}{\lambda v}))}{g} \tag{12}$$

According to the requirement of the skill, knees should be close to the pit of the stomach, so we can regard the athlete as a rigid body. Then we set the maximum kinetic energy of the revolving rigid body as $E_{\max}$,besides, as the athlete should shrink the arm to reduce the moment of inertia, angular velocity can be obtained by best to complete the action, we can set the minimum moment of inertia as J_{min} . Then we obtain the angular velocity:

$$\omega=\sqrt{\frac{E_{max}}{J_{min}}} \tag{13}$$

So we obtain the angle of the revolving is:

$$\sigma=\omega T \tag{14}$$

In conclusion, we obtain the expression of the height $h(R,\theta)$ and the expression of the angle $\sigma(R,\theta)$.

3.4 The solution of the model

On the basis of the model we obtained above, we analyzed the solution of the model we established, and decided to solve it by using Matlab.

3.4.1 The function of T

At first, by consulting the literature, we found that the maximum friction factor of the snowboard course is 0.2, then we assume that the initial velocity (the velocity of the athlete while he is at the bottom of the half-pipe) is 10m/s, the loss rate of the tangential speed is 0.5.Under these circumstance, we substituted them into the function of height, then we plotted the image of the height by Matlab (Figures 1 and 2) as follows:

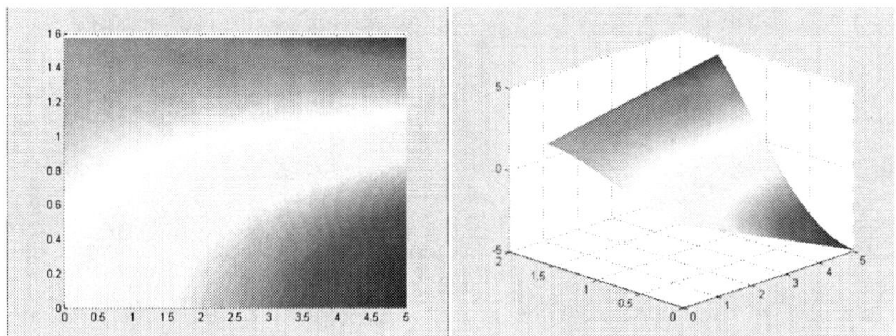

Figure 1: Looking downwards. Figure 2: Maximum height.

These two figures show the same image by different visual angles. The sequence of the height is presented from blue to red as the sequence of the maximum to minimum. We can find out the conclusion obviously by analyzing it. Apparently, in a certain extent of interval, the greatest high emerges in the top right corner (figure 2).

Then, we use the same condition to plot the image of the time. The image is presented as follows (figures 3 and 4):

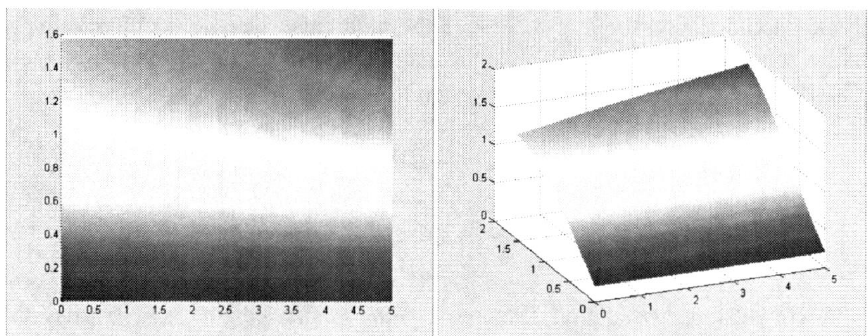

Figure 3: Looking downwards. Figure 4: Maximum time.

The two figures show the same image by different visual angles. Obviously, the sequence of the length of time is presented from red to blue as the sequence of the maximum to minimum. Obviously, we can draw a conclusion from these figures that the longest time emerges in the top right corner.

Now we can draw a conclusion, the best opportunity to lift-off of an athlete is the moment when he is at the top of the arch. As for designing the ideal half-pipe, designer should increase the radius of the half-pipe to guarantee the athlete to move up to the top edge of the arc.

To obtain more data of our analysis, we modify the friction factor from 0.2 to 0.05, and then we plot the superposition image of height by Matlab as follows (figures 5 and 6):

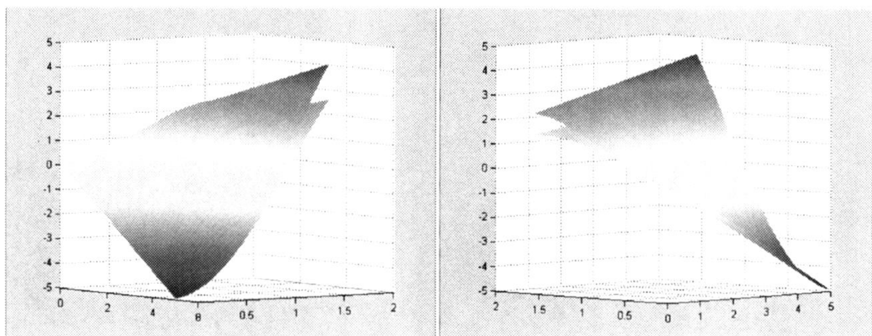

Figure 5: Different friction height. Figure 6: Different friction height.

From the two figures above, we found that the height of the athlete in the air would increase by the reduction of the friction force, which is corresponding with the reality. However, it's obvious that the scope of the increasing height would only be 1m approximately.

We also use the Matlab to describe the superposition of time with the same condition (the friction factor is 0.05):

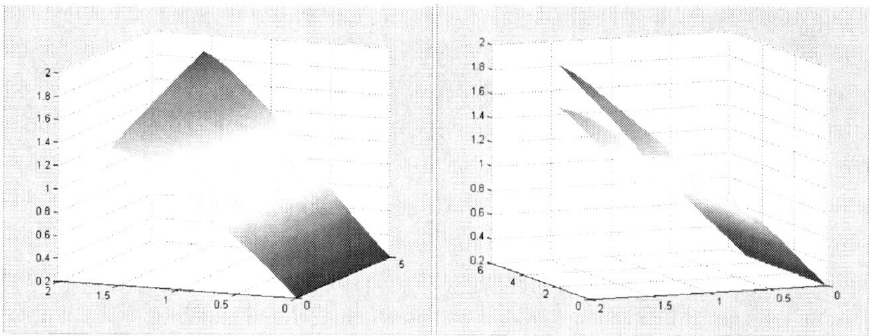

Figure 7: Different friction time. Figure 8: Different friction time.

According to the two figures above, we found that the time of "flying" in the air would increase with the decreasing of the friction factor, which corresponds with the reality. What's more, the time of the athlete "flying" in the air would only increase 0.4s approximately.

We tried to change the initial velocity from 10m/s to 15m/s, and substituted it into the function. The figures below are the equation described by Matlab (figures 9 and 10):

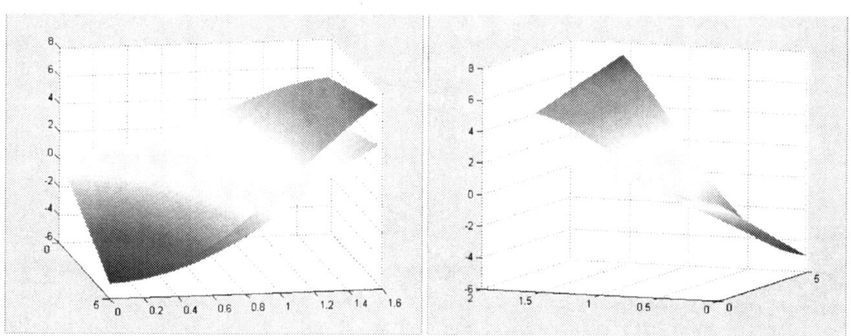

Figure 9: Initial velocity height. Figure 10: Initial velocity height.

Due to the velocity of 15m/s is the greatest controllable velocity of the athletes, we can know that the greatest height of an athlete can only reach to 7m, and it will emerge the drop of 3 meters.

Then we plot the image of time by Matlab with the same condition (initial velocity is 15m/s):

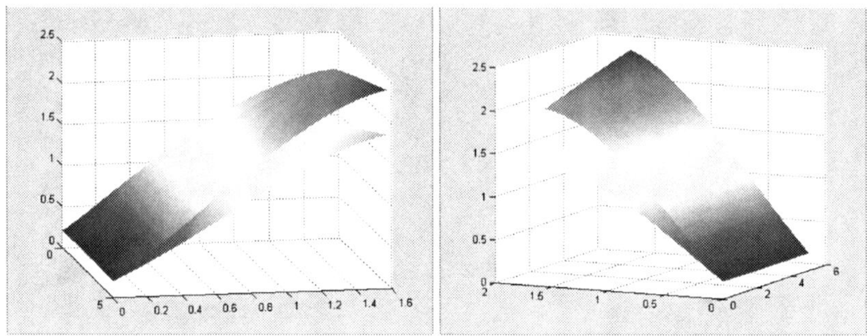

Figure 11: Different initial velocity time.

Figure 12: Different initial velocity time.

From the two figures we obtain the information of the change of time. Because 15m/s is the greatest controllable velocity of the athletes, so the greatest time of athlete "flying" in the air will only reach to 2.5s approximately.

We also tried to modify the loss rate of tangential velocity to gain more details. Firstly, we change the loss rate of tangential velocity from 0.5 to 1. We plotted the image to make a comparison between the loss rate of tangential velocity 0.5 and the condition without any loss:

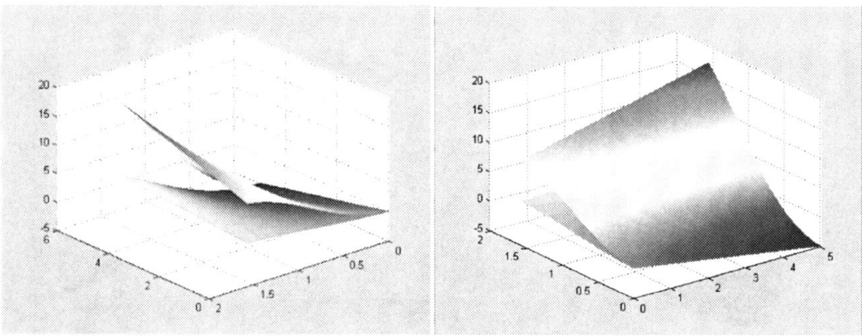

Figure 13: Different loss rate height.

Figure 14: Different loss rate height.

The two images display that the height of "flying" in the air would increase tempestuously without the loss of tangential velocity (from 5m to 15m), which is contrary to the fact. So it confirms the necessity of setting the loss rate of tangential velocity, which makes the model more close to the reality.

Then we plotted the time images of the same condition (the loss rate of tangential velocity is 1):

The two images displays that the time of "flying" in the air would increase tempestuously without the loss of tangential velocity (from 1.5s to 3s), which is also contrary to the fact. It goes a step further to confirm the necessity of setting the loss rate of tangential velocity, which makes the model more close to the reality.

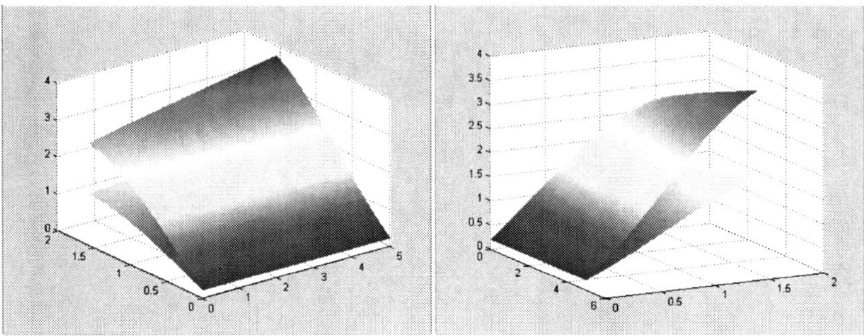

Figure 15: Different loss rate time. Figure 16: Different loss rate time.

4 Model development

In this model, we make an assumption in which said the athlete have the same energy when he join up. But actually, different people have different energy. Maybe their energy distribution will be like the normal distribution. We can do more statistics analysis in this field.

The speed of wind is ignored in this model. However, we can add something to consider such effect.

We can learn more about the action detail in jumping. Some research reported that most athletes can't jump right on the side of the snowboard. We can learn more to help them do the perfect job.

References

[1] Bowden, F.P, Friction on snow and ice. Proceedings of the Royal Society of London, 1953, 217: 462-478.

[2] Colbeck, S. C. The Kinetic Friction of Snow. Journal of Glaciology, 1988, 34(116): 78-86.

www.witpress.com, ISSN 1743-3533 (on-line)

Section 4
Corrosion problems

The effect of different surface topographies on the corrosion behaviour of nickel

A. S. Toloei, V. Stoilov & D. O. Northwood
Department of Mechanical, Automotive and Materials Engineering, University of Windsor, Ontario, Canada

Abstract

The electrochemical and corrosion behaviour of a surface is extremely complicated and depends on various chemical, physical and mechanical factors. In this study the effect of different surface roughnesses on the corrosion resistance of nickel in 0.5 M sulphuric acid was investigated. Open circuit potential, corrosion current density, polarization resistance and corrosion rate were measured for surfaces polished with different grits (120, 240, 400, 600 and 1200) of silicon carbide papers. The surface roughness was measured using a profilometry method both before and after corrosion testing. SEM images were taken and compared for all of the samples before and after corrosion tests. The results showed that surface roughness and surface morphology can considerably change corrosion and corrosion rate. A higher corrosion resistance is obtained for surfaces with lower roughnesses. Finally, the results were compared with specimens where a specific surface patterning was obtained using a laser ablation method.
Keywords: corrosion, nickel, surface roughness.

1 Introduction

There are many different methods of surface modification such as coatings, inhibitors and surface patterning to combat corrosion. Schweinsberg and Flitt [1] have shown that the purity of the anode and the electrolyte, the metallurgical history of the material and pre-treatment of the working electrode (abrasion technique, pre-polarization, time of immersion in the electrolyte) are amongst the most important parameters governing corrosion current density, and the corrosion rate. In different studies [2, 3], grain size, materials composition, mode

WIT Transactions on Engineering Sciences, Vol 77, © 2013 WIT Press
www.witpress.com, ISSN 1743-3533 (on-line)
doi:10.2495/MC130171

of manufacturing, geometry and roughness have been reported to be the most important parameters affecting corrosion potential and current density.

Among the investigated parameters, surface roughness has a major impact both on general corrosion, and the nucleation of metastable pitting and pitting potential [4]. Surface roughness is also known to affect the hydrodynamic and mass-transfer boundary layer, thus influencing the corrosion mechanism and rate [5]. Due to the significant effect of surface roughness on the corrosion resistance, it is possible to meet certain corrosion resistance requirements by specifying a surface finish rather than by upgrading the chosen alloy [6]. It has been shown that an increase in the surface roughness of stainless steels increases the pitting susceptibility and general corrosion rate which is attributed to the passive film breakdown. Behaviour similar to stainless steel has also been observed in copper, aluminium and titanium-based alloys [7–9]. In the case of mild steels, however, a reverse trend has been reported [6]. Also, Alvarez et al. in their work on AE44 magnesium alloy have shown corrosion behaviour consistent with mild steels and opposite to the trend reported in stainless steels [10]. Despite the significant attention that has been paid to the influence of the surface roughness on the pitting corrosion resistance in stainless steels, very little is known of the influence of the surface roughness on the general corrosion resistance [11]. In addition, to the author's best knowledge, there have been no studies on the effect of surface roughness on the corrosion behaviour of nickel and its alloys.

In this study, the corrosion of nickel in dilute sulphuric acid was studied through a potentiodynamic polarization technique to calculate open circuit potential, E_{corr} R_P and I_{corr} values for different surface roughnesses. SEM images were taken and compared for all of the samples and profilometry test measured the roughness of samples before and after corrosion tests. The results were compared with those for samples prepared with specific surface patterning.

2 Experimental procedures

2.1 Sample preparation

Nickel was used as a sample material for this study. All the specimens were cut to 15 × 15 × 1 mm size. Two types of surface modification have been applied, surface roughness obtained through grinding and predesigned surface patterns. The roughness was varied by polishing with SiC papers of different grades, including grit numbers 120, 240, 400, 600 and 1200. A smaller grit number represents a rougher finish. It should be mentioned that the grinding process was performed in one direction only. The 1200 grit surface was further polished with 0.05 μm alumina paste. The letter G, in the used notation, stands for grit, whereas samples with predesigned surface patterns were labelled using DxLy format where D is the hole diameter and L is the inter-hole spacing. The predesigned surface patterns were created using laser ablation. The surface patterns were based on the repetition of holes in the form of equilateral triangles in both X and Y directions. To create the holes, a copper bromide laser was used and single pulse was applied. Nitrogen gas was blown in order to protect the surface from oxidation. To assure repeatability, five sets of specimens were prepared and tested and all reported results are mean values.

2.2 Corrosion testing

Before each electrochemical experiment the specimens were degreased with ethanol and rinsed with deionized water and then immersed in a cell containing the electrolyte. This was followed by leaving the working electrode (WE) at open-circuit potential (OCP) for 30 min before the subsequent corrosion measurements. Corrosion tests were carried out in a 0.5 M H_2SO_4 solution at 23±1°C using a CHI660D, Electrochemical Workstation Beta, (version 11.17). A standard three-electrode cell was used. The counter electrode (CE) was a high-purity Pt electrode, while the reference electrode (RE) was a saturated calomel reference electrode (SCE). The working electrode (WE) was a nickel sample of surface area of 1 cm^2 exposed to the electrolyte. Polarization resistance (R_p), I_{corr} and corrosion rate values of the nickel samples were calculated using the linear polarization method as obtained from eqn (1):

$$R_p = \beta_a\beta_c/2.3i_{corr}(\beta_a + \beta_c) \tag{1}$$

where β_a and β_c are cathodic and anodic Tafel constants and i_{corr} is the corrosion current density obtained from a potentiodynamic curve [12].

2.3 Surface morphology

A Wyko Surface Profiling System NT-1100 was used to measure surface roughness parameters and obtain 3-dimensional topographical map of the tested surfaces. In order to describe the unique features of the surfaces, roughness parameters such as, the average surface roughness (R_a) and the root mean square average roughness (R_q) have been used [13]. In this study we report R_a values.

Scanning electron microscopy (SEM) and energy-dispersive x-ray spectrometry (EDS) were used to compare the surface structure and composition before and after potentiodynamic testing.

3 Results

3.1 Corrosion testing

Table 1 shows the mean calculated values for corrosion parameters of samples with different surface roughness. As can be seen both from Table 1and in Figure 1, which shows the potentiodynamic polarization curves for samples prepared using different grit sizes, the corrosion potential (E_{corr}) shifted towards the noble direction as the surface roughness decreased. Table 1 also shows that the corrosion current (i_{corr}) decreased as the surface roughness decreased. However, no significant difference was observed in cathodic currents among the samples with different surface roughness: (see Figure 1). This suggested that the shift in the E_{corr} and the variation in i_{corr} were primarily due to the anodic behaviour of the alloy. The specimens with the highest surface roughness

showed a sharper increase in the anodic current just above the E_{corr}. This phenomenon can be an indication of pitting [10]. If the pitting potential of all samples is considered at a specific current value (Figure 1), it becomes obvious that in the sample with lowest roughness (G1200) the onset of pitting is shifted to higher potential [9]. In each of the curves in Figure 1, there is a point immediately after the passive region, where the current density begins to rise (Points a, b and c). The breakdown of the protective film begins at that point which is known as critical pitting or breakdown potential. This is where the likelihood of pitting is the greatest and the point is used as a parameter for assessing pitting properties of the tested materials. This point happens at higher potential values for surfaces with lower roughness values.

Table 1: Corrosion parameters calculated for different surface roughnesses.

Sample	E_{corr} (mV)	i_{corr} (μA/cm^2)	β_a (mV)	β_c (mV)	R_P (Ω/cm^2)
G120	-321.8	21.64	109.5	118.3	1.14×10^3
G240	-310.5	19.27	110.0	116.8	1.27×10^3
G400	-313.9	17.26	112.1	113.3	1.41×10^3
G600	-312.6	14.67	123.8	111.1	1.73×10^3
G1200	-272.7	8.43	120.3	105.3	2.9×10^3
D10L20	-327.4	6.01	153.4	128.5	5.04×10^3
D20L20	-303.3	1.16	132.6	173.3	28.07×10^3

Figure 2 illustrates the variation in the corrosion rate with respect to surface roughness (R_a). Clearly, the corrosion rate decreases with decrease in roughness. In addition, it has been observed an increase of the number of pits on the rougher surfaces compared to the smooth surface [14]. This implies that the formation of metastable pits on a smooth surface is more difficult than that on a rougher surface, which corresponds to a lower corrosion rate. The result is consistent with similar studies which show that a higher roughness corresponds to a lower pitting potential (the potential at which metastable pits start to form on the surface) [14].

The patterned samples created by laser ablation, showed significantly lower corrosion rates than the samples with random roughness (Red points in figure 2). The main mechanism behind the significant decrease in the corrosion rate has been shown to be heterogeneous wetting [15, 16]. Heterogeneous wetting results in formation of alternating solid/liquid zones and stable air/vapour pockets. The air/vapour pockets allow the formation of the passive oxide layer but prevent its dissolution by the electrolyte. The heterogeneous interface, formed between the solution and the surface, significantly decreases the overall contact area between the surface and the electrolyte, thus significantly decreasing the corrosion rate [15, 16].

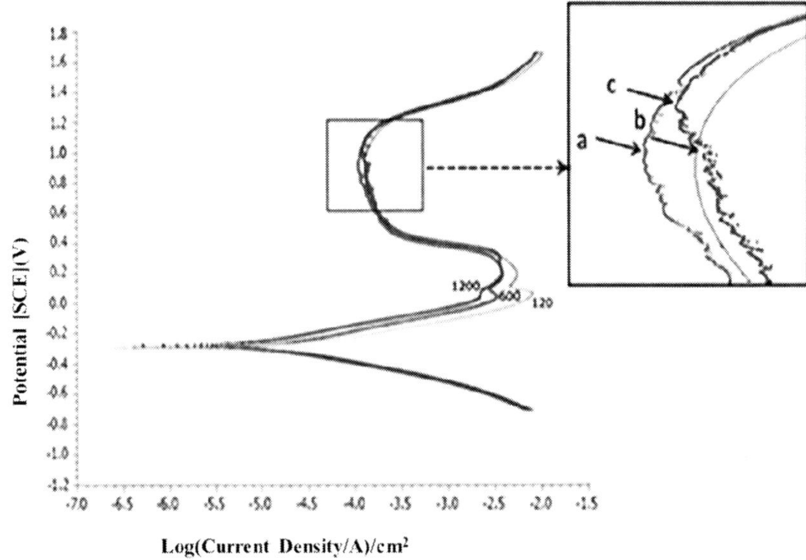

Figure 1: Potentiodynamic polarization curves for ground samples (120, 600 and 1200 grit).

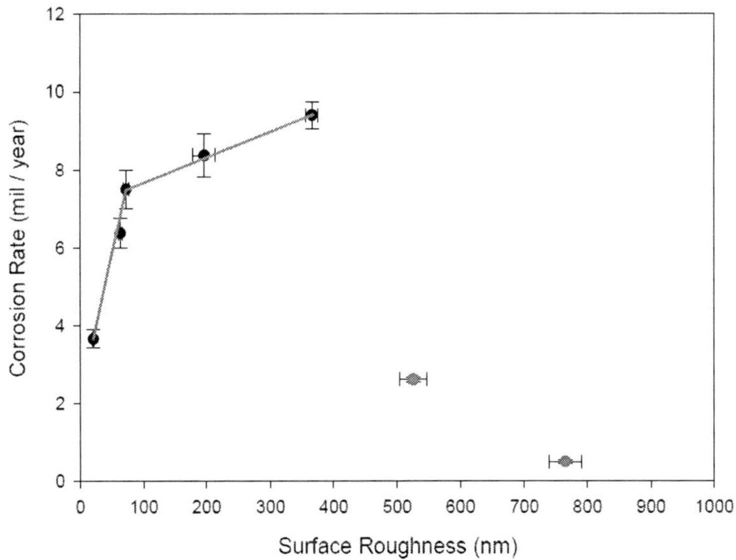

Figure 2: Corrosion rate values for different surface roughnesses.

3.2 Surface roughness measurement

The average roughness value for samples G120 to G1200 and two patterned samples before and after corrosion testing are presented in Figure 3. As it can be seen, for samples G120 through G1200, as grit number increased, the roughness value decreased before corrosion testing. The same trend is almost seen after corrosion testing but with higher roughness values for all samples. In contrast, the samples with predesigned surface patterns, D10L20 and D20L20, had higher roughness values but significantly lower roughness change after corrosion testing compared to the samples with the unidirectional grinding roughness.

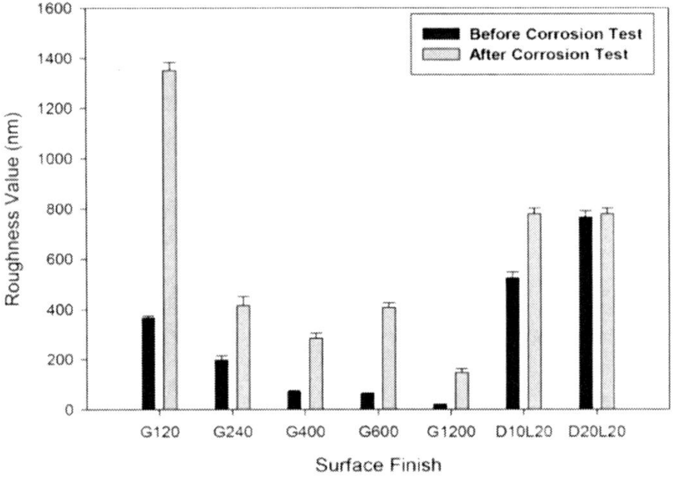

Figure 3: Roughness values before and after corrosion test.

All samples, however, show higher roughness values after corrosion testing. Figures 4(a) and (b) show the 3D topography images for samples G240 before and after corrosion tests. It is obvious that the uniform area has changed to a surface with deeper grooves with higher roughness which is a result of the corrosion process. Diffusion of corrosion products in deeper grooves is limited, hence the solution condition for local dissolution of metal is easily satisfied [4]. This means that on rougher surfaces more metastable pits are formed and the active sites on these surfaces have lower openness (ratio of width to depth at opening of the grooves), hence have a higher possibility to grow larger. Higher depth to width values are reported to indicate more smooth surfaces and it is more difficult to form micro-pits [4].

3D roughness images for the patterned samples showed brighter colour of the surface after corrosion (dark colours show deep point like grooves) which showed surface roughness has increased and no severe corrosion was observed. In these 3D images there was no significant change in the surface appearance (which is in agreement with the SEM images: see section 3.3). It seems that both surfaces, before and after corrosion tests, are almost identical, which reaffirms the results from the potentiodynamic corrosion rate tests.

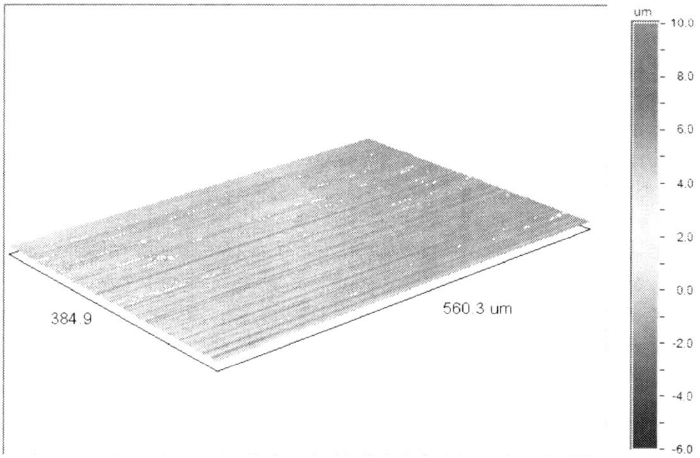

(a)

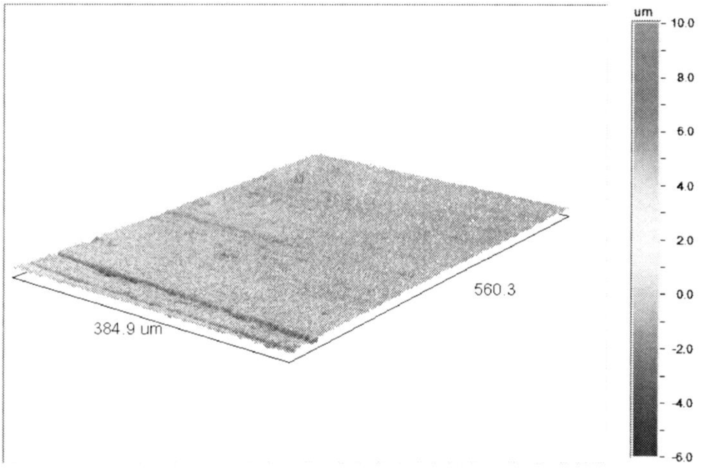

(b)

Figure 4: Sample G240 3D roughness (a) before and (b) after corrosion testing.

3.3 Scanning electron microscopy

Figures 5(a)–(d) are SEM images of samples G120 and G240 before and after corrosion. For G120, (Figures 5 (a), (b)), the corrosion is concentrated along the grooves. By decreasing the roughness from sample G120 to G240, (Figures 5 (c), (d)) less surface degradation is observed after corrosion testing. By comparing this sample before and after the corrosion test it can be observed that the sample has been corroded and the grooves' contrast (depth) is not as visible as before the test. This leads to the conclusion that in this sample the corrosive solution has reached the whole surface including the bottom of the grooves. Corrosion for rougher sample (G120) is more severe compared to the smoother one (G240). The micrograph also shows that the corrosion pits were preferentially aligned along the grooves suggesting that the deep valleys on the ground surface were favourable sites for pit nucleation [13]. The result is in agreement with the measured corrosion potential, corrosion rate and roughness observations. In all cases by increasing the roughness a higher corrosion rate has been recorded.

The surface microstructure for the patterned samples before and after corrosion tests was also analysed and the corresponding SEM micrographs were obtained. Figures 5(e) and (f) illustrate the patterned sample before and after the corrosion tests. In the SEM images there is not a significant change in the surface appearance (good agreement with the 3D profilometry images). This is also in agreement with the measured corrosion rate values. In this case, no severe corrosion has occurred, mainly due to the existence of heterogeneous wetting [15].

3.4 Energy-dispersive x-ray spectrometry (EDS)

The main goal of the EDS examination was to evaluate the change in O concentration on the surface of the samples. In the spectrum, three elements, nickel (Ni), oxygen (O) and carbon (C) were observed. The intensity of the largest peaks (Ni: $L\alpha$=0.851 and O: $K\alpha$=0.523 KeV) was used to calculate the oxygen concentration on the surface (see Table 2).

Table 2: Oxygen content before and after corrosion testing.

Oxygen %wt	Before corrosion	After corrosion
G120	1.37	4.71
G240	1.20	3.6
G400	1.20	2.11
G600	1.48	3.69
G1200	1.49	3.75

All samples had similar oxygen contents before corrosion testing and this value increased for all samples after corrosion testing. In samples with higher corrosion rates, the reason for increased oxygen concentration is the autocatalytic diffusion and corrosion process happened within the grooves which resulted in degradation of metal [6, 7, 9, 13]. In the case of samples with lower corrosion rate the increase of oxygen content was a result of a stable passive layer which was formed on the surface [9].

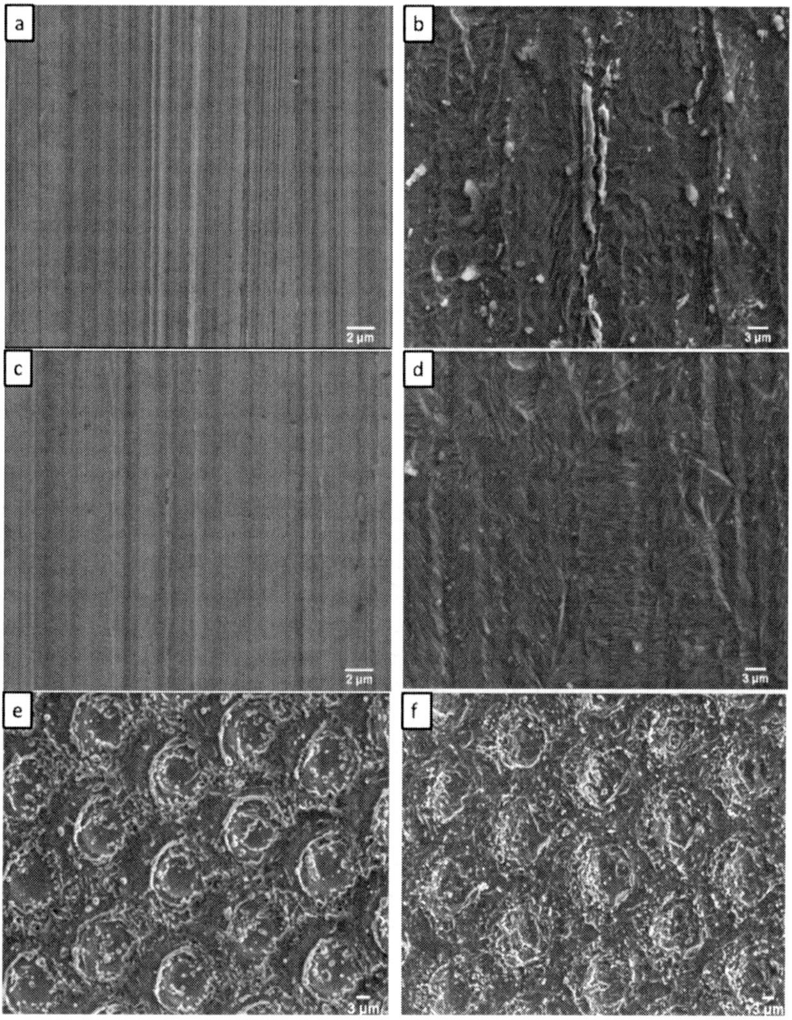

Figure 5: SEM of the sample G120 (a) before and (b) after corrosion, sample G240 (c) before and (d) after corrosion and sample D10L20 (e) before and (f) after corrosion.

4 Discussion

Based on the measured corrosion potentials and corrosion rates, it can be concluded that the corrosion rate of nickel increased with an increase of surface roughness. The reason for this phenomenon is that by increasing the roughness the active sites increased and deeper grooves played an active role as suitable places for localized corrosion. The depth of the valleys that influenced the diffusion of active ions during corrosion and IR drops in the deep valleys seem to be important parameters that affect the corrosion rate. On the rougher surfaces, the grooves trap the corrosion products, which results in more pitting [13]. In a smooth surface the frequency of metastable pitting is reduced because of the number of available metastable pit sites is reduced and consequently the corrosion rate is decreased [17]. Also, in a smooth surfaces weak points are less " activate " than on rough surfaces and the formation of stable passive film on smooth surfaces is more likely to occur. All the EDS results of nickel samples in this study showed an increase in oxygen content. For surfaces with low roughness values this increase in oxygen content can be related to the oxide layer formation. The presence of a protective oxide film lowers the activity of substrate/bulk interface. On rougher surfaces, compared to smooth surfaces, diffusion processes at nucleation sites in small scratches are hindered and there is a higher probability of accumulation of aggressive species and the geometry of small scratches make it easier to maintain a high potential drop [9]. But when the surface is smooth, the pit will survive more due to the formation of lace cover on the pit mouth maintaining the diffusion process which will result in less corrosion on the surface [6].

Rougher surfaces of a metal exhibit lower pitting potentials because by increasing the roughness less-open pit sites are maintained during the early stages of growth as metastable pits. Therefore, there is more restricted diffusion of metal cations during propagation which will result in the transition from metastable to stable pit growth to be made at lower potentials and consequently lower the pitting potential. It is said that deeper, less-open pit site has a greater probability of generating a pit, than a shallower, more-open one. Since the sites on the rougher surfaces are more occluded than those of smoother surfaces, it follows that pits growing on the rougher surface have a greater chance of survival to the stable growth stage, and thereby show a lower pitting potential [7, 8]. On a passive metal, rougher surfaces are more susceptible than smoother surfaces to localised forms of corrosion such as pitting and crevice corrosion. This effect can be related to the surface nucleation of metastable pitting preceding propagation of pitting. Although a higher number of nucleation events take place on a smoother surface as compared to a rough surface [8], propagation of the pits and formation of micropits does not occur as readily [4, 8]. On a rougher surface, several of these nucleation events will lead to propagation of pits and thereby corrosion [3].

In patterned samples, the formation of the passive oxide layer inside the pattern holes prevents the electrolyte reaching the bottom of the patterned hole. In other words, the contact between the electrolyte and the metal surface is heterogeneous wetting–alternating solid/liquid zones and stable air/vapour pockets. The air/vapour pockets allow the formation of the passive oxide layer but prevent its dissolution by the fluid. The heterogeneous interface, formed between the solution and the surface, decreases the overall contact area between the surface and the electrolyte, thus significantly decreasing the corrosion rate [15].

5 Conclusions

This study demonstrates that surface morphology makes a significant contribution to corrosion. For materials and environments where a passive film formed, as surface roughness decreased, both the corrosion potential and the pitting potential shifted to more noble values. The observed trends include: the rough surfaces limit diffusion out of the formed grooves; more available active sites on the rough surfaces; and there is fast formation of a stable oxide film on the smoother surfaces. When the material has the ability to form a passive film quickly, as is the case with nickel, aluminium and stainless steel, the rougher surface would be easily pitted because the smooth surface has fewer places for pit nucleation and can quickly form a passive film preventing pit nucleation. SEM and EDS results supported the conclusion of more corrosion on rougher surfaces and the formation of oxide layer on the surface. Specific surface patterning can further reduce the corrosion rate through a process of heterogeneous wetting.

References

[1] Schweinsberg, D. and Flitt, H.J., Some practical considerations relevant to the recording of corrosion data. *Corrosion & Prevention,* **45**, pp. 541-553, 2005.

[2] Alvarez, R., Martin, H., Horstemeyer, M. and Chand, M., Corrosion relationships as a function of time and surface roughness on a structural AE44 magnesium alloy. *Corrosion Science,* **52**, pp. 1635-1648, 2010.

[3] Hilberta, L., Bagge-Ravnb, D., Koldc, J. and Gram, L., Influence of surface roughness of stainless steel on microbial adhesion and corrosion resistance. *International Biodeterioration & Biodegradation,* **52**, pp. 175-185, 2003.

[4] Zuo, Y., Wang, H. and Xiong, J., The aspect ratio of surface grooves and metastable pitting of stainless steel. *Corrosion Science,* **44**, pp. 25-35, 2002.

[5] Asma, R.N., Yuli, P. and Mokhtar, C., Study on the effect of surface finish on corrosion of carbon steel in CO2 environment. *Journal of Applied Sciences,* **11(11)**, pp. 2053-2057, 2011.

[6] Abosrra, L. Ashour, A., Mitchell, S.C. and Youseffi, M., Corrosion of mild steel and 316L austenitic stainless steel with different surface roughness in sodium chloride saline solutions. *WIT Transactions on Engineering Sciences,* **65**, pp. 161-172, 2009.

[7] Sasaki, K. and Burstein, G., The generation of surface roughness during slurry erosion-corrosion and its effect on the pitting potential. *Corrosion Science,* **38(12)**, pp. 2111-2120, 1996.

[8] Burstein, G. and Vines, S., Repetitive nucleation of corrosion pits on stainless steel and the effects of surface roughness. *Journal of The Electrochemical Society,* **148(12)**, pp. B504-B516, 2001.

[9] Suter, T., Muller, Y., Schmutz, P. and Van Trzebi, O., Microelectrochemical studies of pit initiation on high purity and ultra high

purity aluminium. *Advanced Engineering Materials,* **7(5)**, pp. 339-348, 2005.

[10] Walter, R. and Bobby Kannan, M., Influence of surface roughness on the corrosion behaviour of magnesium alloy. *Materials and Design,* **32**, pp. 2350-2354, 2011.

[11] Shahryari, A., Kamal, W. and Omanovic, S., The effect of surface roughness on the efficiency of the cyclic potentiodynamic passivation (CPP) method in the improvement of general and pitting corrosion resistance of 316LVM stainless steel. *Materials Letters,* **62**, pp. 3906-3909, 2008.

[12] Roberge, P., *Corrosion Engineering Principles and Practice*, McGraw-Hill, New York, 2008.

[13] Mok Lee, S., Gyu Lee, W., Ho Kim, Y. and Jang, H., Surface roughness and the corrosion resistance of 21Cr ferritic stainless steel. *Corrosion Science,* **63**, pp. 404-409, 2012.

[14] Li, W. and Li, D., Influence of surface morphology on corrosion and electronic behaviour. *Acta Materialia,***54**, pp. 445-452, 2006.

[15] Toloei, A., Stoilov, V. and Northwood, D., The effect of creating different size surface patterns on corrosion properties of Nickel. *ASME International Mechanical Engineering Congress & Exposition, IMECE2012*, Houston, 2012.

[16] Bigdeli Karimi, M. Stoilov, V. and Northwood, D., Improving corrosion performance by surface patterning. WIT Transactions on Engineering Sciences, **72**, pp. 85-93, 2011.

[17] Burstein, G. and Pistorius, P., Surface roughness and the metastable pitting of stainless steel in chloride solutions. *Corrosion,* **51(5)**, pp. 380-385, 1995.

Self-healing of steel corrosion in a model alkaline medium: electrochemical response and surface analysis

D. A. Koleva[1], J. Hu[2], H. Kolev[3], K. & Van Breugel[1]
[1]*Department of Materials and Environment,*
Faculty of Civil Engineering and Geosciences,
Delft University of Technology, The Netherlands
[2]*School of Materials Science and Engineering,*
South China University of Technology, P.R. China
[3]*Bulgarian Academy of Sciences, Institute of Catalysis, Sofia, Bulgaria*

Abstract

The application of polymeric nano-particles for corrosion control in reinforced cement-based systems was studied. As an initial investigation, steel electrodes (St37) were tested in modified cement extract (CE), which is a model pore solution of pH 12.9 and chemical composition: 201 mg/l Ca; 3.85 mg/l K; 1.33 mg/l Na; 4 mg/l Al, Fe < 1 mg/l. Corrosion initiating factor was 5% NaCl. PEO_{113}-b-PS_{780} vesicles at 0.0024 wt.% in the CE were employed. These are nano sacs of 220 nm, enclosing a volume with a thin membrane, thus able to effectively entrap water-soluble compounds. Ion binding (bulk liquid medium), steel surface adsorption and/or participation in oxide layers formation were the expected mechanisms of particles' involvement. Both "empty" and "Ca-containing" vesicles were studied to evaluate "barrier" effects only vs. self-healing mechanisms. Increased steel corrosion resistance in the presence of both vesicle types was observed, however, it was more pronounced when Ca-containing ones were involved. CVA and XPS analysis, substantiate the self-healing phenomena, through a direct evidence of pitting formation and propagation and consecutive healing (and no further pit formation) when Ca-containing vesicles are present.

Keywords: corrosion, cement extract, vesicles, self-healing.

WIT Transactions on Engineering Sciences, Vol 77, © 2013 WIT Press
www.witpress.com, ISSN 1743-3533 (on-line)
doi:10.2495/MC130181

1 Introduction

The alkaline nature of concrete (pH=12.6-13.5) generally provides a physicochemical barrier against corrosion of the embedded steel and assures the formation of a stable passive layer on the steel surface [1]. However, corrosion is initiated due to carbonation (pH drop < 9) or chloride contamination (pH drop < 5) [2–8], resulting in corrosion propagation, accelerated damage and rapid failure [9]. Various protection methods and techniques have been investigated and are successfully applied, e.g. protective coatings/sealers [10, 11]; polymer coatings and cathodic protection (CP) (the latter combination leading to increased current demand/de-bonding [12] and possible hydrogen embrittlement or bond-strength degradation [13–18]); concrete re-alkalization [19]; corrosion inhibitors (quickly exhausted and with none or even negative effects on the bulk cement-based matrix [19–21]). Well known is that each method/technique has its limitations and side effects. Moreover, there is no technique that considers both steel reinforcement and cement-based bulk matrix, aiming at improved properties of both materials and superior global performance of reinforced concrete as a system. To this end, in the frame of a novel approach and (initially) within the IOP Self-Healing project SHM08743, executed in TU Delft and financially supported by Agentschap, NL, tailor-made polymeric nano-materials were studied. The aim was to increase steel corrosion resistance via self-healing of corrosion damage, along with improved microstructural characteristics.

As far as the cement-based bulk matrix properties were concerned, the suggested hybrid core-shell nano-particles already proved their positive effect, for example: in the presence of polyethylene oxide $(PEO)_{113}$-b- polystyrene $(PS)_{218}$ micelles (50 nm size) of only 0.025 wt.% per cement weight, the porosity of the micelle-modified matrix was reduced two times, whereas water permeability was reduced with six orders of magnitude; a more uniform distribution of calcium-silicate-hydrate (C-S-H) and enhanced mechanical properties, but un-disturbed water balance was evident from nano-indentation tests [22–25]. For establishing the effects on corrosion resistance, the aforementioned micelles, as well as stabilized (kinetically frozen) PEO_{113}-b-PS_{760} vesicles (220 nm size), "empty" and Ca-containing ones, were studied [26–28]. The aim of these studies was to investigate the possibility for maintained positive effects on the cement-based microstructure, along with increased corrosion resistance and self-healing (or repair) of corrosion damage, resulting from Ca-core release. Initial tests were performed in model pore solutions, prior to studies in reinforced mortar and concrete. This paper briefly presents some main outcomes from the investigation on corrosion resistance of St37 steel electrodes in cement extract, modified with the above defined vesicles [28]. Further, this work substantiates improved corrosion resistance and evidence of self-healing effects through correlating electrochemical performance and steel surface analysis.

2 Materials

Polymeric vesicles: vesicles were prepared via dialysis from PEO_{113}-b-PS_{760} block-polymer [24–28]. In order to distinguish between "barrier" effects and "self-healing" due to Ca-release, both "empty" and Ca-containing vesicles were used. Their hydrodynamic radius was approximately 220–250 nm, identified by dynamic light scattering (DLS) measurement and transmission electron microscopy (TEM) (Fig. 1).

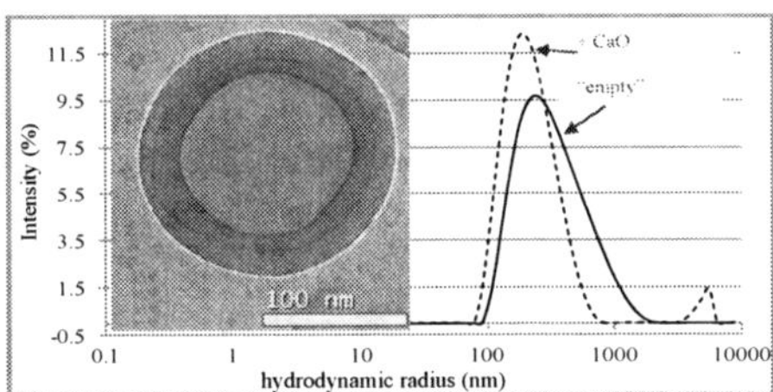

Figure 1: DLS and TEM of vesicles

Cement extract (CE): as a model pore solution was prepared by mixing Ordinary Portland cement CEM I 42.5N and tap water in 1:1 ratio; after 24h rotation the suspension was filtrated and this solution was used as a model medium. The pH of CE is 12.6–2.9, the chemical composition, derived via Inductive coupled plasma spectrometry (ICP) was: Ca – 201 mg/l; K – 3.85 mg/l; Na – 1.33 mg/l; Al – 4 mg/l and Fe < 1 mg/l; NaCl was corrosion-inducing factor, added in a concentration of 10 g/l to the CE. The aqueous solution of PEO_{113}-b-PS_{760} vesicles was added to CE, thus the vesicles concentration was 0.0024 wt. % in chloride-free and chloride-containing CE.

Steel electrodes: low-carbon steel St37 with surface area of 4 cm^2 was used. All electrodes were equally treated (grinding, polishing, acetone cleaning and water rinsing) prior to treatment in the relevant solutions and electrochemical measurements. Three replicates per condition were tested. Sample designation is as follows: group CE – steel in CE only; CEn – steel in chloride-containing CE; CEV and CEVC stand for steel in CE modified with "empty" and "Ca-containing" vesicles, whereas CEVn and CEVCn stand for the corroding cells respectively.

3 Methods

Electrochemical measurements: Autolab-Potentiostat PGSTAT 30, combined with FRA2 module, using GPES and FRA interface was used. Linear polarization resistance (LPR), Potentio-dynamic polarization (PDP), Impedance spectroscopy (EIS) and Cyclic voltammetry (CVA) were employed, using a common 3-electrode cell arrangement (saturated calomel electrode, SCE, as reference electrode). EIS was performed at open circuit potential (OCP) in the frequency range of 50 kHz to 10 mHz by superimposing an ac voltage of 10 mV; LPR used a polarization window of ± 20 mV vs. OCP; PDP was in the range of

-0.2V to +1.0V vs. OCP at a scan rate 0.5 mV/s; CVA scans were in the potential region of −1.4V to 0.6V at a scan rate of 5 mV/s (all electrochemical measurements are as generally employed for the systems under study [22–27]).
Surface analysis: morphology and composition of the product layers were investigated using scanning electron microscopy (SEM), coupled with energy dispersive X-ray (EDX) analysis as a qualitative and semi quantitative technique, using ESEM Philips XL 30. X-ray photoelectron spectroscopy (XPS) was carried out in the UHV chamber of an electron spectrometer ESCALAB-MkII VG Scientific with a base pressure of about 1×10^{-10} mbar (during the measurement 1×10^{-9} mbar). The photoelectron spectra were obtained using unmonochromatized Al Kα (hν = 1486.6 eV) radiation. Passing through 6 mm slit (entrance/exit) of a hemispherical analyzer, electrons with energy 20 eV are detected by a channeltron. The instrumental resolution measured as the full width at a halfmaximum (fwhm) of the Ag3d5/2, photoelectron peak is about 1 eV. The energy scale is corrected to the C1s (peak maximum at 285 eV) for electrostatic sample charging. The fitting of the recorded XPS spectra was performed, using a symmetrical Gauss-Lorentzian curve fitting after Shirley-type subtraction of the background. Except scanning the steel surface after 7 days i.e. product layers formed on the steel electrodes in each relevant condition, four-times consecutive mechanical polishing of equal pressure and time duration was performed in depth of the layer in the vacuum chamber of the spectrometer. This technique was chosen to avoid influence on the chemistry of the formed layers, which would otherwise be relevant if Ar sputtering was used. The XPS measurements thus resulted in quali- and quantifying of the composition of the product layers in depth i.e. for each specimen, four in depth scan results were recorded.

4 Results and discussion

4.1 Electrochemical measurements

4.1.1 OCP evolution and polarization resistance

A general perception is that the evolution of open circuit potential (OCP) determines the time to corrosion initiation. For reinforced concrete, and alkaline medium as the here used CE, the steel surface is considered to be passive if OCP is equal or more anodic than -270 mV SCE. When chlorides (or other corrosion-inducing factors) are present, passivity is lost and localised corrosion initiated – OCP readings shift to more cathodic (more negative) values in this case. Figure 2 depicts OCP evolution for the control (non corroding) cases (dotted notation), Fig. 3 shows these values for the corroding cases. Additionally, Figs 2 and 3 depict the derived (via linear regression) polarization resistance values (*Rp*) (*Rp* in global terms corresponds to corrosion resistance). As can be observed, OCP values fall in the region of passivity and shift towards more noble potentials at the end of the test. These correspond to the gradual increase of *Rp* values, the highest being for the steel treated in “empty” vesicles containing CE. Since there is no corrosion accelerator in the medium for the control samples, only gradual passivation is expected rather than any pronounced effect of the Ca-rich inner "core" of the Ca-containing vesicles. However, adsorption of both vesicle types on the steel surface and their incorporation in the passive layer could result in possible alterations in compactness and/or speed of layer formation. Obviously, barrier effects in terms of diffusion limitations from the adsorbed "empty" vesicles are relevant for specimen CEV, since OCP values are more cathodic, whereas *Rp* values are the highest. This does not necessarily mean a better corrosion resistant layer, but possibly limited oxygen access and hence locally impeded growth of a more inhomogeneous layer. In contrast, the CE and CEVC specimens exhibit similar *Rp* values.

Apparently the Ca-containing vesicles either act as "nucleation sites" on the steel surface, adsorb on active spots, or both and therefore do not result in diffusion limitations as in the case of CEV specimens.

For the corroding cases (Fig. 3), a shift to more active (cathodic) OCP values is as expected and as observed for specimens CEn (steel treated in vesicles-free solutions). This corresponds to the decreasing *Rp* values with time of treatment. In contrast, specimens CEVn and CEVCn depict OCP values in the passive range at the end of the test (more anodic than -200 mV) and higher *Rp* values. The influence of Ca-containing vesicles is here already well pronounced: after initially lower values, and even lowest after 4 days treatment, the *Rp* values for

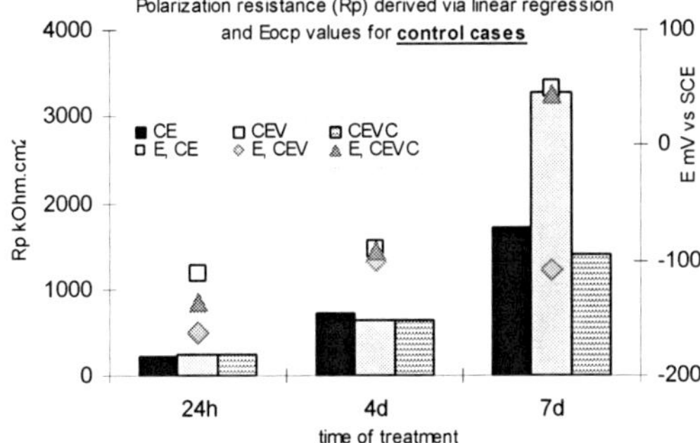

Figure 2: OCP and *Rp* values for control cases.

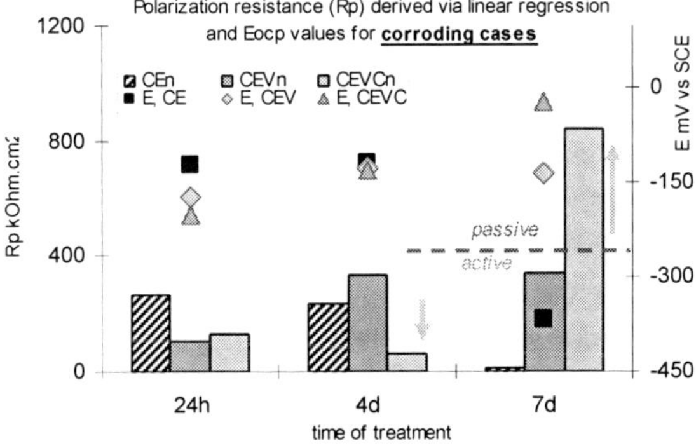

Figure 3: OCP and *Rp* values for corroding cases.

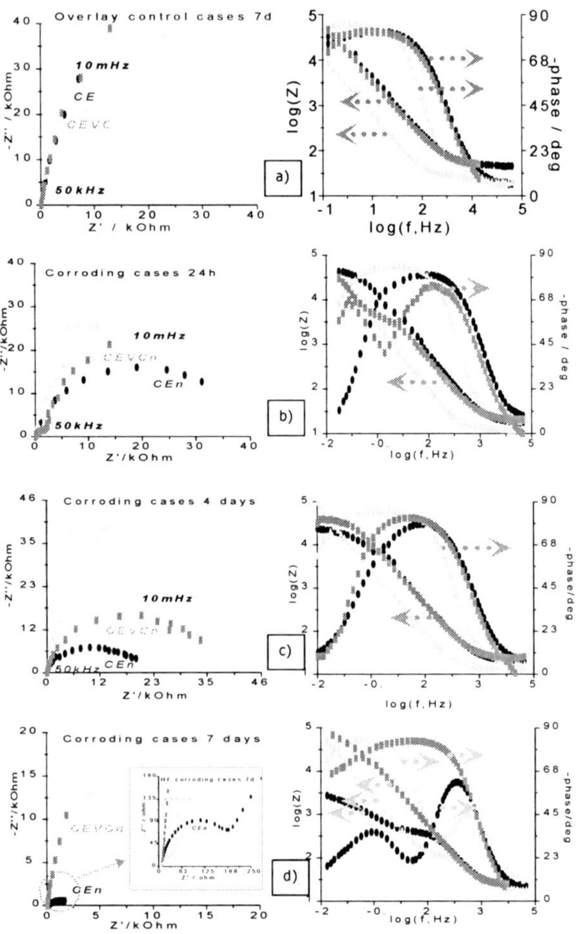

Figure 4: 7d EIS response for CE, CEV and CEVC (a) and CEn, CEVn, CEVCn after 24h (b), 4d (c) and d (d).

specimens CEVCn depict a significant increase at the end of the test. Since the steel electrodes, treated in CE containing “empty” vesicles do not show this trend (specimens CEVn), the only plausible explanation is that the Ca-containing vesicles exert a “self-repair” or ‘self-healing” of the already induced corrosion damage on later stages of treatment. These observations are actually confirmed by EIS, PDP and CVA tests and supported by surface analysis.

4.1.2 EIS, PDP and CVA results

The EIS response is not subject to elaborated discussion in this paper, but is presented as a direct visualisation of the electrochemical behaviour of the steel electrodes. Figure 4(a) presents the response for the control specimens after 7days. Figure 4(b)–(d) depicts the response of all corroding cases after 24h (b), 4d(c) and 7d (d) treatment. As seen from the plots, the magnitude of impedance

|Z| is higher for the steel treated in vesicles containing solutions at all time intervals, accompanied by a mostly constant phase angle of approx. 80°, whereas for the vesicles-free cases the phase angle drops to approx. 65° at the end of the test. The EIS results clearly show the "self-repair" of the product layer on the steel surface when Ca-containing vesicles are involved: specimens CEVCn depict a response with two time constants after 24h treatment, meaning a pronounced contribution of product layer alterations in parallel to charge transfer resistance phenomena. Later on, the magnitude of |Z| slightly drops, whereas the phase angle increases, which can only be denoted to modified product layer morphology and composition, rather than increase in charge transfer. Further, after 7 days treatment, both magnitude of |Z| and phase angle increase, denoted to improved and more corrosion resistant steel surface (one time constant only at 7d). The only barrier, rather than "self-healing" effect in specimens CEVn ("empty" vesicles involved) is seen from the gradual decrease of impedance and phase angle and pronounced contribution of a second time constant at the end of the test. These observations are in line with the electrochemical behaviour with external polarization – PDP and CVA response. Fig. 5(a) depicts PDP curves for all corroding cases as an overlay of 4d and 7d response; Fig. 5(b) presents the response of specimens CEVCn only for all time intervals.

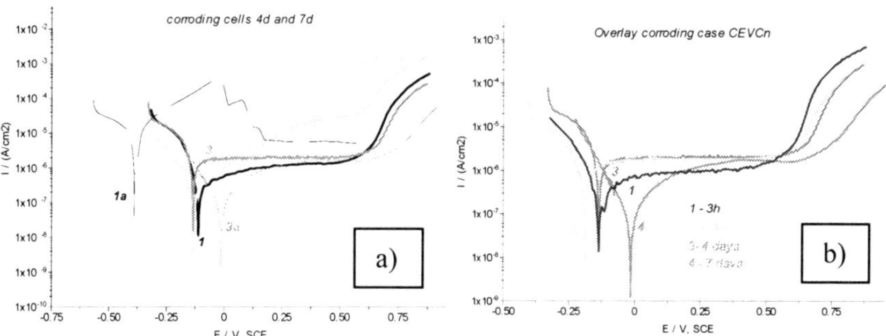

Figure 5: (a) PDP curves for specimens: CEn at 4d and 7d (curves 1 and 1a); CEVn at 4d and 7d (curves 2 and 2a); CEVCn at 4d and 7d (curves 3 and 3a); (b) PDP response for specimen CEVCn at 3h, 24h, 4d and 7d (curves 1, 2, 3 and 4 respectively).

After 4 days treatment the corrosion current density for specimens CEn and CEVn are similar (curves 1 and 2, Fig. 5a)), the anodic current with external polarization for specimens CEVn is even higher than that for CE. This proves the previously discussed diffusion limitations in specimen CEVn and barrier effects of the "empty" vesicles, which do not result in a stable and homogeneous product layer, since anodic currents are not limited but increasing with polarization. Specimens CEVCn present slightly higher corrosion current density, but a significant anodic limitation with external polarization is relevant (curve 3, Fig. 5a). The latter observation is in line with the previously discussed attempts to "repair" of initially induced corrosion damage. After 7d treatment,

the CEn specimens (curve 1a, Fig. 5a) exhibit the highest anodic and corrosion currents (as expected) and significant shift of corrosion potentials to more cathodic values. In contrast, specimens CEVCn (curve 3a, Fig. 5a) present a drop of corrosion and anodic current densities, a shift to more noble corrosion potentials and maintained anodic currents limitations.

After 7d treatment, specimens CEVn present similar to the 4d behaviour of active dissolution with external polarization (curves 2a and 2, Fig. 5a). The self-healing effect of the "Ca-containing" vesicles, present in the medium of CE+ NaCl are well seen if a comparison is made of the response for specimens CEVCn with time – Fig. 5(b): after initially low currents (curve 1, 3h), corrosion is initiated (curve 2, 24h); the process is maintained until 4days, where slightly higher corrosion current is recorded, most likely corresponding to product layer transformations and Ca-content release and/or nucleation sites effect due to the "Ca-containing" vesicles on the steel surface. The result with prolonged treatment is product layer "repair", evident by a significant drop of corrosion and anodic currents (curve 4, 7d) and further anodic limitations.

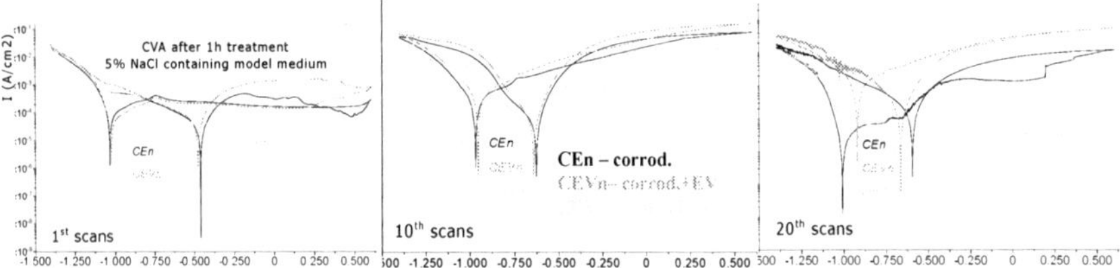

Figure 6: CVA scans for specimens CEn, CEVn and CEVCn

CVA tests strongly support all above observations. Fig. 6 presents the 1st, 10th and 20th scans, recorded after treatment in the relevant solutions. As seen from the plots, pitting initiates for specimens CEn and CEVn (at approx. -325 mV) with the 1st scan, whereas no pitting was observed for specimens CEVCn, the latter exhibiting the lowest current densities in the reverse scan and most noble re-passivation potential (Fig. 6). After 10 scans, all specimens, including CEVCn show pitting formation at around -500 mV (Fig. 6) and shift towards higher currents in both anodic and cathodic sweeps. Finally, within 20 scans (Fig. 6) the self-healing effect of the Ca-containing vesicles is evident from the recorded significantly low currents for specimens CEVCn in both anodic and cathodic sweeps, accompanied by the lack of pitting formation. In contrast, for specimens CEn and CEVn, pitting propagation is still relevant, accompanied by already largely increased currents in both anodic and cathodic sweeps.

4.2 Surface analysis

The previously discussed possible characteristics of the formed on the steel surface product layers with respect to electrochemical response were actually as observed via ESEM, EDX and XPS analysis. Fig. 7 presents ESEM micrographs and the relevant EDX analysis for the corroding cases (b-d); for a comparison,

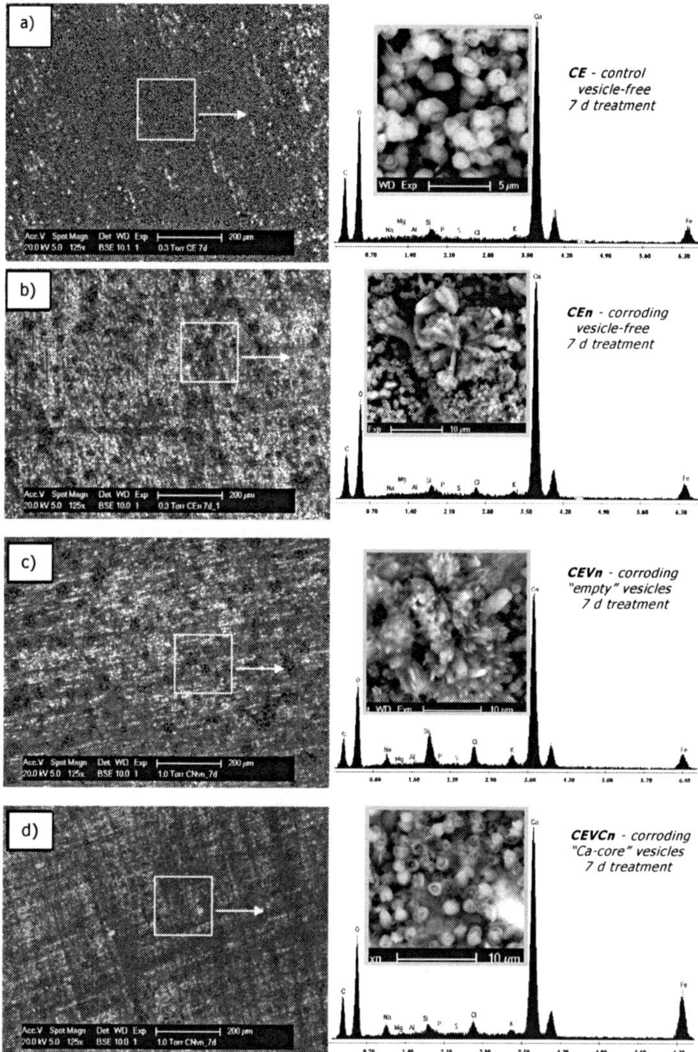

Figure 7: ESEM micrographs and the relevant EDX analysis for control (a) and corroding cases CEn (b), CEVn (c) and CEVCn (d).

the control case (CE) is depicted in Fig. 7a) (the layers at higher magnification are given as inlets in the EDX patterns).

As seen from Fig. 7 the morphology and composition of the product layer for corroding specimen CEVCn (Fig. 7d) resembles the one observed for the control specimen CE (Fig. 7a) i.e. a more compact layer, depicting mainly calcium carbonate was observed. Should be noted that for all specimens, $CaCO_3$ was detected by both ESEM/EDX and XPS (Fig. 8) (the XPS and ESEM/EDX investigations were performed after removing the specimens from the model solutions and drying on air). Since the specimens were in immersed condition, carbonation of the solutions during treatment is minimal. Moreover, if a substantial concentration of CO_2 in the solutions was present, $FeCO_3$ will form, which would be either detected by XPS and ESEM, or would rather rapidly transform when the specimens are taken on air to more stable iron oxides ($FeCO_3$ or "green rust" is not stable in oxygen containing environment). Therefore, the detected surface $CaCO_3$ is a result of the reaction of Ca-containing compounds (as CaO and $Ca(OH)_2$ previously deposited on the steel surface during treatment) with atm. CO_2 upon drying of the specimens on air. Consequently, the presence and amounts of $CaCO_3$ on the steel surface can serve as an indirect indication for the presence of adhered Ca-containing compounds during treatment in immersed condition. To this end, the control cases and specimens CEVCn present mainly carbonate-rich product layers, whereas for specimens CEn and CEVn (Fig. 7(b), (c) except $CaCO_3$, chloride-containing iron oxides/hydroxides are relevant. This is in line with the lower corrosion resistance, detected via electrochemical measurements. In specimens CEn and CEVn localised corrosion is basically observed, evidenced by the deposition of voluminous, needle-shape corrosion products (Figs. 7(b),(c).)

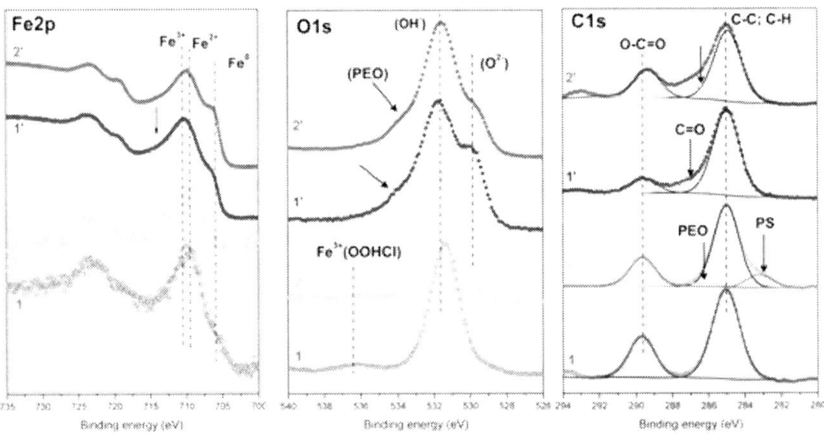

Figure 8: XPS iron (Fe2p), oxygen (O1s) and carbon (C1s) ionization for specimens CEn (curves 1 and 1') and CEVCn (curves 2 and 2').

An elaborated discussion on the otherwise extensive XPS analysis is not subject to this paper and will be separately reported. What needs to be mentioned are only main observations and features, supporting the above results and discussion.

As previously defined, XPS was performed on the surface of the as-treated specimens (after 7 days) and in depth of the (7 days-old) product layers. Fig. 8 presents the ionizations of Fe, O and C on the "as-treated" surface (scans 1 and 2) and in depth of the products (scans 1' and 2'). With respect to iron oxides and hydroxides, all specimens exhibit a mixture of these with a larger portion of the latter in the outermost sections of the layers (shape and features of the scans 1 and 2 in Fe2p, pronounced peak at 531.6 eV in O1s, corresponding to OH^-) and a larger portion of the former (oxides) in depth of the layers (hematite/magnetite based layers as evident from the characteristic features of Fe2p, 1' and 2' and the peaks at lower binding energy ~ 529 eV in the O1s for 1' and 2'). As also observed, metallic iron appears in dept of the layers as well (~706 eV), which is as expected within product layer removal. The presence of adsorbed in the product layers vesicles for specimen CEVCn is confirmed by the characteristic binding energies in the C1s ionizations i.e. 283 eV for PS, 286.5 eV for PEO and 534 eV for PEO in the O1s ionizations. Iron oxy/hydroxy chlorides are detected only in the outermost layers for (mainly) specimens CEn (peak at 536.1 eV in O1s), which is actually supported by the morphological observations on the surface of CEn (Fig. 7b). In other words, morphological and chemical alterations in the product layers, formed on the steel surface in the presence of Ca-containing vesicles and chloride result in a "self-repair" or "self-healing", evidenced by a more compact and uniform product layer, featuring iron oxy/hydroxides and calcium carbonate only (similar to the layers, formed in the control, chloride-free cases). In contrast, for steel treated in vesicles-free or "empty" vesicles containing solutions, voluminous iron-oxy/hydroxy chlorides and lower amount of Ca-containing compounds are present.

5 Conclusions

In the frame of a novel approach to corrosion control in reinforced concrete, the application of polymeric nano-particles was investigated. In order to evaluate the difference between "barrier" effects only and self-healing mechanisms, resulting from the "Ca-core" release, the effect of both "empty" and Ca-containing vesicles was studied. In the presence of both vesicle types corrosion resistance was increased, but the effect was significantly more pronounced when Ca-containing vesicles were involved. The concentration of 0.0024 wt.% vesicles in the model solutions is minimal and therefore self-repair or self-healing of the product layer solely due to the Ca-component is not realistic. The most plausible mechanism is the nature of incorporation of the Ca-containing vesicles in the product layer, enhanced chloride binding effects and adsorption on active sites on the steel surface. The result is a more uniform and stable steel surface layer, initial pitting formation and propagation, but consecutive healing, as evidenced by surface analysis and electrochemical response i.e. largely reduced anodic and corrosion currents and no further pit propagation when Ca-containing vesicles are present.

References

[1] Conway, BE, Theory and principles of electrode processes, 170-272, The Ronald Press Company, NY (1965)

[2] Broomfield J (1997) Corrosion of steel in concrete, understanding, investigating and repair. Spon, London

[3] Sahmaran M (2007) J Mater Sci 42:9131. doi:10.1007/s 10853-007-1932-z

[4] Elsener B (2002) Cem Concr Compos 24(1):65

[5] Garces A, Andrade MC, Saez A, Alonso MC (2005) Corros Sci 47(2):289

[6] Uhlig HH (2000) In: Revie RW (ed) Uhlig's corrosion handbook., p 583

[7] Bertolini L, Elsener B, Pedeferri P, Polder R (2004) Corrosion of steel in concrete: prevention, diagnosis, repair, 1st edn. Wiley, Weinheim

[8] Petterson K (1994) In: Swamy RN (ed) Corrosion and corrosion protection of steel in concrete. Academic Press, Sheffield, p 461

[9] Pedeferri P, Constr. Build. Mater., 10, 391 (1996)

[10] Bassi R., Davies H, Testing anti-carbonation coatings for concrete, BRE Information paper IP7/96, 1996

[11] Buenfeld N. Buenfeld, J.Z. Zhang, Cem. Concr. Res. 28 (1998) 665-674

[12] Castle J.E., Cathodic disbondment, Proceedings of the AIP Conference p. 165, (1996)

[13] Bird C.E, Nature, 194:798 (1962)

[14] Herrera MJ, Martinez M and Perez JT, Evaluation of Electrochemical behavior of Two Inhibitors for Galvanized Rebars in Alkaline Solutions, paper 831, 210th meeting ECS, Cancun, MX (2006)

[15] Intern. Lead Zinc Research Organisation. Galvanized reinforcement for concrete II. Res. Triangle Park NC, USA, 1981

[16] Obada K, Stephen R. Yeomans, Cem.Concr.Comp.22, 459-467, (2000)

[17] Andrade C, Holst JD. Coating protection for reinforcement. Comite Euro-International Du Beton. State of the Art Report. London: Thomas Telford Publication; 1995. p. 51

[18] Mietz J, Electrochemical Rehabilitation Methods for Reinforced Concrete Structures: A State of the Art Report, in, Institute of Materials, 1 Carlton House Terrace, SW 1 Y 5 DB, London, UK, 1998

[19] Andrade C., Keddam M, Nóvoa X. R., Pérez M. C., Rangel C. M., Takenouti H., Electrochim. Acta, 46, 3905-3912 (2001).

[20] Gaidis M., Cement and Concrete Composites, 26 (2004) 181-189

[21] Wombacher F., Maeder U., Cem. Concr. Comp., 26 (2004) 209-216

[22] Koleva D.A., van Breugel K., Ye G., Koenders E., ACI Mater. J. 267 (2009)101-110

[23] Koleva D.A., Jie Hu, Van Breugel K., J. Intern. Sci. Publ.: MMT5 (2011) 63

[24] Hu J., Koleva D.A., Petrov P., van Breugel K., Corr. Sci. 65 (2012) 414-430

[25] Hu J., Koleva D.A., Ma Y., Schlangen E., Petrov P., van Breugel K., Cem. Concr. Res. 42 (2012) 1122-1133

[26] Hu J., Koleva D.A., van Breugel K., J. Mater. Sci. 47 (2012) 4981-4995

[27] Hu J., Koleva D.A., et. al. J. Electrochem. Soc. 158 (2011) C76

[28] Hu J., Koleva D.A., van Breugel K. ECS Transactions, 41 (16) 1-9 (2012)

Corrosion resistance of Au/Ni thin films coated stainless steel used for a PEFC separator

Y. Kimura[1] & M. Hirano[2]
[1]*Department of Environmental and Energy Chemistry, Kogakuin University, Japan*
[2]*Graduate school, Kogakuin University, Japan*

Abstract

Recently, as a new clean and renewable energy much attention has been paid to the PEFC (Polymer Electrolyte Fuel Cell) because of energy and environmental problems, such as depletion of petroleum, global warming and so on. PEFC has the characteristics of a simple structure, easy to maintain and a low cost. One of the parts of the PEFC is called the 'separator'; metal separators have a major issue regarding compatibility with corrosion resistance and electrical conductivity, in addition to low cost. Using current technology, we have selected a stainless steel AISI316L with an Au/Ni skin film. Therefore, there is a need to make a thinner skin film especially an Au top film for suppressing the PEFC total cost. When thinking about skin film, defects in the Au top film increases gradually as the thickness of Au film decreases. As a result, intensive corrosion is generated due to defects in the Au/Ni skin film. However, there are few reports evaluating correlation between the corrosion characteristics of stainless steel with an Au/Ni skin film under a PEFC environment and existing defects in the Au/Ni skin film. Therefore in this study, the pinhole defect ratio especially in Au film was evaluated using the CPCD method. Au/Ni double layered coating was formed on AISI316L stainless steel employing the sputtering method because of its high adhesion strength. As the quasi-PEFC environment 1M H_2SO_4 aqueous solution was selected and then the pinhole defect ratio was evaluated by the CPCD method. As a result, the pinhole defect area ratio of 80nm thick Au film was obtained to be about 1.1% in 1M H_2SO_4 aqueous solution.
Keywords: separator of PEFC, AISI316L stainless steel with Au/Ni skin film, corrosion, pinhole defect area ratio evaluation, CPCD method.

WIT Transactions on Engineering Sciences, Vol 77, © 2013 WIT Press
www.witpress.com, ISSN 1743-3533 (on-line)
doi:10.2495/MC130191

1 Introduction

Recently, as a new clean and renewable energy much attention has been paid to the PEFC (Polymer Electrolyte Fuel Cell) because of energy and environmental problems, such as depletion of petroleum, global warming and so on. PEFC has the characteristics of a simple structure, easy to maintain and a low cost. One of the parts of PEFC, called 'separator', metal separators have a major issue regarding compatibility with corrosion resistance and electrical conductivity, in addition to low cost [1, 2]. Using current technology, we have selected stainless steel AISI316L with Au/Ni skin film. Therefore, there is a need to make a thinner skin film especially Au top film for suppressing PEFC total cost [3]. When thinking about skin film, defects in Au top film increases gradually as the thickness of Au film decreases. As a result, intensive corrosion is generated due to defects in Au/Ni skin film. However, there are few reports evaluating correlation between corrosion characteristics of stainless steel with Au/Ni skin film under PEFC environment and existing defects in Au/Ni skin film. Therefore in this study, the pinhole defect ratio especially in Au film was evaluated using the CPCD method [4–7]. And also, Corrosion resistance of Au/Ni thin films coated stainless steel under quasi-PEFC environment was investigated.

2 Experimental procedures

2.1 Evaluation of pinhole defect area ratio in Au skin film

The substrate material used in this study was pure Niplate specimens whose size of 11×11×1mm. The surface of substrate was polished until #2000 emery paper and then made mirror finish by buff polishing. For making Au film with four different kinds of thickness from 20 nm to 120nm DC sputtering system (L-332S-FH, Canon ANELVA) was employed (Table 1) [8]. The pinhole defect ratio especially in Au film was evaluated using the CPCD (Critical Passivation Current Density) method. The CPCD method was based on the critical passivation current density, just before corrosion state shifted from active to passive with the potential sweep toward noble direction, is proportional to ratio of the pinhole defect area through coating film.

At first, for determining optimal concentration of H_2SO_4 in aqueous solution for obtaining larger current density data from 1M to 5M H_2SO_4 solutions were employed and anodic polarization measurement of substrate Ni specimen was conducted under the conditions shown in Table 2. When evaluating the pinhole defect area ratio of Au coating film, it is necessary to make sure the suppression of exfoliation of formed Au film. The corroded morphologies of coating film surface were examined in detail by optical microscope, scanning electron microscope (SEM) and atomic force microscope (AFM).

Electrochemical measurement by three electrodes methods was conducted in H_2SO_4aqueous solution after deaeration treatment of more than one hour using N_2 gas.

Table 1: Details of specimen used for Au pinhole defect ratio evaluation.

Specimen	Remarks
Substrate : Pure Ni	#2000 emery paper polish and buff finish
Au coating	Au film thickness: 20, 50, 80, 120nm made by sputtering

Table 2: Conditions of anodic polarization measurement for Au pinhole defect area ratio evaluation.

Potential sweep range	−500mV ~ 500mV(SCE)
Sweep rate	60mV/min
Temperature	303±0.5 K
Solution	1, 3, 5M H_2SO_4

And then the defect area ratio R[%]of Au thin film was evaluated from the following equation.

$$R = \frac{1}{fs} \times \frac{i_{crit}\,(\text{Coating/Substrate})}{i_{crit}\,(\text{Substrate})} \times 100[\%] \qquad (1)$$

where, i_{crit} (Coating/Substrate) is the critical passivation current density of Au thin film coated Ni specimen and i_{crit} (Substrate) is that of non-coated substrate Ni specimen. The value of the shape factor of corrosion pit fs is evaluated to be 2, because the morphology of corrosion pit is expected to be semielliptical.

2.2 Evaluation of corrosion resistance of Au/Ni thin films coated stainless steel under quasi-PEFC environment

For evaluating corrosion resistance of Au/Ni coated stainless steel AISI 316L stainless steel under quasi-PEFC environment AISI 316L (chemical composition were shown in Table 3) plate specimens whose size of 11×11×2.5 mm was selected as substrate material because of having good corrosion resistance, low hydrogen embrittlement susceptibility and actual results using as separator. The surface of substrate was finished as the same as Ni palate described in previous section. Au/Ni double layered coating film was formed under the conditions as shown in Table 4. Then for quasi-PEFC environment 0.5M H_2SO_4 aqueous solution at 353K was employed and static immersion test of 24hrs was conducted (Table 5) using corrosion cell shown in Fig. 1.

Table 3: Chemical composition of material used in this study (wt. %).

Material	Ni	Fe	C	Cr	Mo
AISI 316L	12.2	Bal	0.01	16.3	2.10

Table 4: Details of Au/Ni coated SUS316L specimen for H_2SO_4 corrosion test.

Specimen	Au film thickness
Au / Ni coated AISI 316L (Ni thickness:100nm)	20, 50, 65, 80nm

Table 5: Conditions of immersion corrosion test in H_2SO_4aqueous solution.

Immersion time	24H
Solution temperature	353±1 K
Solution	0.5M H_2SO_4aqueous solution

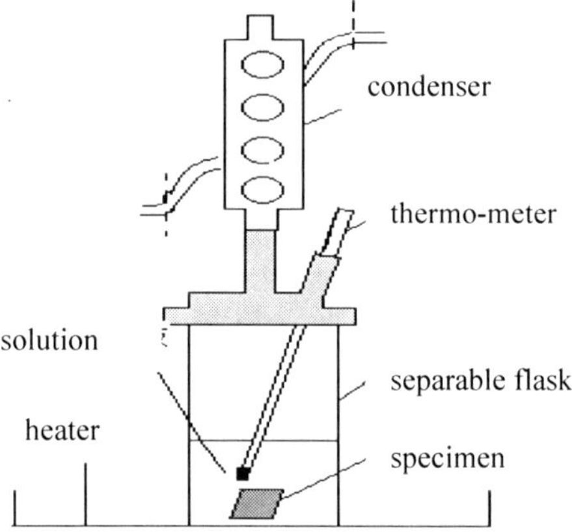

Figure 1: Setup of immersion corrosion test in H_2SO_4aqueous solution.

2.3 Evaluation of corrosion resistance of Au/Ni thin films coated stainless steel under quasi-PEFC environment by anodic polarization measurement

Electrochemical measurement was conducted using three electrode method (HZ-3000, Hokutodenkou, Japan) composed of specimen (working electrode), Pt (counter electrode) and the saturated calomel electrode (S.C.E.) (reference electrode). All tests were performed in 0.5 H_2SO_4 aqueous solution of 348K. Potentio-dynamic corrosion test was conducted with the sweep rate of

www.witpress.com, ISSN 1743-3533 (on-line)

60mV/min from –200 mV to 1500 mV (vs. S.C.E.) considering real potential range under actual usage (100~800mV SCE) [11]. Through considering separator's actual operating conditions anodic polarization measurement were conducted under the conditions summarized in Table 6. In this case, specimens with 200nm thick Ni intermediate film whose Au top film thickness were 50, 65 and 80 nm, were also prepared for understanding the effects of Ni intermediate film thickness upon corrosion resistance of Au/Ni/AISI 316L specimen (Table 7).

Table 6: Conditions of anodic polarization measurement.

Potential sweep range	−200mV–1500mV(SCE)
Sweep rate	60mV/min
Solution temperature	348±0.5 K
Solution	0.5M H_2SO_4aqueous solution

Table 7: Details of Au/Ni/AISI 316L specimen for anodic polarization measurement.

Ni intermediate film thickness	Au top film thickness
100nm	50, 65, 80, 120nm
200nm	50, 65, 80nm

3 Experimental results and discussions

3.1 Pinhole defect area ratio in Au skin film

At first, to determine optimal concentration of H_2SO_4 in aqueous solution for highly accurate measurement from 1M to 5M H_2SO_4aqueous solutions were employed and anodic polarization measurement of substrate Ni specimen was conducted under the conditions shown in Table 2. From the obtained results shown in Fig. 2, 1M H_2SO_4aqueous solution was selected as the optimal concentration of H_2SO_4 aqueous solution for CPCD method because of showing maximum current density within the potential range at which no anodic dissolution of Au was generated.

Then, the pinhole defect area ratio in Au top skin film was evaluated employing CPCD method [4–7].Results obtained by anodic polarization measurements of specimens with different Au top skin film thickness on Ni substrate were shown in Fig. 3. Critical passivation current density ($i_{crit.}$) of 20nm thick Au film coated specimen showed almost the same level as that of substrate pure Ni specimen. Other specimens with thicker Au coating showed relatively smaller $i_{crit.}$ as the thickness of the film became larger. Drastic suppression in critical passivation current density ($i_{crit.}$) was recognized in specimen with Au film thickness between 50nm and 80 nm.

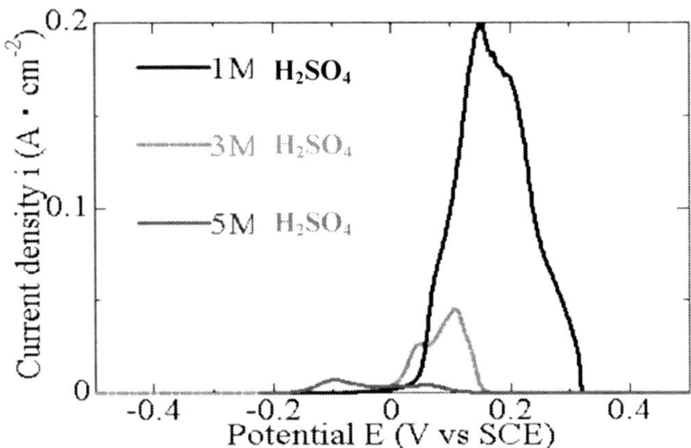

Figure 2: Results of anodic polarization measurements in aqueous solutions with different H_2SO_4concentration.

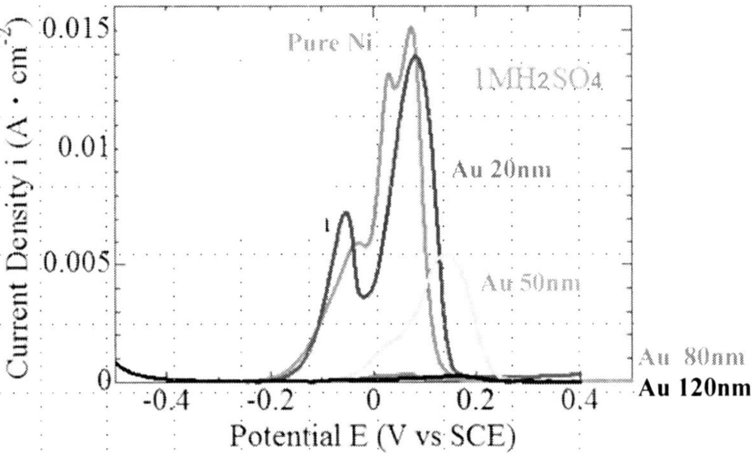

Figure 3: Results of anodic polarization measurements with different Au skin film thickness on Ni substrate.

In the case of 20nm thick Au film coated specimen, almost all of the Au film was exfoliated after polarization measurement. Therefore, the Au top film defect area ratio R_{Au} in this case was not evaluated. Then, the defect area ratio R_{Au} with thicker Au film was evaluated by using equation (1) and shown in Table 8. At the same time, an Au pinhole defect area ratio evaluation was also conducted by microscopic surface observation.

Table 8: Results of Au film defect area ratio evaluation.

Au film thickness [nm]	Au pinhole defect ratio: R_{Au} [%] (CPCD method)	Au pinhole defect ratio [%] (Surface observation)
20	–	–
50	18. 0	6. 34
80	1. 12	0. 84
120	0. 854	1. 45

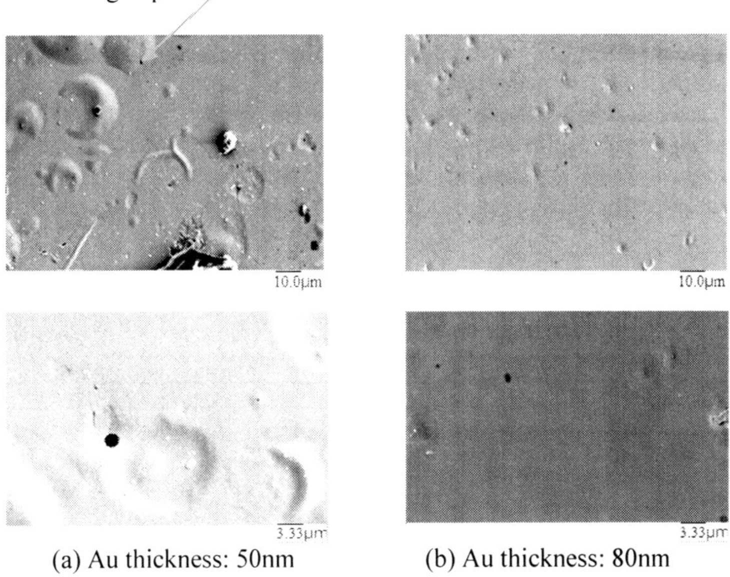

(a) Au thickness: 50nm (b) Au thickness: 80nm

Figure 4: Surface morphology after polarization measurement.

Also in this Table 8, Au pinhole defect ratio [%] evaluated by surface observation was summarized. From these results, clear difference in pinhole defect area ratio values was recognized between data obtained by CPCD method and microscopic surface observation for 50nm thick Au film. In case of specimens with thicker Au films than 50 nm, there were no large difference between evaluated values both by the CPCD method and microscopic surface observation. Therefore, coherency of evaluated results between two different evaluating methods was recognized. For making clear the reason of the deference in evaluated results obtained in 50nm thick Au film coated specimen, surface morphology after polarization measurement was investigated by SEM and shown in Fig. 4. As a result, surface morphology of specimen with 50nm thick Au film showed clear deference to specimens with thicker Au film. In case of specimen with 50nm thick Au film some trace of Au film swelling [9, 10] was recognized and shown by circle in Fig. 4(a). In this case, anodic dissolution of Ni substrate was expected to generate and extended inside of pinhole. For this reason some excess current density may be detected in this case.

3.2 Corrosion resistance of Au/Ni thin films coated stainless steel under quasi-PEFC environment

Then for quasi-PEFC environment 0.5M H_2SO_4 aqueous solution at 353K was employed and static immersion test of 24hrs was conducted using corrosion cell shown in Fig. 1 under the conditions shown in Table 5. From these tests effect of Au film thickness upon corrosion resistance of Au/Ni thin films coated AISI 316L stainless steel under quasi-PEFC environment was evaluated. Then, corrosion rate was calculated and shown in Table 9. As a result, both of evaluated corrosion rates were suppressed. From these results, sufficient corrosion resistance of Au/Ni thin films coated AISI 316L stainless steel under static quasi-PEFC environment of 0.5M H_2SO_4 aqueous solution at 353K was obtained within thin Au top film thickness range of 20nm and over.

Table 9: Calculated corrosion rate.

Au/Ni thickness(nm)	Corrosion rate(mg/dm^2· day)	Corrosion rate(mm/year)
20/100	9.30 x 10^{-3}	4.27 x10^{-4}
50/100	9.38 x 10^{-3}	4.29 x10^{-4}
65/100	9.72 x 10^{-3}	4.45 x10^{-4}
80/100	15.8 x 10^{-3}	7.22 x10^{-4}

3.3 Corrosion resistance of Au/Ni thin films coated stainless steel under quasi-PEFC environment evaluated by anodic polarization measurement

For evaluating corrosion resistance of Au/Ni thin films coated AISI 316L stainless steel under quasi-PEFC environment potentio-dynamic corrosion test was conducted with the sweep rate of 60mV/min from –200 mV to 1500 mV (vs. S.C.E.) considering real potential range under actual usage (100~800mV S.C.E.) [11].

Obtained anodic polarization curves were summarized in Figs 5 and 6.

In Fig. 5, results obtained from the Au/Ni/AISI316L specimen with100 nm thick intermediate Ni film were indicated. On the contrary, those obtained from specimens whose intermediate Ni film thickness is 200 nm were shown in Fig. 6.

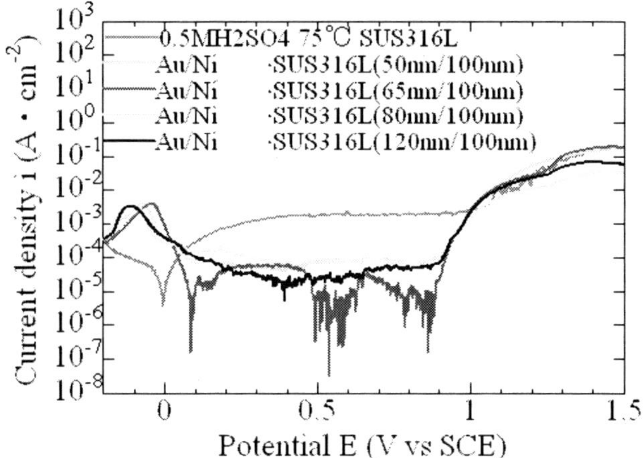

Figure 5: Results of polarization measurement obtained from Au/Ni/AISI316Lspecimen whose intermediate Ni film thickness is 100 nm.

Current density of Au/Ni thin films coated AISI 316L stainless steel was suppressed until about 1/100 levels comparing to that detected in bare substrate AISI 316L stainless steel.

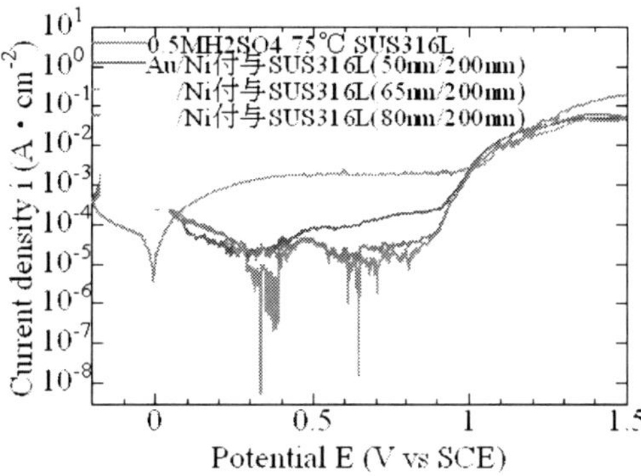

Figure 6: Results of polarization measurement obtained from Au/Ni/AISI316Lspecimen whose intermediate Ni film thickness is 200 nm.

And also, under the potential range of actual usage (100~800mVSCE) double layers coated AISI316L specimen indicated to stay still in passive state. Current density of Au/Ni thin films coated AISI 316L stainless steel whose intermediate Ni film thickness was 200 nm showed furthermore suppressed passivation current density compared with that obtained from the same system specimen with thinner intermediate Ni thickness of 100 nm. And also, improvement in

corrosion resistance due to increase of Au top film thickness was more remarkable in case of specimens with thicker intermediate Ni film.

From abovementioned results, improvement of corrosion resistance was recognized in all the double layered Au/Ni coated AISI 316L stainless steel. Contribution of intermediate Ni film thickness increase upon improvement in corrosion resistance became more remarkable when Au top film thickness became larger. Adhesion strength improvement due to Ni film thickness increase may result in promotional effect of dense Au film formation together with Au film thickness increase effect itself. Best corrosion resistance in Au/Ni coated AISI 316L stainless steel specimen was obtained when Au/Ni is 65 nm/200 nm. These results may be brought about due to optimization of adhesion strength [10]improvement and fill up of pinhole in Au top film by thicker Ni film formation. For this reason increase of intermediate Ni film thickness results in remarkable corrosion resistance improvement together with suppressing top Au film thickness.

From these results, it was made clear that CPCD method was applicable to Au top film defect area ratio evaluation of Au/Ni double layered coated AISI 316L stainless steel. And through establishing adhesion strength improvement by intermediate Ni film Au/Ni double layered coated film has sufficient corrosion resistance under quasi-PEFC environment. At the same time, outlook of specimen after anodic polarization measurement was investigated and typical results of these observations were indicated in Fig. 7. As shown in Fig. 7(a), specimen with thin top Au coating film resulted in remarkable exfoliation of coated layers after anodic polarization measurement under the conditions shown in Table 6. However, this specimen indicated sufficient corrosion resistance under the conditions of 24hrs static immersion test in 0.5M H_2SO_4 aqueous solution of 353K, as shown in Table 9.

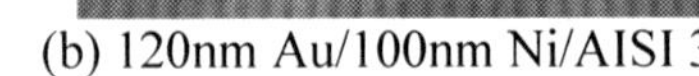

(a) 50nm Au/100nm Ni/AISI 316L (b) 120nm Au/100nm Ni/AISI 316L

Figure 7: Outlook of specimen surface after anodic polarization measurement.

On the contrary, in case of 120 nm thick Au top coated specimen no damage was recognized in this specimen as shown in Fig. 7(b) even after anodic polarization measurement under the condition shown in Table 6.

4 Conclusions

There are few reports evaluating correlation between corrosion characteristics of stainless steel with Au/Ni skin film under PEFC environment and existing defects in Au/Ni skin film. Therefore in this study, the pinhole defect area ratio especially in Au top film was evaluated using the CPCD method. And also, corrosion resistance of Au/Ni thin films coated AISI 316L stainless steel under quasi-PEFC environment was evaluated.

Results obtained are summarized as follows:

1. CPCD method was applicable to Au top film defect evaluation of Au/Ni double coated AISI 316L stainless steel. The pinhole defect area ratio in Au thin film depends upon thickness of film itself.
2. The coherency of evaluated results of Au top film defect between CPCD method and surface morphology observation after polarization measurement was recognized.
3. Improvement of corrosion resistance was recognized in all the double layered Au/Ni coated AISI 316L stainless steel. Current density of Au/Ni thin films coated AISI 316L stainless steel suppressed until about 1/100 levels compared with that detected in bare substrate AISI 316L stainless steel. And also, under the potential range of actual usage double layered coated AISI316L specimen whose top Au film thickness is larger than 65 nm indicated to stay still in passive state.
4. Best corrosion resistance in Au/Ni double layered coated AISI 316L stainless steel specimen was obtained when Au/Ni is 65 nm/200 nm under the condition of polarization measurement. However, specimen whose top Au film thickness is less than 50nm indicated sufficient corrosion resistance under the conditions of 24hrs static immersion into 0.5M H_2SO_4 aqueous solution of 353K.

References

[1] Takagi, S., Development of New Metallic Separator Material for PEFC "NANOCLAD", Electric Furnace Steel, **77(4)**, pp. 321-326, 2006.

[2] Ueda, M., *et al.* Development of Metallic Separator for Polymer Electrolyte Membrane Fuel Cell (PEFC) (Report 1), pp. 2-6, Kurimoto Technical Report, No.55, 2006.

[3] Ueda, M., *et al.* Development of Carbon Coated Stainless Separators and Their Polarization Performance for Polymer Electrolyte Cell, J. of Japan Inst. of Metals, **71(7)**, pp. 545-552, 2007.

[4] Sugimoto, K., 95th JSCE symposium, pp. 1-11, JSCE, 1993.

[5] JSME standard S-010 "Standard Method for Evaluating the Defects in the Coatings Made by Dry Processing", JSME, 1996.

[6] Nitta, S. and Kimura, Y., Evaluation of Defects in Ceramic Coating Films by Various Electrochemical Techniques, Trans. Japan Soc. Mech. Engineers, **61(589)**, pp. 1914-1920, 1995.

[7] Nitta, S. and Kimura, Y., Evaluation of Defects in TiN Thin Films by Various Electrochemical Techniques, Advances in Surface Engineering, Volume II: Process Technology, Edited by P. K. Datta and J. S. Burnell-Gray, pp. 39-47, The Royal Society of Chemistry 1997.
[8] The Surface Finishing Society of Japan, "Basics and applications of PVD and CVD Coatings", Tokyo, Makishoten, 1994.
[9] Nitta, S. and Kimura, Y., Morphologies and Quantitative Evaluation by Electrochemical Methods of the Defects in Ceramic Coating Films, Journal of Material Testing Research Association of Japan, **42(1)**, pp. 4-14, 1997.
[10] Ohata, M. and Kimura, Y., Correlation between Localized Corrosion Morphology and the State of Interface of Coated TiN Thin Film, Surface Treatment IV, Editors, C. A. Brebbia *et al.* pp. 23-32, WIT Press, 1999.
[11] Chiba, H., *et al.* Influence of Mechanical and Structural Factors on Durability of PEFC with Gel Electrolyte Membranes, Electrochemistry, **75(2)**, pp. 238-242, 2007.

PFA-linings for avoidance of dewpoint corrosion in heat-exchangers and ducts of waste-incineration power plants

M. Lotz
Quadrant EPP AG, Switzerland

Abstract

The combustion of wastes of diverging qualities leads to the formation of gases with a high content of sulphur, chlorine and fluorine compounds. When the dewpoint of these gases is underrun in heat exchangers, housings and ducts in order to re-cycle more heat out of the combustion gases, this leads to a heavy condensation of hydrochloric, hydrofluoric and sulphuric acid on the involved equipment and in consequence to one of the strongest types of corrosion known, called dewpoint corrosion. The latter even corrodes the most resistant steel grades and thus requires other technical solutions. A lining of the condensing heat-exchangers, their housings and the flue gas ducts with the fully-fluorinated thermoplastic PFA (tetrafluoroethylene-perfluoroalkylvinylether) renders possible a service of the heat exchangers under condensing conditions and consequently increases the total power production efficiency of the waste-incineration plant, even having a positive environmental effect due to a positive influence on the overall energy mix. Although the suitability of this technical principle had already been demonstrated in multiple successful examples in fossil- and waste-fired power plants, constructive and technical details can have an influence on the performance of the lined equipment, such as material properties, welding of the liner, installation and sealing against the unlined parts as well as constructive aspects of the heat exchanger pipes and tubes. This will be discussed in the following.
Keywords: waste incineration, dew point corrosion, PFA, linings, fix points, PTFE.

WIT Transactions on Engineering Sciences, Vol 77,
www.witpress.com, ISSN 1743-3533 (on-line)
doi:10.2495/MC130201

1 Introduction

1.1 Dew point corrosion in power plants

Combustion of fossil fuels or waste leads to the formation of gases containing sulphur, chlorine and fluorine. The dewpoint of the flue gases typically is reached if the gas temperature drops below approx. 125°C, which however depends on the exact combustion gas composition (Oil & Gas Journal [1]). Then, droplets of sulphuric acid, hydrochloric acid and hydrofluoric acid form on the surfaces of the heat exchangers and the surrounding equipment. Accordingly, these heat exchangers are called condensing heat exchangers. The condensation of the acid droplets on the metal surfaces leads to the so called dew point corrosion (Huijbregts and Leferink [2], Lewandowski [3], McKetta [4]). If at all, only high nickel alloys are able to resist the extremely corrosive conditions, whereas regular steels and stainless steels are destroyed (Cox *et al.* [5], de Weijer and Huijbregts [6], Agarwal and Grossmann [7]).

In order to increase the total efficiency of the power production process, dewpoint temperatures are underrun. For this, condensing heat exchangers are installed in addition to the known air pre-heaters.

The efficiency increase of the total power production process, which can be achieved with condensing heat exchangers depends on the total setup of the power plant, moisture (in case of coal fired plants) and the flue gas composition. The efficiency increase typically is in a range of 0.5 to 1.5% in Europe. In American and Asian countries the potential is even higher (up to 5%). In consequence, fossil fuels and emissions can be saved, if condensing heat exchangers are used in fossil-fired power plants and waste incineration plants.

1.2 PFA

PFA is a semi-crystalline, fully fluorinated thermoplastic. The strong chemical bond between the carbon backbone of the plastic molecules can hardly be broken by chemical attack of the compounds contained in the flue gas stream, and the fluorine side chains of the carbon backbone shield the latter against chemical attack. In combination, this confers an extreme chemical and temperature resistance to the PFA material.

Unlike sintered fluoropolymers like PTFE and modified PTFE, PFA forms a liquid melt and hence can be extruded into foils and pipes. In contrast, foils from PTFE and modified PTFE are skived from sintered blocks and hence vary in their quality alongside the cooling radius of the sintered drums. Furthermore, PTFE and modified PTFE can have micro-pores in regions of imperfect sintering, which then also lead to a weakening of the protective effect of these materials. Both effects, quality variations due to diverging temperature regimes during the cooling phase of the sintering and micro-pores are not found in PFA due to the technically completely diverging melt-extrusion process.

The ability of PFA to flow in the molten state furthermore renders possible to weld PFA sheets with PFA weld rods, and to achieve weld seam qualities, which are close to the strength of the parental, non-welded material (Fig. 1). Such weld seam qualities cannot be reached using sintered fluoropolymers, since the melt of sintered polymers does not flow. Besides its excellent thermoforming-properties, the ability of PFA to form strong weld seams is a crucial prerequisite to yield long life times of the lined equipment.

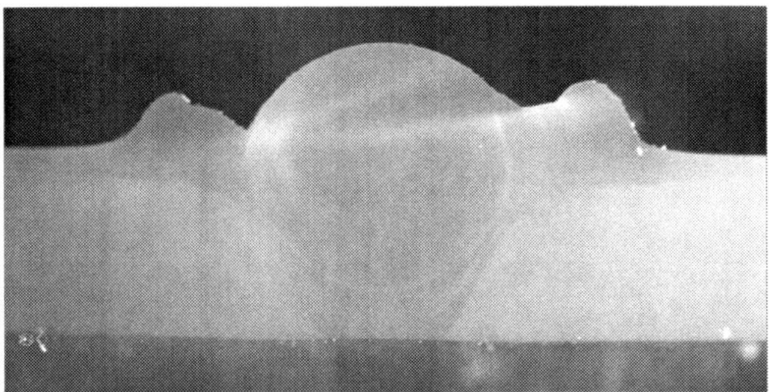

Figure 1: Picture of a typical PFA weld seam (cross-section). The molten material of the plate and the weld rod (round, in the middle) coalesce and hence lead to weld strengths close to the non-welded material.

2 Condensing heat exchangers and housings

2.1 Location of the condensing heat exchangers in the flue gas stream

Condensing heat exchangers are located after the boiler, air pre-heaters and electrical precipitators (Fig. 2). The condensing heat exchangers are furthermore upstream of the FGD (flue gas desulphurisation). Upstream of the condensing heat exchangers the gas temperatures are high and no flue gas condensation occurs. Therefore, metals are the appropriate material for these regions in the flue gas stream. In the FGD section, the flue gases are less corrosive and the gas temperatures lowered to a point, where often cheaper materials like polypropylene (usable to approx. 100°C) can be used. Hence, the usage of PFA linings and condensing heat exchangers is attributed to a particular part in the flue gas stream, where the corrosion is extremely high and the temperatures still too high to use other plastic materials. Gas/liquid (gas: flue gas stream, liquid: heat transfer media) condensing heat exchangers are hold in a steel frame, lined with PFA natural foil in thickness 1.5 mm and 2.3 mm; both thicknesses often are mixed depending on the mechanical load of the foil. For example in a system of the company Flucorrex AG, the liquid runs through steel pipes, coated with enamel and an additional PFA tube in order to protect the tubes against the chemically extremely aggressive environment (Fig. 3).

In gas/gas (gas 1: flue gas stream, gas 2: heat transfer media) heat exchangers, the frame also consists of PFA lined steel, but the tubes are made from glass (Fig. 4).

www.witpress.com, ISSN 1743-3533 (on-line)

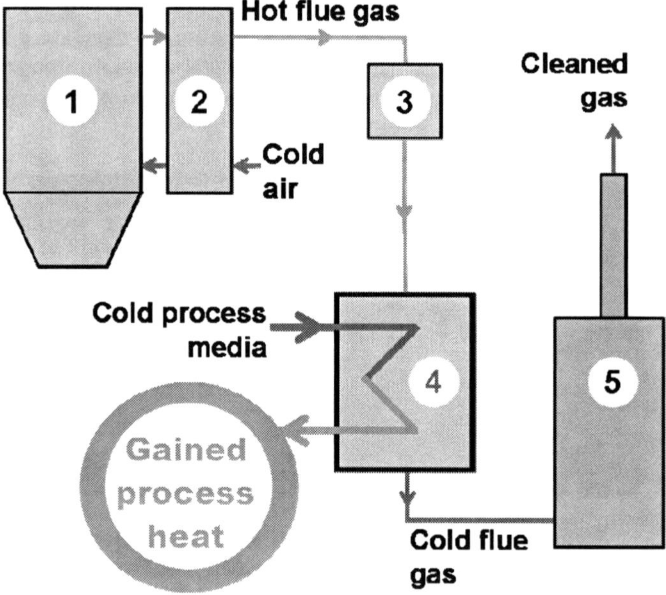

Figure 2: Scheme of flue gas stream. (1) Boiler, (2) air pre-heater, (3) electrostatic precipitator, (4) PFA-lined condensing heat exchanger/flue gas cooler, (5) FGD scrubbers.

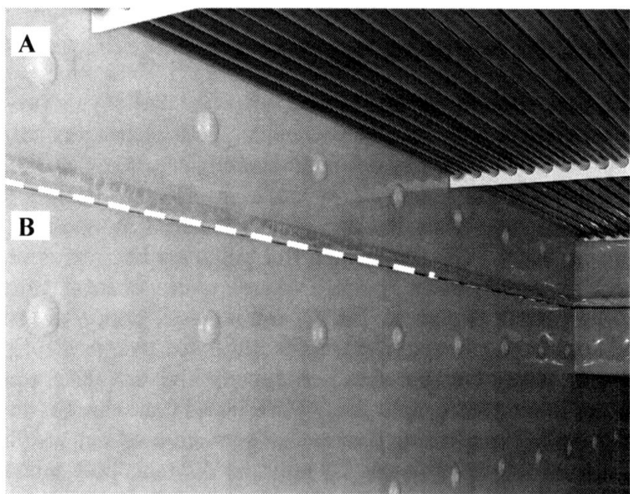

Figure 3: Gas/liquid condensing heat exchanger. The PFA fix point lining resists application temperatures of 260°C. (A = above dashed line.) Heat exchanger element. The visible tubes were made of carbon steel, coated with enamel and lined with PFA. (B) PFA-lined inspection room between to exchanger elements. Courtesy Flucorrex AG, CH-Flawil.

Figure 4: Gas/gas condensing heat exchanger. The gaseous heat transfer media runs through glass pipes. Courtesy Panergetic AG, FL-Schaan.

2.2 Flue gas ducts and housings

Likewise for the heat exchangers, PFA linings are applied onto the metal surfaces of the ducts and housings in order to protect them against dew point corrosion (Fig. 5). Thereby, the seals in the transition zones towards the following sections of the flue gas ducts are a critical part for the performance of the lining and require a high level of experience of the companies carrying out the lining works (Fig. 6). The duct walls downstream of the PFA lining thereby often are coated with resin containing glass flakes (flake coating), which resists the flue gas conditions following the condensing heat exchangers.

Other critical parts can be for example the floor regions: for cleaning, the condensing heat exchangers are flushed from time to time with water, which afterwards runs along the floor. If the workmanship of the weld seams in the floor region is not sufficient or the weld seam design is not chosen appropriately, this can lead to leakage problems with the cleaning water.

2.3 Liner installation

The occurring high temperatures in the flue gas stream have to be considered for the way of liner installation in the piece of equipment. Bonded installations using resins and adhesives as known for PFA linings for liquid media or wet fumes in chemical industry cannot be used, because a bonded connection cannot stand temperatures above 160°C (in FRP; material and fabrication dependent) and 120°C (in steel; material and fabrication dependent). Therefore, a mechanical fixation without resins or adhesives is chosen for the PFA foils in condensing heat exchangers, flue gas ducts and housings. Then, the full temperature range of

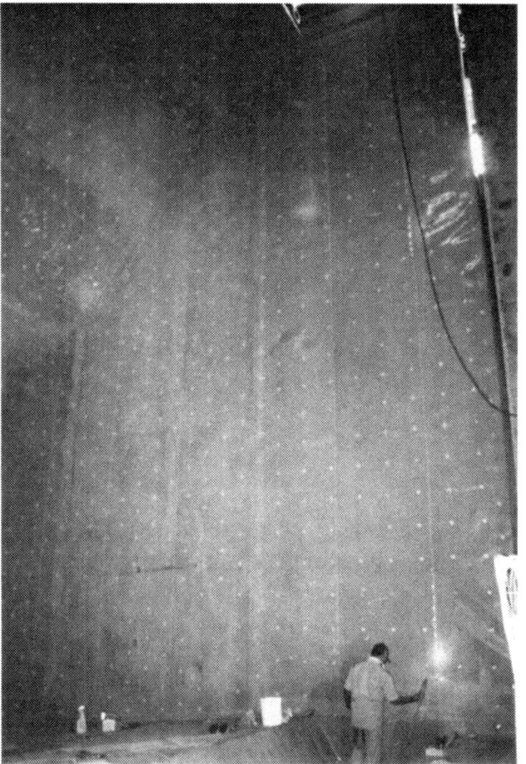

Figure 5: PFA-lined flue gas duct of a waste incineration plant. Courtesy Plasticon Europe, D-Dinslaken.

the PFA material of up to 260°C can be used, which is sufficient for the flue gas temperatures in the condensing heat exchanger region.

Two methods of mechanical fixation are used. For the first method, the PFA lining foils are brought into the shape of the equipment piece, which has to be lined, by tailoring, thermo-forming and welding. The fixation of the liner is only done over the flanges, where it is 'clamped' (loose or clamped liner; Fig. 7). An advantage is a relatively quick installation and an easy replacement of the liner, when required. Disadvantages are a limitation to comparatively small work pieces; if high walls have to be lined, the PFA foil has to be supported due to its relatively high weight (approx. 3.4 kg/m^2 for a natural foil of thickness 1.5 mm). Furthermore, a loose lining cannot be used, if significant negative pressures have to be expected.

For the second method, PFA linings are applied using an appropriate number of screws, washers and bolts per area for the fixation (fix point linings). Fix point linings are common for the lining of condensing heat exchangers and housings, where large areas have to be lined and changing flue gas pressures can occur. The number of fix points usually is in a range of 8–12 per square meter, but can vary significantly if lower or higher mechanical loads are expected (Fig. 7).

Figure 6: Examples for PFA linings fabricated on a high technical level. Transition zone (upper) and lining of a housing with two diverging fix point application methods (lower). Courtesy Plasticon Europe, D-Dinslaken.

The fix points are covered with welded PFA caps in order to seal the surface completely against the flue gasses. Thereby, either caps made from thermo-formed PFA foils or injection-moulded PFA caps can be used (Fig. 8). Injection-moulded PFA caps have the advantage that they can be produced in the shape of washer and screw, which leads to a relatively flat assembly. In addition, the time-consuming thermo-forming step can be saved. On the other hand, thermo-formed PFA caps can be produced from the same PFA foils as used for the lining, which is advantageous for welding (identical melt flow index).

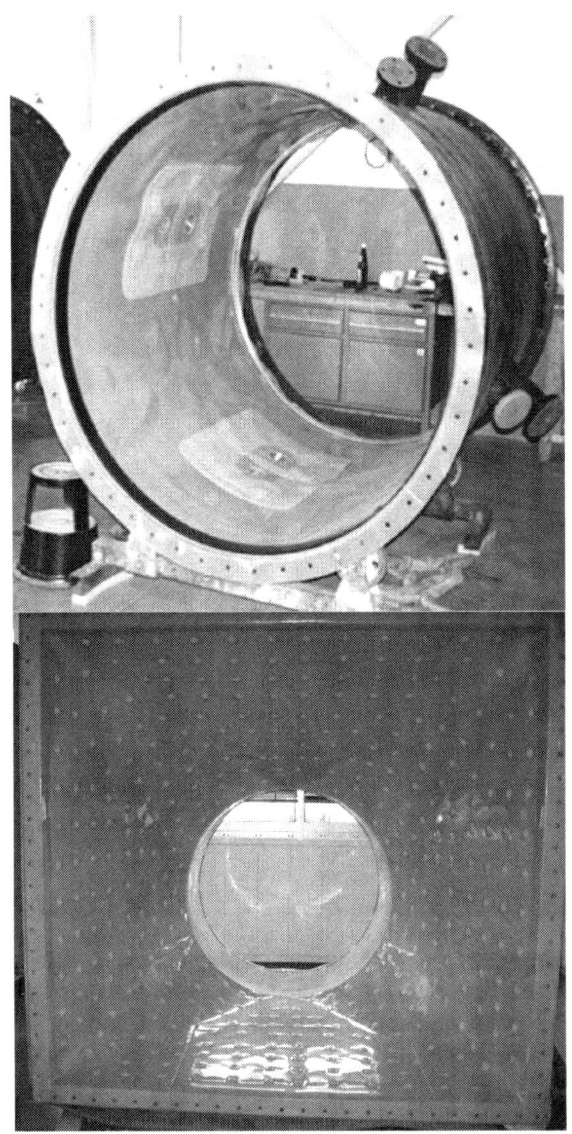

Figure 7: Examples mechanically fixated PFA linings for condensing flue gas applications. Clamped loose lining (upper) and fix point lining of a heat exchanger hood (lower). The hood is exposed to high mechanical loads due to the gas stream velocities. Accordingly, the foil is supported by a comparatively large number of fix points. Courtesy Plasticon Europe, D-Dinslaken (upper) and Flucorrex AG, CH-Flawil (lower).

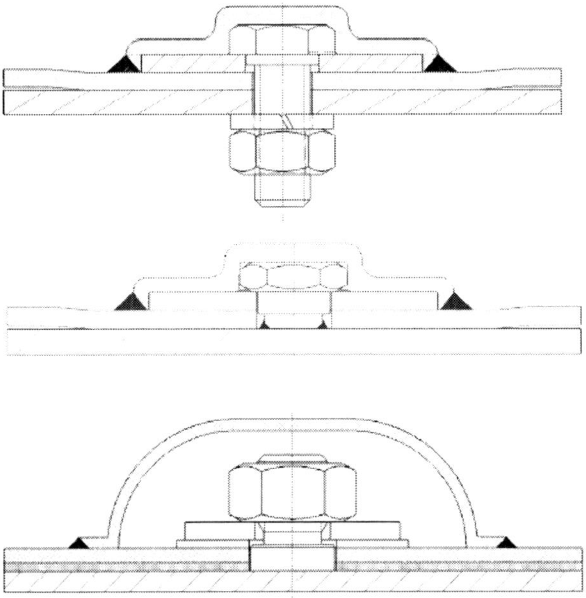

Figure 8: Drawings of typical fix point constructions. Injection-moulded PFA cap and hexagonal screw (upper), injection-moulded PFA cap and welded bolt (middle) as well as thermo-formed PFA cap and hexagonal nut (lower).

The fix points at the same time protect the PFA weld seams against the influences of the thermal expansion of the foil: if the fix points are placed close enough to the weld seams, the influence of the thermal expansion on the weld seams can be kept low.

2.4 Thermal retraction after heating of PFA foils

Besides a good quality of the PFA weld rods (typically in thickness 3.5 mm or 4 mm), the retraction after heat behaviour of the used PFA foils (typically delivered in dimensions of 1.5 mm x 1250 mm x 20000 mm or 2.3 mm x 1250 mm x 10000 mm) is important for the liner design and performance.

Retraction after heat in this respect means that the PFA foils shrink to some extend after they had been heated. The shrink thereby is caused by crystallisation events in the material and is a normal behaviour of all semi-crystalline thermoplastic materials. The effect becomes more significant, the closer the application temperatures of the materials come to the melting temperature of the semi-crystalline thermoplastics (Peterson [8]).

Accordingly, the phenomenon usually can be technically neglected in fluoropolymer-lined equipment for wet media as it is used for example in chemical industry, since the application temperatures do not come close to the melting temperatures (for PFA: 290–310°C).

www.witpress.com, ISSN 1743-3533 (on-line)

However, in hot flue gas applications a significant shrink in the foil can occur as the temperatures go up to 260°C. This on the one hand has to be taken into account when the lined equipment is designed. On the other hand, PFA foils manufactured from the suppliers with a minimum shrink should be chosen to minimise the effect and hence to reduce the risk of heat shrink problems in the lined equipment.

Noteworthy, the retraction after heat values of diverging PFA foil manufacturers can differ. Whereas some suppliers define the values for 200°C, others provide heat shrink values at 250°C temperatures. Since the heat shrink becomes more severe, if the temperatures come close to the melting point of PFA, the heat shrink at 250°C in our opinion should be observed, if retraction after heat can be critical.

3 Conclusions

Lining with the fully fluorinated polymer PFA is a powerful tool to handle dew point corrosion in fossil-fired and waste incineration plants. The linings can be used to gain more energy from the combustion process and to protect FGD equipment, respectively. The lining system technology, workmanship and quality of the foil however play a key role for the performance of the lined equipment and should be chosen appropriately.

The author likes to thank the companies Plasticon Europe, Flucorrex AG and Panergetic AG for providing the images of the lined equipment.

References

[1] Oil & Gas Journal. New correlation predicts dewpoints of acidic combustion gases. 22 February 2010.
[2] W. M. M. Huijbregts, R. G. I. Leferink, Anti Corrosion Methods and Materials Volume 51, No 3, (2004), 173-188.
[3] David A. Lewandowski. Design of Thermal Oxidation Systems for Volatile Organic Compounds, 1st Edition (2000). CRC Press. ISBN 1-56670-410-3.
[4] John J. McKetta. Encyclopedia of Chemical Processing and Design, Volume 61, 1st Edition (1997). CRC Press. ISBN 0-8247-2612-X.
[5] W. Cox, W. Huijbregts, R. Leferink, Components Susceptible to Dew-Point Corrosion. ASM Handbook Volume 13C, Corrosion: Environments and Industries (2006). ASM International. 491-496.
[6] T. de Weijer, W. Huijbregts. Severe corrosion in a waste incinerator plant due to flue gas and steam leakage. Anti-corrosion methods and Materials Volume 50 (2003). 334-340.
[7] D. C. Agarwal and G.K. Grossmann, Case Histories on the Use of Nickel Alloys in Municipal and Hazardous Waste Fueled Facilities. CORROSION. (2001), March 11 - 16, 2001, Houston, Tx. Copyright NACE International.
[8] J. M. Peterson. An energy-driven mechanism for retractive forces in semicrystalline polymers. Polymer letters Volume 7 (1969). 231-236.

Section 5
Cementitious materials

Application of the X-ray absorption technique for monitoring hydration and setting processes in cementitious composites

A. J. Klemm & P. Baker
Glasgow Caledonian University, UK

Abstract

This paper presents an application of the X-ray absorption technique for monitoring hydration and setting processes in cementitious composites. The method, which is a relatively new approach in concrete research, is based on the measurement of attenuation of a finely collimated beam of X-rays by a sample. The absorption of X-rays by sample is proportional to the material density. Changes of density in a function of time can be related to the absorption and moisture transport within porous media. Although interpretation of the results may be very problematic, chemical and physical reactions associated with hydration may be qualitatively analysed by this technique. This paper is focused on a setting process of cement mortars modified by superabsorbent polymers.
Keywords: X-ray absorption, hydration process, cementitious materials, superabsorbent polymers.

1 Introduction

1.1 Background

Application of the X-ray absorption technique for monitoring hydration and setting processes in cementitious composites is a relatively new approach in concrete research. The method is based on the measurement of attenuation of a finely collimated beam of X-rays by a sample. The absorption of X-rays by sample is proportional to the material density. Changes of density in a function of time can be related the absorption and transport of moisture in porous materials. Experimental results may be used in simulation models for predicting moisture movement at material inter phases and other macroscopic features [1].

WIT Transactions on Engineering Sciences, Vol 77, © 2013 WIT Press
www.witpress.com, ISSN 1743-3533 (on-line)
doi:10.2495/MC130211

Some attempts were also made to monitor progress of hydration process [2-4]. Although interpretation of the results may be very problematic chemical and physical reactions associated with hydration may be examined by this technique. Recorded alterations in densities of cementitious systems resulting from the formation of hydration products and microstructural changes, especially at early ages, are valuable information in the hydration process analysis.

1.2 Hydration process of Portland cement

The vast majority of cements used in the construction industry are based on Portland cement. In order to obtain its bonding properties Portland cement is mixed with water. It hardens with time and form hardened cement paste. Starting point for the reactions is the moment of water – cement contact. During the hydration process volume of pores is gradually reduced by newly created products of reaction. In the first few minutes after cement is mixed with water, the cement grains are wetted and $CaSO_4$ and C_3A are hydrated [5]. The aluminate phase reacts with gypsum, and on the surface of the grains, ettringite is formed, preventing hydration for a certain period of time. At the same time, a coat of CSH gels is formed on the surface of C_3S grains. Then hydration reactions slow down for few hours (1-6 hours) depending on the conditions. When the concentration of Ca^{2+} ions reaches the state of oversaturation the process of CSH gel formation begins. In the next 8–14 hours C_3S reacts quickly at the formation of CSH gel and $Ca(OH)_2$, which is precipitated in the form of crystals of portlandite. A basic solid structure is formed by C-S-H gel, crystalline phases of aluminates and ferrites, and by ettringite (Figure 1).

Figure 1: Hydration products in immature cement pastes.

Simultaneously, reduction of porosity takes place and plastic structure turns into solid structure. Final set is obtained in this stage. The concentration of SO_4^{2-} ions decreases by consuming $CaSO_4$ in cement and ettringite begins to react with unreacted C_3A to form monosulphate $3CaO.Al_2O_3.CaSO_4.12H_2O$. After all ettringite has transformed to monosulphate, no $CaSO_4$ is available for the reaction with aluminoferrite phase. The products of hydration of cement are chemically the same as products of hydration of the individual compounds under similar conditions [6, 7].

1.3 Superabsorbent polymer (SAP)

Application of superabsorbent polymers (SAP) significantly changes water management of the system and the kinetics of hydration process. SAP is a general term referring to polymers that are able to absorb and retain a large amount of water when compared to their own weight and slowly release it as the hydration process continues preventing self-desiccation [8, 9]. The importance of SAP particles is based on a process of formation of a system of fine, evenly distributed pores, which are filled with water in fresh or young concrete. These air-filled pores may be to a great extent empty in the hardened concrete, performing similarly to air-entrained pores or in a collapsed state of SAP may be filled with hydration products [10–12]. Figure 2 shows superabsorbent polymer during the water absorption phase [13].

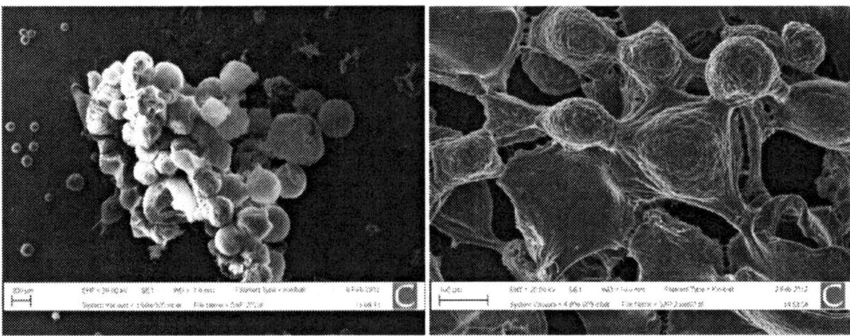

Figure 2: Superabsorbent polymers.

The paper will present results of the X-ray absorption analysis of hydration process and early curing in cementitious mortars modified by the superabsorbent polymers.

2 Experimental method

X-ray adsorption chamber is used to measure the attenuation of a finely collimated beam of X-rays by a sample. The apparatus consists of an X-ray source (up to 70 keV) and a cadmium–zinc–telluride (CZT) high resolution detector with a range of collimators to define the spatial resolution of the system to less than 1 mm. These are supported on a movable gantry within a climate controlled chamber. A dedicated PC and software allows fully-automated measurement, data collection and processing. The PC also controls the precise location of the gantry to within 0.05 mm as well as the chamber temperature and humidity. A sample is positioned between the X-ray source and the detector. The working space in the chamber is 790 mm along the horizontal-axis, 419 mm along the vertical-axis and 368 mm between the X-ray source and detector. Figure 3 shows operation schema of the X-ray apparatus.

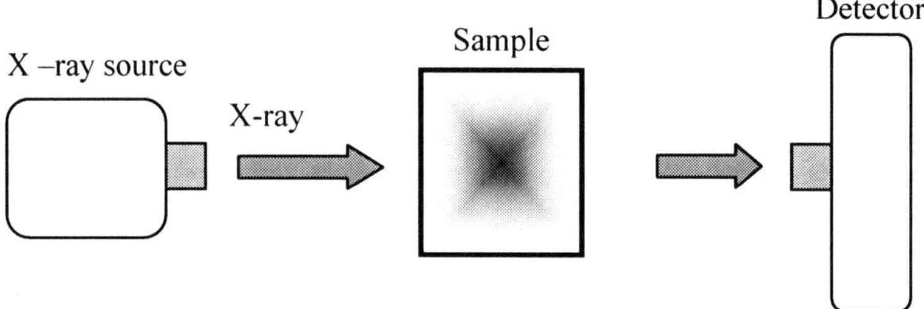

Figure 3: Operation schema of the X-ray apparatus.

It allows for a large sample size or a number samples to be tested in the same sequence. The location of the sample within the chamber may be determined precisely relative to the detector gantry. The movement of the gantry may then be programmed to enable the scanning of the sample at precise locations.
The main advantages over other available moisture measurement systems (such as Nuclear Magnetic Resonance), include [1]:

- rapid, precise measurement of the two-dimensional distribution of moisture in samples large enough to be representative of the macroscopic features;
- very high spatial resolution, allowing the collection of data for fundamental research on transport coefficients and processes at material interfaces;
- multi-tasking – the measurement of a number of samples sequentially.

The attenuation of X-rays in an absorbing medium is governed by Beer's law:

$$I = I_0 e^{-\mu\rho x} \tag{1}$$

where:
I_0 is the incident X-ray intensity;
I is the transmitted X-ray intensity;
μ is the mass attenuation coefficient;
ρ is the density of the medium;
x its thickness.

The attenuation coefficient can be determined experimentally by measuring the intensities I_0 and I of the X-ray beam reaching a detecting device without and with an absorber of thickness x interposed.

In an attempt to analyse changes that occur during curing, water movement in cementitious specimens has been continuously monitored in the X-ray environmental chamber. Since the X-ray absorption is proportional to the density of the materials through which the X-rays are passing, a low w/c ratio (denser) paste absorbs more X-rays than a higher w/c ratio paste that is associated with the reduced counts throughout the thickness of the specimen. Similarly, a paste, which has dried out absorbs less X-rays than the same paste under its initial saturated [2].

3 Materials tested

For the purpose of this research cement was mixed with fine sand at 1:1 and 2:1 sand/cement ratio (s/c by weight). Throughout the investigation an even total water to cement ratio of 0.45 was maintained for all mixes: reference samples (R1 and R2) and samples with SAP additions (A1, B1, C1, A2, B2, C2). Three types of SAPs were used in concentration of 0.25 % by weight of cement content. The effective water to cement ratio $(w/c)_{eff}$ for mixes containing SAPs was lower than total water to cement ratios (w/c) due to water absorption by SAPs. Detailed information about mix composition is presented in Table 1.

Table 1: Cement mortar composition.

Mix code		A1	B1	C1	R2	A2	B2	C2
s/c	1:1	1:1	1:1	1:1	2:1	2:1	2:1	2:1
$(w/c)_{tot}$	0.45	0.45	0.45	0.45	0.45	0.45	0.45	0.45
$(w/c)_{eff}$	0.45	0.42	0.43	0.37-0.39	0.45	0.42	0.43	0.37-0.39
SAP [%]	0	0.25	0.25	0.25	0	0.25	0.25	0.25

Portland fly ash cement CEM II/B-V 32.5R was used in this study. All polymers have absorption capacity of 200-250 ml/g in demineralised water. However, their absorption characteristics in cement paste solution differ significantly; approximately 10 g/g for SAP A, 5 g/g for SAP B and 25-30 g/g for SAP C [13].

4 Experimental results and analysis

The measurements in the X-ray absorption chamber commenced approximately 15 minutes after water addition and the intensities I_0 (incident X-ray intensity) and I (transmitted X-ray intensity) were measured. The location of the sample within the chamber was determined precisely by a series of 'dry' runs. The sealed transparent containers (70 x 70 x 70 mm) used for the mortar tests were located on a stand within the chamber, separated by metal spacer and a scanning sequence was carried out. The scans across the samples were taken at approximately half-way up the container: The minimum achievable interval between the start of successive scans was 20–30 minutes. Figure 4 shows samples positions inside the X-ray absorption facility.

The natural logarithm of the detector count has been used as a metric which is inversely proportional to the density multiplied by the mass attenuation coefficient, assuming constant incident X-ray intensity during the test. The increase in the detector count represents a reduction in density or the attenuation coefficient.

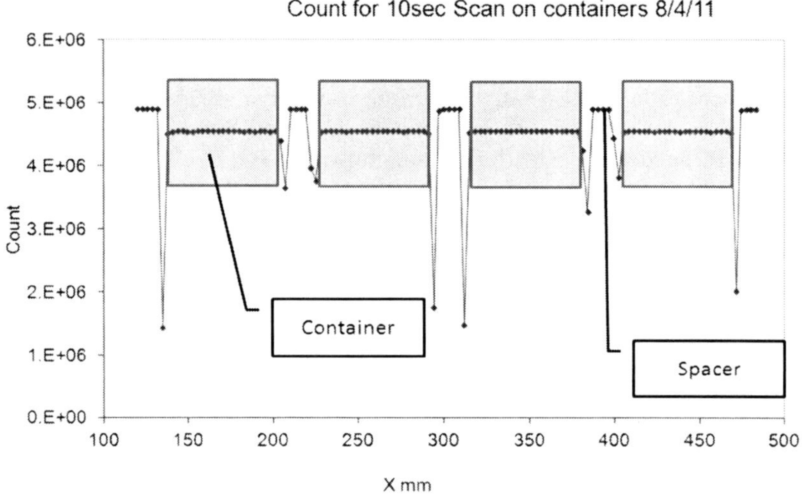

Figure 4: Samples position in the X-ray absorption chamber.

The profiles of measured attenuations of X-ray beam by samples with sand to cement ratio 1:1 are shown in Figure 5. Results represent recordings from the initial 24 hours for each sample approximated by polynomials' of fourth degree.

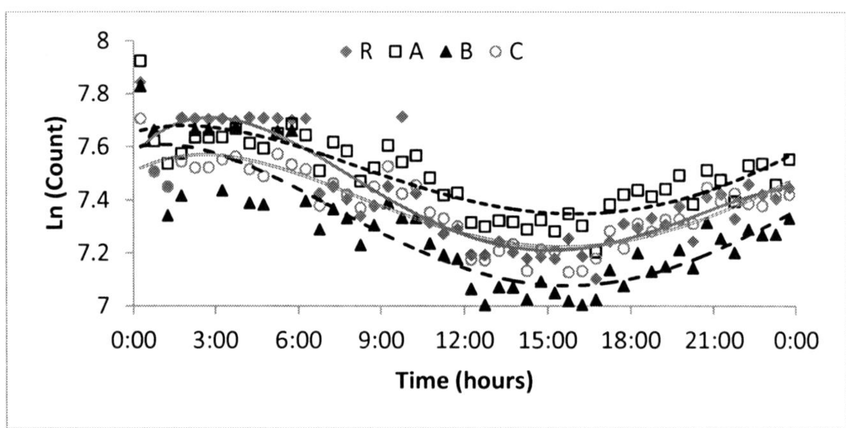

Figure 5: X-ray absorption results for mortar mixes with sand to cement ratio 1:1 for initial 24 hours.

Due to the fact that samples were completely sealed during the measurements and there was no evaporation from their surfaces, desorption of SAPs was probably delayed and not as intense as in normal curing conditions. By the time of first readings (15 min) a significant part of the mixing water could have been absorbed by the superabsorbent polymers. In the first fifteen hours after water addition densification has been observed. This trend has changed in the following hours. The results obtained for sample A have indicated lower density and for samples B higher density than the reference sample. During the first day of hydration mix containing SAP C has followed the pattern of reference sample in first day of hydration. Almost instantaneous absorption of high amount of water was responsible for the similarities in behaviour. This water was gradually released leading to formation of a denser CSH network.

Results shown in Figure 6 represent daily average of recordings for each sample during the first six weeks. Rather insignificant changes have been recorded during testing, however some densification has been observed from the fourth week for all samples. Between second and third week sample C has started to more resemble sample A, while sample B shown some similarities to the reference sample R. This could be attributed to delayed water desorption of SAP C in cementitious composites.

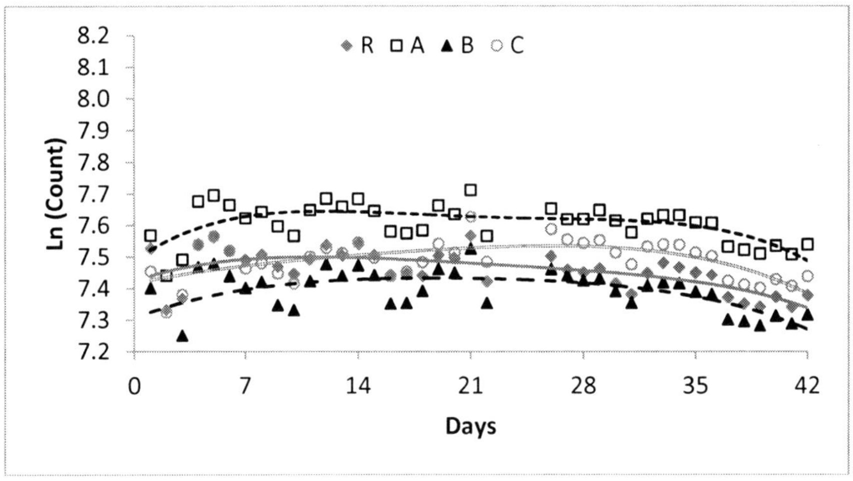

Figure 6: X-ray absorption results for mortar mixes with sand to cement ratio 1:1.

Figure 7 presents the X-ray absorption results for mortar mixes with sand to cement ration 2:1 for the initial 24 hours. The higher values have given an indication of less denser microstructure than for samples with 1:1 sand to cement ratio. This may be associated with lower percentage of cement paste in the mixes. Moreover, the change in densification trend has been observed, similarly to samples with 1:1 ratio, after about fifteen hours. Since cement used in the study contained substantial fly ash addition, the large amounts of aluminates appeared in the mixes. Thus, when aluminates have started to react with calcium hydroxide, this could have inverted the values trend.

Similarly to samples with s/c 1:1, the resemblance of the reference samples R and B is well pronounced. After the period of 12 hours significant increase in densities has taken place in all samples. Lower densities have been recorded for sample B, which absorbs water in the smallest quantities and then slowly release

www.witpress.com, ISSN 1743-3533 (on-line)

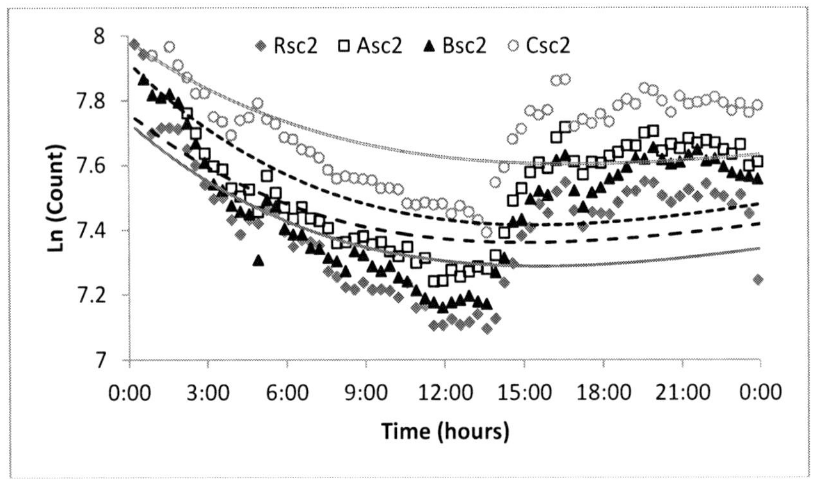

Figure 7: X-ray absorption results for mortar mixes with sand to cement ratio 2:1 for initial 24 hours.

it. This makes it similar to the reference sample not containing any SAPs. As anticipated samples C characterised by the high water absorption capacity and relatively long period of water release gave the highest values, indicating more uniform and denser structure. This feature is more evident in samples with lower cement content.

The results of X-ray absorption for mortar mixes with sand to cement ratio 2:1 for 6 weeks are presented in Figure 8. The smallest differences have been

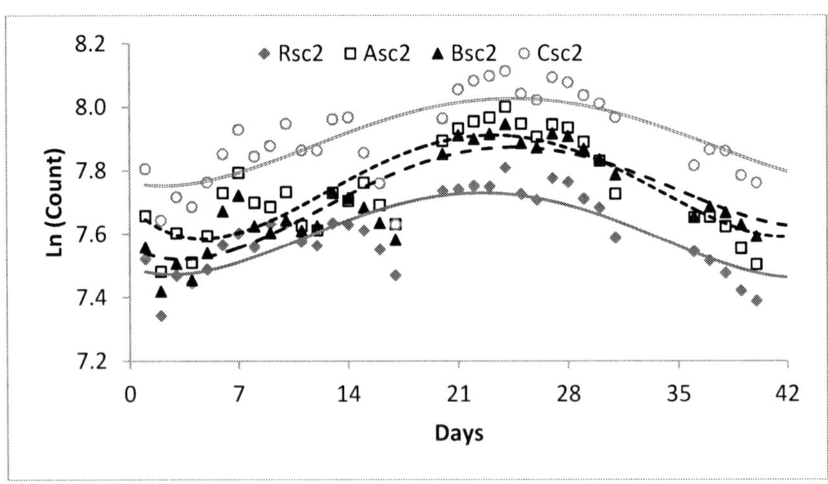

Figure 8: X-ray absorption results for mortar mixes with sand to cement ratio 2:1.

observed for reference sample referring to highest density of mix. An increase in densities has been observed for samples modified by SAPs, although of much bigger magnitude and earlier for sample A.

Results obtained after 6 weeks of curing for samples A and B is comparable; however sample C has shown less dense structure. This phenomenon has been recorded at every point for sample with sand to cement ratio 2:1, indicating different absorption/desorption characteristics than SAP A and SAP B.

5 Summary

The recorded values refer to the sealed system, comprising; newly created hydration products, water absorbed by the polymers, free water, pores and some unreacted cement particles. Prolonged period of time leads to more water being consumed for creation of smaller volume of solid phase. Therefore, after the initial densification of internal structure, a steady decrease has been observed. The process is very complex, particularly in case of blended cements and its understanding requires further thorough investigation.

Undoubtedly absorption/desorption of water by SAPs has a strong impact on the kinetics of hydration process and consequently on formation of microstructure.

Even though the quantitative assessment of densities is not possible a qualitative analysis can be successfully performed by the X-ray absorption technique.

The most pronounced differences between samples have been observed during the first 24 hours when the water desorption processes are the most intense. This applies to both cement to sand ratios, however the results were more consistent for mixes with lower cement content. High cement content and hence higher percentage of superabsorbent polymers makes the water management more complex and more difficult to analyse.

The characteristic feature, clearly identifiable from X-ray absorption, was the similarity between samples R and B as well as samples A and C. The latter contained SAPs with higher water absorption capacities and hence the stronger effect on microstructure formation.

X-ray absorption may also find an application in the assessment of water desorption of SAPs in cementitious matrices. A sudden increase in X-ray absorption gives an indication of the intensified CSH formation and thus increased water release.

References

[1] Baker, P. H., Bailly, D., Campbell, M., Galbraith, G. H., McLean, R. C., Poffa, N., Sanders, C. H., The application of X-ray absorption to building moisture transport studies. Measurement, 40, 951-959 (2007).

[2] Bentz, D. P., Hansen, K. K., Preliminary observations of water movement in cement pastes during curing using X-ray absorption. Cement and Concrete Research, 30, 1157-1168 (2000).

[3] Benz D.P., Hansen, K.K., Madsen H.D., Vallee F., Griesel E.J., Drying/hydration in cement pastes during curing. Materials and Structures, 24, 557-565, (2001).
[4] Hu J., Stroeven P., X-ray absorption study of drying cement paste and mortar, Cement and Concrete Research, 33, 397-403, (2003).
[5] Cerny R., Rovnanikova P., Transport processes in concrete, London: Spon Press (2002).
[6] Bogue R..H., Lerch W., Hydration of Portland cement compounds, Industrial Engineering Chemistry, 837-847 (1934).
[7] Neville A.M., Properties of concrete, Harlow: Longman Group Ltd. (1995).
[8] Jensen O. M., Hansen P. F., 'Water-entrained cement-based materials I. Principles and theoretical background', Cement and Concrete Research, 31, 647-654 (2001).
[9] Jensen O. M., Hansen P. F., Water-entrained cement-based materials II. Experimental observations, Cement and Concrete Research, 32, 973-978 (2002).
[10] Reinhardt H. W., Assmann A, Enhanced durability of concrete by superabsorbent polymers, Proc. Int. Symp. Brittle Matrix Composites 9, Warsaw (2009).
[11] Mönnig S., Lura P., 'Superabsorbent Polymers – An Additive to Increase the Freeze-Thaw Resistance of High Strength Concrete, Advances in Construction Materials, V, 351-358 (2007).
[12] Klemm A.J., Sikora K., The effect of cement type on the performance of mortars modified by superabsorbent polymers, Conference Proceedings the 3rd ICCRRR, Cape Town, South Africa, (2012).
[13] Sikora K., The effect of superabsorbent polymers on the properties of cementitious mortars containing fly ash, PhD Thesis, Glasgow Caledonian University, (2013).

Overview about the use of Fourier Transform Infrared spectroscopy to study cementitious materials

M. Horgnies, J. J. Chen & C. Bouillon
Lafarge Centre de Recherche, St Quentin-Fallavier, France

Abstract

Fourier-Transform Infrared (FT-IR) spectroscopy shows several advantages that make it interesting to investigate cementitious materials: from clinker or hydrated phases to the bulk or the surface of hardened concrete. The FT-IR analyses in Transmission mode need only a few milligrams of material to provide its composition while other techniques (such as thermo-gravimetric analysis, X-ray diffraction, X-ray fluorescence) need a few grams. Moreover, FT-IR analyses are rapid and give results after a few minutes while other methods need at least a few hours to study one sample. The analyses done in Attenuated Total Reflection mode allow studying the surface of materials without any specific sampling methods (such as cleaning with solvent or storage under vacuum), which can alter the final composition. Clinker and anhydrous cement phases were already studied in 70s to establish the specific peaks of alite, belite and calcium aluminate. The study of major hydrates has been largely performed for the last two decades. FT-IR spectroscopy highlights easily the presence of portlandite ($Ca(OH)_2$), which is detected by a thin unique peak. Moreover, the shift of the peaks assigned to Calcium Silicate Hydrate (C-S-H) can be detected using transmission mode, in order to study the polymerization of the silicates according to the conditions of cure and ageing. AFt and AFm phases (such as ettringite, sulfoaluminate or hemicarbonate) can also be studied using FT-IR spectroscopy. Finally, this method is also able to detect organic demolding agents or $CaCO_3$ efflorescences at the surface of hardened concrete and can be used to study the interface between the hydrated paste and the polymers or coatings.
Keywords: FT-IR, clinker phases, hydrates, C-S-H, afwillite, cement paste, concrete.

doi:10.2495/MC130221

1 Introduction

This work deals with the description of the advantages and limits of the FT-IR spectroscopy to analyze the cementitious materials. Contrary to organic solutions, polymers and coating, the Fourier-Transform Infrared (FT-IR) spectroscopy is not usually used to study cementitious materials. Indeed, researchers prefer to characterize these materials using other methods such as thermo-gravimetric analysis (TGA), X-ray diffraction (XRD), X-ray fluorescence (XRF)...etc [1]. However, FT-IR spectroscopy was already used in 70' to characterize the clinker phases and shows several advantages that make it useful to study a large range of cementitious materials.

In this paper, several applied results are described in parallel of a short bibliographic review. First, the main analytical modes available on a FT-IR spectrometer are rapidly described: (i) transmission mode (Tr); (ii) Attenuated Total Reflection mode (ATR); and (iii) diffuse reflectance spectroscopy (DRIFTS). The advantages and limits of FT-IR spectroscopy to study the cementitious materials are then specified. In a second part, a brief overview of the scientific literature is presented in parallel of selected FT-IR results for each topic (clinker phases, anhydrous cement, synthetic hydrates, hydrated cement paste, bulk of mortar and concrete surface).

2 Presentation of the FT-IR spectroscopy

2.1 Main modes of analyses

The FT-IR results shown in this work were obtained using a Nicolet iS10 spectrometer (Thermo Fisher Scientific Inc.). It was equipped with a deuterated triglycine sulfate (DTGS) detector and controlled by OMNIC software.

In Transmission mode (Tr), our sampled pellets were manufactured by mixing 250 mg of potassium bromide (KBr) with 3 mg of material (clinker or cement powder, or powder directly sampled into the bulk of hardened samples). 16 scans were recorded over the range 4,000–400 cm^{-1} with a spectral resolution of 4 cm^{-1}. The background spectrum was collected using a pure KBr pellet and the spectra of samples were corrected with a linear baseline.

In Attenuated Total Reflexion (ATR) mode, our hardened samples were directly analyzed over a thickness of a few µm and a sampling area of about 1 mm^2. The crystal used was made of diamond. 16 scans were routinely recorded over the range 4000–650 cm^{-1} with a spectral resolution of 4 cm^{-1}. The background spectrum was collected at ambient atmosphere before analyzing each sample. The spectra of hardened samples were corrected with a linear baseline.

The diffuse reflectance infrared Fourier transform spectroscopy (DRIFTS) allowed measuring a range between 400 and 4000 cm^{-1}. The incident beam was reflected off the ground sample towards an overhead mirror upon which the diffusely scattered rays were collected and measured in the detector.

2.2 Advantages and limits compared to other methods

2.2.1 Main advantages

First, the analyses in Transmission mode need only a few milligrams of material to provide its composition while other techniques need a few grams. This specific light sampling is particularly relevant when only a small mass of hydrates can be synthesized. It is also easy to sample the cement paste into the bulk of mortar/concrete without any contamination from sand or aggregates.

Second, FT-IR results are directly obtained a few minutes after analyses without any complex treatment while other methods (TGA, XRD, XRF…etc) need at least a few hours to study one sample.

Third, the analyses in Attenuated Total Reflection (ATR) mode allow studying the surface of the cementitious materials without any specific sampling methods (such as cleaning with solvent or storage under vacuum, which can modify the composition of the material).

2.2.2 Main limits

FT-IR spectroscopy cannot be considered such as a quantifying method. Even if the intensity of certain specific IR bands, such as the one of portlandite, can vary as a function of their content in the sample (semi-quantification), it is not possible to make the relation between the intensity or the area of an IR band and a specific ratio.

Moreover, traces of chemical compounds can be under-estimated or erased due to the high intensity of certain IR bands, such as those of silica or calcium carbonate.

Finally, the presence of a large amount of capillary water into the hydrated sample can mask certain chemical compounds and disturb the interpretation of the spectra.

3 Selection of results obtained on cementitious materials

3.1 Analyses of anhydrous clinker phases

Clinker and anhydrous cement phases were already studied in the 70s in order to characterize the IR bands of alite, belite and calcium aluminate. Bensted and Varma [2], and Ghosh and Chatterjee [3, 4] published in 1974 certain of the first main papers about the FT-IR analyses applied in the cement industry. Ghosh and Handoo [5] wrote one of the first review papers in 1980. These first papers were mainly devoted to the analyses using transmission mode (only [4] used ATR mode).

Table 1 and Figure 1 show our own analyses done in Transmission mode concerning the specific peaks of synthetic clinker phases. Indeed, pure Ca_3SiO_5 and Ca_2SiO_4 phases were prepared by successive heating (at 1600°C) of a mixture of finely divided calcium carbonate (from Sigma-Aldrich, Germany) and silica (Aerosil 200 from Degussa, France) with appropriate stoichiometric proportions. All the spectra are conformed to those found in literature [2–5].

www.witpress.com, ISSN 1743-3533 (on-line)

A double peak at 995-900 and 938-883 cm^{-1}, assigned to Si-O asymmetric stretching vibrations, characterizes mainly C2S and C3S phases, respectively. The detection of ferrite phase is difficult due to the absence of resolved and fixed peaks but aluminates phase (C3A) is easier to detect due to a large set of thin peaks assigned to Al-O bond.

Table 1: FT-IR bands of the synthetic clinker phases.

Clinker phases	Wave-number (cm^{-1}) with intensity of the band (s: strong; m: medium; l: low)						
	Si-O	Al-O	Si-O	Al-O	Si-O		
C3S - Ca_3SiO_5	938 s 883 s		812 l		522 s		430 s
B-C2S - Ca_2SiO_4	995 s, 900 s		844 s, 810 l		518 s		
C3A - $Ca_3Al_2O_6$		898 s	817 s	786 l, 739 s, 704 s, 588 l, 521 s		456 s	
C4AF – $Ca_4Al_2Fe_2O_{10}$			*Poorly resolved bands*				

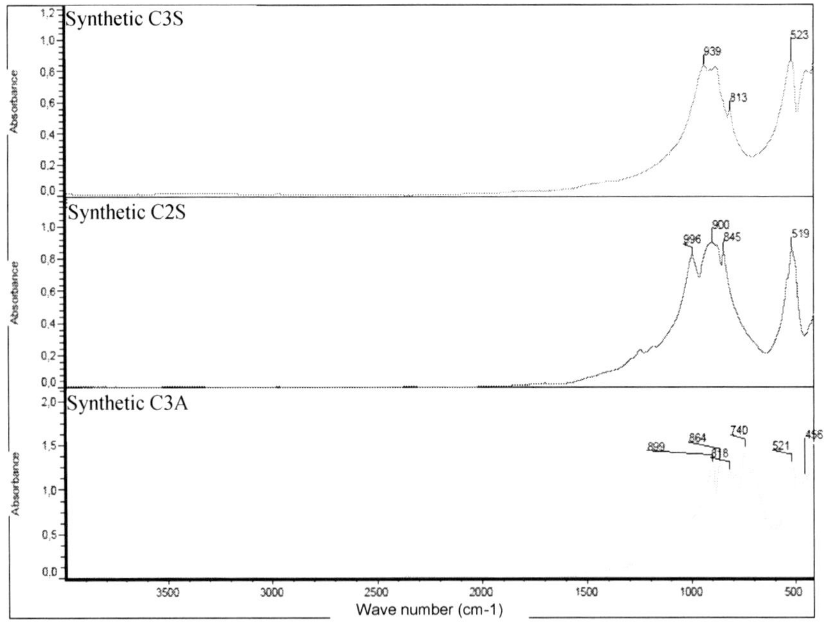

Figure 1: IR spectra of synthetic C3S, C2S and C3A.

3.2 Analyses of hydrated cement phases

3.2.1 Detection of the main hydrated phases

FT-IR spectroscopy highlights easily the presence of portlandite ($Ca(OH)_2$), which is well detected by a resolute unique peak at 3640 cm^{-1}. This method is also used to study the shift of the bands assigned to Calcium Silicate Hydrate (C-S-H) according to the conditions of cure and ageing.

Table 2: FT-IR bands of the main hydrated phases.

Main hydrated phases	Wave-number (cm^{-1}) with intensity of the band (s=strong; m=medium; l=low)						
	O-H	O-H; H_2O capil.	Si-O (asym. stretching vib.)		Si-O	Si-O (out-of plane vib.)	Si-O (in-plane vib.)
Afwillite (crystalline micro-structure; Ca/Si ratio=1.5)		3352 l, 1660 l	985 m, 963 s	911 s	860 m, 781 m	617 m, 520 s, 490 s	450 s
C-S-H (colloidal gel like structure; Ca/Si ratio=1.5)		3356 l, 1640 l	1000 s, 950 s	950 s	814 m	667 l, 496 s	456 s
Portlandite – $Ca(OH)_2$	3642 s						

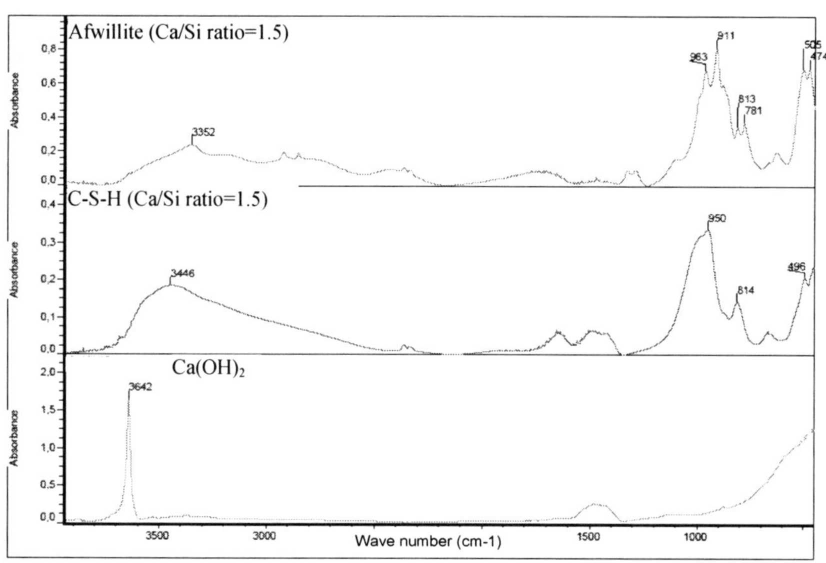

Figure 2: IR spectra of synthetic $Ca(OH)_2$, C-S-H (Ca/Si=1.5) and afwillite.

Table 2 and Figure 2 provide our own analyses done in Transmission mode on synthetic hydrated phases. The $Ca(OH)_2$ (portlandite) was provided directly by a supplier (Sigma-Aldrich) while the calcium silicate hydrates (C-S-H, a colloidal gel like structure [6]) were prepared by mixing CaO (Sigma-Aldrich) and pure silica (Aerosil 200 from Degussa) with a calcium/silicon ratio of 1.5 and 0.9. A crystalline calcium silicate hydrate (afwillite) was also manufactured using a grinding process with ball milling and a calcium/silicon ratio of 1.5 [7]. The main IR bands of our powder samples made of C-S-H or afwillite are conformed to those detected in literature [8].

3.2.2 Shift due to the silica polymerization of C-S-H

The main characteristic peaks of C-S-H are located in the range between 1100 and 900 cm^{-1}. During ageing and de-calcification process, these IR bands shift according to the process of polymerization of the silica [9].

García-Lodeiro *et al.* [10, 11] and Bhat and Debnath [12] studied the hydration of C-S-H gel by FT-IR (in transmission mode). An interesting de-convolution of Si-O asymmetric stretching bands has been done [11].

Moreover, Ylmén *et al.* [13, 14] performed FT-IR analyses by DRIFTS to study different types of C-S-H according to the hydration time.

Hughes *et al.* [15] summarized the different assignment of IR bands (detected by DRIFTS) concerning the cement before and after hydration (clinker phases, sulfate-based and hydrated/carbonated products). The authors highlighted that the peak area of O-H of $Ca(OH)_2$ was proportional to the peak area of $Ca(OH)_2$ measured by TGA.

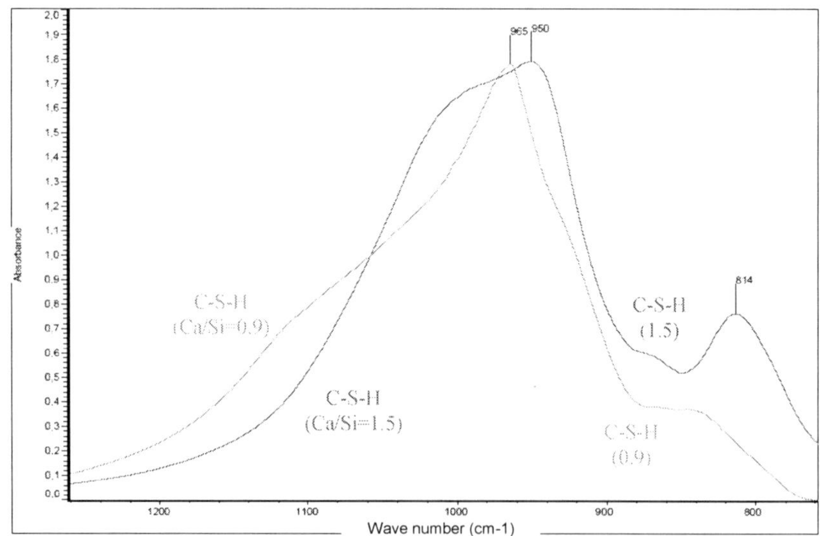

Figure 3: Main IR bands of C-S-H (1.5) and C-S-H (0.9).

Yu *et al.* [16] investigated the structure of different synthetic C-S-H. The FT-IR spectra of their C-S-H were compared to those of tobermorite in order to establish that the IR bands of C-S-H (between 980 and 1080 cm^{-1}) varied according to the calcium/silicon ratio. Indeed, the 1100–950 cm^{-1} region (assigned to Si-O stretching) shifted to lower wave-numbers due to the decreasing polymerization.

Figure 3 shows the IR bands of our synthetic C-S-H samples in the 1300–700 cm^{-1} range. The spectrum of C-S-H 1.5 (calcium/silicon ratio of 1.5), which is usually found in young hardened cement paste, is then compared to the one of aged C-S-H (calcium/silicon ratio of 0.9). A shift of the bands to higher wave numbers is then highlighted after ageing, due to the polymerization of silica during the de-calcification process.

3.2.3 Detection of the Aft/AFm phases

A part of the actual studies done using FT-IR spectroscopy is devoted to the investigation of Aft and AFm phases (such as ettringite, sulfoaluminate, hemicarbonate, monosulfoaluminate, C4AH13…etc), which can be responsible for an anticipated ageing of the hydrated cement paste. For example, precipitation of secondary ettringite can be responsible for the swelling, the fissuring and the decreasing of the mechanical resistance of concrete [17].

Certain works are focused on the evolution of the hydration when silica fumes and fly ash are added into the cement paste. Hidalgo Lopez *et al.* [18] showed by FT-IR in transmission mode certain phases formed in the hydrated pastes, such as Stratlingite (C2ASH8), hydrogarnet (C3AH6), gibbsite ($Al(OH)_3$) and hydrated tetracalcium aluminate (C4AcH11).

We performed FT-IR analyses in transmission mode on different synthetic Aft/AFm phases (created using specific mixes made of pure $AlSO_4$, CaO, $Ca(OH)_2$, $CaCO_3$, C3A and/or gypsum and processed at distinct temperatures). Table 3 provides the assignment of IR bands of the spectra of ettringite, mono-carbo-aluminate, hemi-carbo-aluminate, mono-sulfo-aluminate, C4AH13, C3AH6 (hydrogarnet) and C2ASH8 (stratlingite). The contaminations due to the process of synthesis were deduced by using XRD analyses and were removed (e.g. ettringite in sulfo-aluminate or portlandite in hemi-carbo-aluminate). All these phases show complex spectra (as shown in Figure 4) due to the presence of aluminate, sulfate and carbonate units. We notice that both the IR spectra of hemi- and mono-carbo-aluminates are equivalent.

3.3 Analyses of the bulk and surface of hardened pastes

FT-IR spectroscopy is mainly known for its efficiency to characterize polymer or organic compounds [19]. This specific ability can be useful to: (i) investigate the bulk of lightweight mortar containing aggregates made of polymer waste; (ii) detect residues of demolding agents (e.g. based on vegetable base oil) [20]; (iii) highlight the presence of $CaCO_3$ efflorescences [21]; (iv) study the interface between concrete and protective coating [21].

Table 3: FT-IR bands of Aft/AFm phases (contamination due to the creation process deduced).

AFt/AFm phases	Wave-number (cm^{-1}) with intensity of the band (s=strong; m=medium; l=low)									
	O-H			O-H; H_2O capil.	C-O; $[CO_3]^{2-}$	S-O; $[SO_4]^{2-}$	Si-O	Al-O		
Ettringite $Ca_6Al_2(SO_4)_3(OH)_{12}$ / $26H_2O$		3637s		3431m, 1680-1640m		1115s, 617m		857l		537m
Mono-Carbo-aluminate $Ca_4Al_2(CO_3)(OH)_{12}$ / $5H_2O$	3676m	3624m	3543m	3363m, 3005m, 1650l	1363s			953m	669l	535s
Hemi-Carbo-aluminate $Ca_4Al_2(CO_3)_{0.5}(OH)_{13}$ / $5.5H_2O$	3676m	3642m, 3624m	3544m	3367-3007m, 1645l	1364s			954m	671l	537s
Mono-sulfo-aluminate $Ca_4Al_2(SO_4)(OH)_{12}$ / $6H_2O$	3672m		3549m	3423s, 1650m	1380l	1150m			579m	525s
C4AH13 ; $Ca_4Al_2(OH)_{14}$ / $6H_2O$	3673m		3541m	3400m, 1641l	1383l				667l	522s
C3AH6 (Hydrogarnet) ; $Ca_3Al_2(OH)_{12}$	3660s							810l		537s
C2ASH8 (Stratlingite) ; $Ca_2Al_2SiO_2(OH)_{10}$ / $2.25H_2O$	3669l			3442m, 1652l			1050s, 452s	951s		524s

Indeed, FT-IR spectroscopy in transmission mode was used to study the bulk of lightweight mortar containing polyamide waste powder. The polymer waste powder was then recycled as aggregates to substitute 50% of the sand (in volume ratio) contained in the mortar. Table 4 describes the IR spectra obtained 28 days after demolding. Specific bands assigned to polyamide are perfectly detected (data in italic). Several samplings performed into the bulk show a homogeneous dispersion of the polymer powder into the mortar mix. Assignment of the other bands did not show any specific disturbance of hydration: portlandite was detected at 3640 cm^{-1} (O–H stretching vibration); C–S–H were characterized by the major bands at 1030-970 cm^{-1} (Si-O asymmetric stretching vibration); the S-O stretching vibration of $[SO_4]^{2-}$ was also detected at 1150-1100 cm^{-1}.

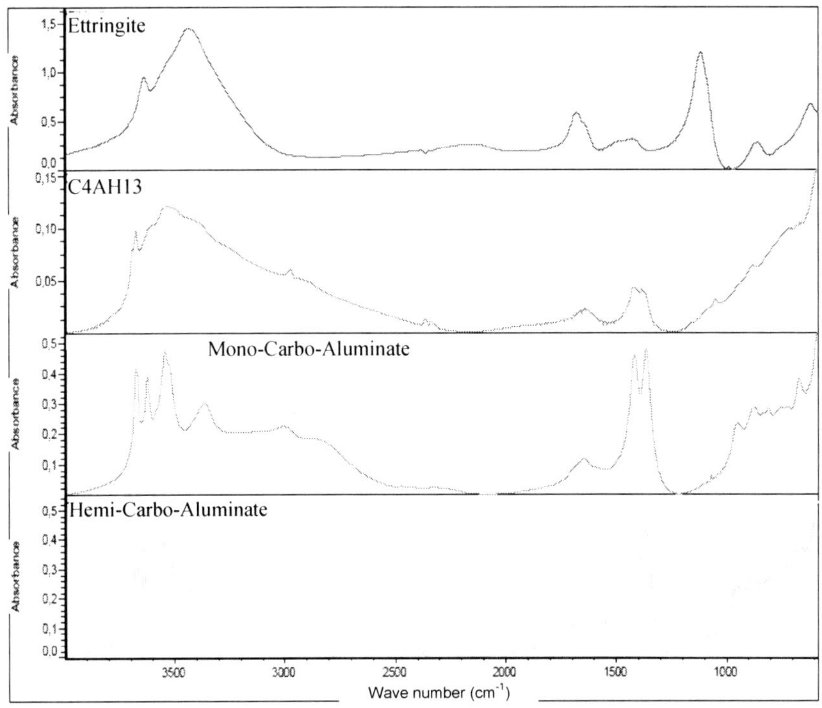

Figure 4: IR spectra of the different AFt/AFm phases studied.

Table 4: FT-IR bands of the bulk of mortar containing polymer recycled waste.

Bulk of mortar containing polyamide waste powder	Wave-number (cm^{-1}) with intensity of the band (s=strong; m=medium; l=low)							
	O-H	O-H; H_2O capil.	*N-H*	*CH_2 / CH_3*	C-O ; $[CO_3]^{2-}$	*C=O*	*C-N*	S-O; $[SO_4]^{2-}$	Si-O (asym. stretching vib.)
	3640m	3400 m, 1650 l	*3300s, 3100m, 1460m*	*2920s, 2850s*	2510l, 1410s, 872m, 710s	*1640s*	*1560s*	1100s	1030s - 970s

A concrete sample, protected by a polyurethane based resin, was also studied by FT-IR. The coating was sprayed on the concrete surface 28 days after demolding. The ATR mode was used to investigate the interface between concrete surface and coating. Table 5 provides the assignment of the main IR bands detected from the concrete. Portlandite (at 3640 cm^{-1}) and C-S-H (1020-970 cm^{-1}) were well detected. A double peak at 935–900 cm^{-1} was assigned to the presence of anhydrous alite. The FT-IR spectroscopy highlighted the presence of $CaCO_3$ (3 distinct bands at 1410, 870 and 710 cm^{-1}). The detection of $CaCO_3$ was assigned to the carbonation process or to the limestone filler used into the concrete mix. On the other hand, specific peaks of PU-based coating were well detected (data in italic in Table 5). No specific band attributed to the isocyanates group (at 2270 cm^{-1}) was detected, confirming a complete crosslinking of the resin.

Table 5: FT-IR bands of the interface between concrete and PU resin.

Interface between concrete and a PU based resin	Wave-number (cm^{-1}) with intensity of the band (s=strong; m=medium; l=low)									
	O-H	OH; H_2O capil.	*N-H*	*CH_2 / CH_3*	C-O; $[CO_3]^{2-}$	*C=O*	S-O; $[SO_4]^{2-}$	Si-O (asym. stretching) vib.	Si-O	Si-O
	3640m	3400m, 1650l	*3335s, 1450m*	*2950s, 2850s*	2510 l, 1410 s, 872 m, 710 s	*1690s*	1110s	1020s, 970s	935l, 900l	797m, 777m

4 Conclusions

FT-IR spectroscopy is a useful and convenient method to study the cementitious materials. This technique of characterization benefits certain advantages such as a small amount of sampling material, a short time of interpretation and a simplified sampling process compared to other usual methods of characterization. Moreover, specific modes of analysis such as ATR can directly investigate the surface of the hardened cementitious samples. However, FT-IR spectroscopy does not give any quantifying interpretation and is limited to detect the traces of certain compounds. Concerning the different applications, clinker phases and anhydrous cement can be well characterized by FT-IR in order to highlight the presence of C3S, C2S and C3A. Synthetic hydrated powders are easily investigated to detect the presence of major hydrates such as portlandite and C-S-H. The shift of the position of the major bands of C-S-H is studied to obtain specific data about the polymerization of silica and the decalcification process that occur during the ageing of the hydrated cement paste. New developments are under progress to study the presence of phases such as secondary ettringite, hemicarbonate, monosulfoaluminate, sulfoaluminate, C4AH13, etc. Finally, FT-IR spectroscopy is also interesting to characterize the mortar or concrete samples, containing or covered by polymer compounds, such as organic coatings.

References

[1] Taylor, H.F.W., Cement Chemistry, Thomas Telford Publishing, London, 2nd Edition, 1997.

[2] Bensted, J. and Varma, S.P., Some applications in Infrared and Raman spectral studies in cement industry (Part 2: Portland cement and its constituents). *Cement Technology*, pp. 378-382, July/August 1974.

[3] Ghosh, S.N. and Chatterje, A.K., Absorption and reflection infrared spectra of major cement minerals, clinker, and cements. *Journal of Materials Science*, **9**, pp. 1577-1584, 1974.

[4] Ghosh, S.N. and Chatterje, A.K., Attenuated total reflectance spectra of Portland cement. *Journal of Materials Science*, **10**, pp. 1454-1456, 1975.

[5] Ghosh, S.N. and Handoo, S.K., Infrared and Raman spectral studies in cement and concrete (review). *Cement and Concrete Research,* **10**, pp. 771-782, 1980.

[6] Allen, A.J., Thomas, J.J. and Jennings, H.M., Composition and density of nanoscale calcium-silicate-hydrate in cement, *Nature Materials*, 6, pp. 311-316, 2007.

[7] Kantro, D.L., Brunauer, S. and Weise, C.H., The ball mill hydration of tricalcium silicate at room temperature. *Journal of Colloid Science*, **14**, pp. 363-376, 1959.

[8] Ryskin, Y.I., The structure and infrared spectra of acid silicates. *Inorganic Materials*, **7**, pp. 331-344, 1971.

[9] Puertas, F., Goni, S., Hernandez, M.S., Varga, C. and Guerrero, A., Comparative study of accelerated decalcification process among C3S, grey and white cement pastes. *Cement and Concrete Composites*, **35**, pp. 384-391, 2012.

[10] García Lodeiro, I., Macphee, D.E., Palomo, A. and Fernández-Jiménez, A., Effect of alkalis on fresh C–S–H gels. FTIR analysis. *Cement and Concrete Research,* **39**, pp. 147-153, 2009.

[11] García Lodeiro, I., Palomo, A., Fernández-Jiménez, A. and Macphee, D.E., Compatibility studies between N-A-S-H and C-A-S-H gels. Study in the ternary diagram Na_2O–CaO–Al_2O_3–SiO_2–H_2O. *Cement and Concrete Research,* **41**, pp. 923-931, 2011.

[12] Bhat, P.A. and Debnath, N.C., Theoretical and experimental study of structures and properties of cement paste: the nanostructural aspects of C-S-H. *Journal of Physics and Chemistry of Solids,* **72**, pp. 920-933, 2011.

[13] Ylmén, R., Jäglid, U., Steenari, B-M. and Panas, I., Early hydration and setting of Portland cement monitored by IR, SEM and Vicat techniques. *Cement and Concrete Research,* **39**, pp. 433-439, 2009.

[14] Ylmén, R., Wadsö, L. and Panas, I., Insights into early hydration of Portland limestone cement from infrared spectroscopy and isothermal calorimetry. *Cement and Concrete Research,* **40**, pp. 1541-1546, 2010.

[15] Hughes, T.L., Methven, C.M., Jones, T.G.J., Pelham, S.E., Fletcher, P. and Hall, C., Determining cement composition by Fourier Transform Infrared Spectroscopy. *Advanced Cement Based Materials*, **2**, pp. 91-104, 1995.

[16] Yu, P., Kirkpatrick, R.J., Poe, B., McMillan, P.F. and Cong, X., Structure of calcium silicate hydrate (C-S-H): near, mid and far-infrared spectroscopy. *Journal of American Ceramic Society*, **82**, pp. 742-748, 1999.

[17] Pajares, I., Martinez-Ramirez, S. and Blanco-Varela, M.T., Evolution of ettringite in presence of carbonate and silicate ions. *Cement and Concrete Composites*, **25**, pp. 861-865, 2003.
[18] Hidalgo Lopez, A., Calvo, J.L.G., Olmo, J.G., Petit, S. and Alonso, MC., Microstructural evolution of calcium aluminate cements hydration with silica fume and fly ash additions by SEM, and mid and near-infrared spectroscopy. *Journal of American Ceramic Society*, **91**, pp. 1258-1265, 2008.
[19] Almeida, E., Balmayore, M. and Santos, T., Some relevant aspects of the use of FT-IR associated techniques in the study of surfaces and coatings, *Progress in Organic Coatings*, **44**, pp. 233-242, 2002.
[20] Chollet, M. and Horgnies, M., Analyses of the surfaces of concrete by Raman and FT-IR spectroscopies: comparative study of hardened samples after demoulding and after organic post-treatment, *Surface and Interface Analysis,* **43,** pp. 714-725, 2011.
[21] Horgnies, M., Willieme, P. and Gabet, O., Influence of the surface properties of concrete on the adhesion of coating: characterization of the interface by peel test and FT-IR spectroscopy, *Progress in Organic Coatings,* **72**, pp. 360-379, 2011.

Study of the microstructure and pores distribution of lightweight mortar containing polymer waste aggregates

V. Calderón[1], S. Gutiérrez-González[1], A. Rodriguez[1] & M. Horgnies[2]
[1]*Departamento de Construcciones Arquitectónicas e Ingenierías de la Construcción y del Terreno, Universidad de Burgos, Spain*
[2]*Lafarge Centre de Recherche, Saint Quentin-Fallavier, France*

Abstract

The study of mortars containing polymer wastes is a key-topic to create new construction materials for a sustainable development. This work deals with the characterization of the microstructure of lightweight mortars containing polymer waste powder, which are recycled, such as aggregates. The observations of cross-sections and fractured samples using Scanning Electron Microscopy (SEM) highlight that the particles of polymer are homogeneously dispersed into the mortar matrix. Numerous capillarity pores are detected into the hydrated cement paste. Calcium Silicate Hydrates covering the particles of polymer are also detected in environmental mode. The porosity is analyzed using two complementary techniques: (i) Mercury Intrusion Porosimetry (MIP) revealing the micro-pores sized from 170–200 µm up until 5–10 nm; and (ii) X-ray computerized axial tomography providing the macro-pores sized larger than 170 µm. The distribution of the total intruded volume is quantified by MIP according to the ratio sand/particles of polymer. The distortion of the pores distribution increases as a function of the content of polymer residues, excepted when the particles of polymer are replacing 25% of sand. In this specific case, the measurements by MIP and tomography show a homogeneous distribution of the larger pores, close to the one of the reference mortar.
Keywords: lightweight mortar, waste polymer, microstructure, tomography, MIP, SEM.

WIT Transactions on Engineering Sciences, Vol 77, © 2013 WIT Press
www.witpress.com, ISSN 1743-3533 (on-line)
doi:10.2495/MC130231

1 Introduction

New construction materials and an efficient recycling of polymer-based wastes are both expected to promote a sustainable development. Numerous studies have been performed on mortars and concretes containing polymer particles or waste polymer residues to obtain improved properties in terms of thermal or acoustic insulation [1–4]. The relations between the final porosity and the properties of hardened mortars were studied by several authors. Bouvard *et al.* [5] investigated the microstructure of lightweight concrete containing millimeter-size expanded polystyrene spheres by X-ray tomography. Following previous studies [6, 7] dealing with the substitution of sand by residues of polyurethane foam, this work focuses on the characterization of hardened mortars containing waste polyamide powder. First, the interface between the particles of polymer and the hydrated cement paste is investigated by SEM and Energy-Dispersive X-ray spectroscopy (EDS). Second, the porosity of hardened mortars is measured using two complementary techniques: MIP and X-ray tomography, which provide results about the micro-porosity and macro porosity, respectively.

2 Material and methods

2.1 Materials

2.1.1 Polymer Waste Particles (PWP)

Particles made of polyamide were obtained from waste raw material generated in an industrial Laser Sintering process. Its granulometric size was determined through laser diffraction analysis using a HELOS 12kv SYMPATEC analyzer (as shown in Figure 1). The real density of the PWP is 1.07 g/cm^3, which was measured with the Pycnometer method, using isopropyl alcohol [8].

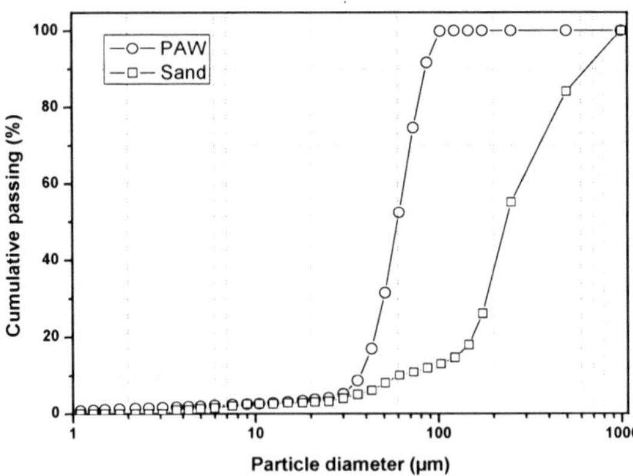

Figure 1: Laser granulometry of the polymer waste particles and sand.

2.1.2 Lightweight mortar samples

The mortar mix-design was traditional, containing cement, sand and water. Ordinary Portland cement (type CEM I 42.5 R) with a density of 3065 kg/ m^3 and river sand, sieved between 0 and 2 mm with a density of 2600 kg/m^3, were used. The lightweight mortar samples were manufactured by replacing different percentages of sand with the polymer waste particles. The particle size distribution of river sand is compared to the one of the PWP in Figure 1.

The lightweight mortar was produced using a 1/3 relation cement/(sand + PWP). Although the initial dosages were considered by weight, the sand was substituted by an equivalent quantity of polymer particles in volume (Table 2). The amount of water was adjusted to ensure good workability, fluency and plastic state, in accordance with Standard EN 1015-3 [9].

Fresh density and occluded air in fresh mortar samples were measured at 20 (± 1)ºC and 50 (± 1)% of relative humidity, according to the standards [10, 11]. The occluded air increases according to the amount of polymer waste particles introduced into the mix (replacing the sand) except for the P25 and P50 samples, which show respectively a minor and a similar content of occluded air compared to the one of the reference sample. As expected by the intrinsic low density of the polyamide waste, the fresh density of the samples decreases proportionally as a function of the replacement of sand by the particles of polymer.

Table 1: Mix proportions and properties of the mortar samples.

Samples	Sand replaced by PWP (% in volume)	Water/Cement	Fresh density (kg/m^3)	Occluded air (%)
R	0	0.71	2123	4.7
P25	25	0.74	1910	4.2
P50	50	0.75	1690	4.5
P75	75	0.77	1440	6.3
P100	100	0.89	1200	7.4

2.2 Experimental methods

2.2.1 SEM and EDS

Selected mortar sample was cut, impregnated in resin and then polished before being characterized in Back Scattering Electron Diffraction (accelerating voltage of 15 keV; current intensity of 1 nA), using a high-resolution field-effect gun scanning electron microscope (SEM FEG Quanta 400 from FEI Company, USA). EDS spectrometry was used to map the distribution of specific elements such as carbon and chlorine, which characterized the polished cross-section. Certain fractured specimens were not polished and were analyzed in environmental mode using a pressure of 3.2 Pa (to limit the dehydration).

2.2.2 MIP

The porosity of mortar samples was investigated by MIP technique (Autopore IV from Micromeritics, USA). The pressure range of the porosimeter was from sub ambient up to 400 MPa, covering the pore diameter range from about 360 μm to 3 nm. Tests were carried out on 10x10x10 mm^3 samples cut from the core of the mortar. The samples were dried in an oven at 45°C overnight before being tested. The advancing/receding contact angle was assumed to be 130°, as recommended for ordinary cement pastes [12].

2.2.3 X-ray computerized axial tomography

X-ray tomography allows the non-invasive characterization of the microstructure of numerous types of materials evaluation [13, 14], including concrete [15]. The mortars were cut and drilled after being cured to obtain cylindrical samples. The scans were made with a transmission type X-ray tube Yxlon working at 225 Kv/30mA. The sample was placed on a platform rotating 360° with a step of 0.3. A 2D scintillator panel detector of 2048x2048 pixels was used to record the images of the samples to reconstruct the internal structure. In these conditions, 1200 projections were taken for each sample and the data were reconstructed into l-CT slices using Mimics 10.0 software.

3 Results

3.1 Observations by SEM and EDS spectrometry

Figure 2 shows a SEM image of a P50 cross-section: numerous particles of polymer are dispersed into the hydrated paste. Numerous small pores (related to the large capillarity pores) are also detected into the hydrated cement paste at high magnification.

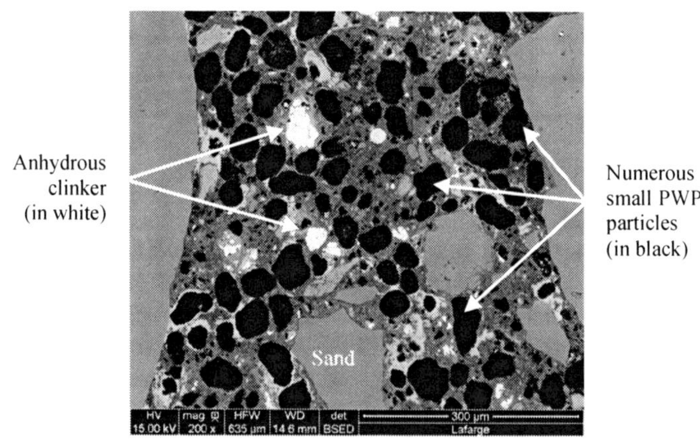

Figure 2: Polished cross-section of the P50 sample observed by SEM.

Figure 3 provides a map acquired by EDS spectrometry, showing the elemental composition of a P50 cross-section. Carbon and chlorine elements are then represented in red and green, respectively. Because chlorine is assigned to the epoxy resin used to impregnate the mortar samples before the polishing, the areas completely green can be assigned to the presence of air voids. Carbon is mainly related to the particles of polymer, which are homogeneously dispersed into the cement paste surrounding the sand aggregates (darker areas).

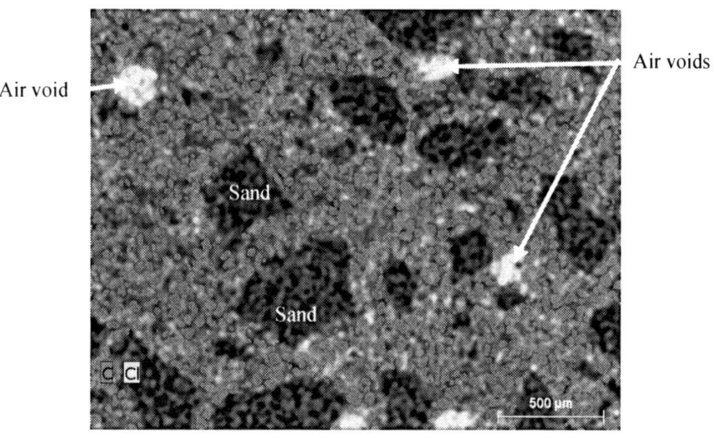

Figure 3: Distribution of carbon (red) and chlorine (green) for a P50 cross-section (acquired by EDS spectrometry).

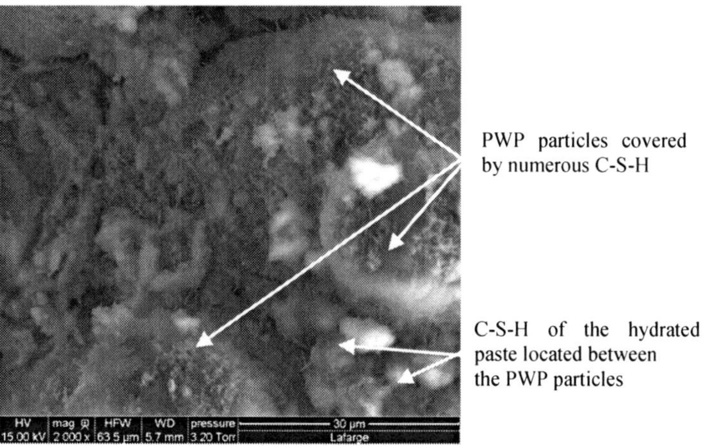

Figure 4: SEM images of C-S-H covering the particles of polymer (P50 sample).

Figure 4 provides a SEM image of the fractured P50 sample acquired in environmental mode: Calcium Silicate Hydrates (C-S-H) are covering the particles of polymer. Moreover, a large amount of C-S-H and some crystals of portlandite are detected into the paste located between the particles confirms that the recycled polymer does not disturb the hydration of cement paste. No crystals of ettringite are highlighted by the SEM observations.

3.2 Study of the pores distribution using MIP

Figure 5 shows the differential distribution curves, according to the ratio of sand replaced by the particles of polymer. These curves were plotted to better estimate the critical pore diameter that corresponds to the maximum peaks in the distribution, above which no connected path could form throughout the sample [16]. The distribution of the pore diameter into the mortar samples becomes more and more disturbed when more than 50% of sand is replaced by the PWP. However, the differential distribution curve of P25 sample is close to the one of the reference mortar sample. These results are conformed to the values of occluded air measured in the fresh mortar mixes.

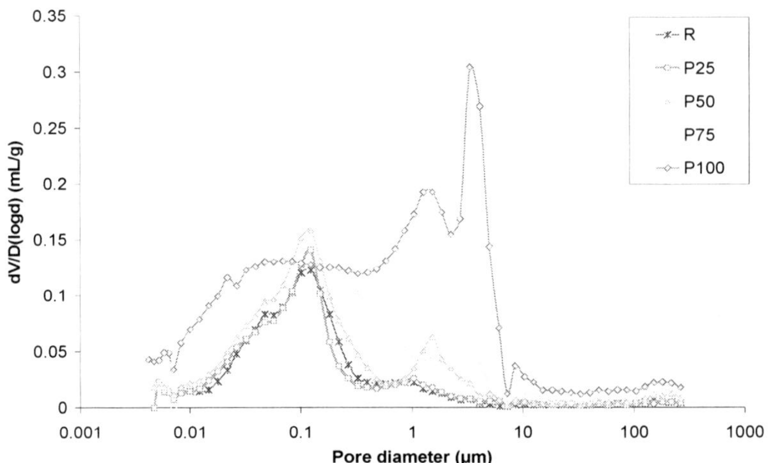

Figure 5: Differential distribution curves according to the ratio of sand replaced by the PWP particles.

Table 1 provides the distributions of the total intruded volume for the different mortar samples. The pore sizes are classified in 4 distinct ranges [17, 18]: (i) d>10 µm: air voids; (ii) 0.05<d<10 µm: large capillarities; (iii) 0.01<d<0.05 µm: medium capillarities; (iv) <0.01 µm: small capillarities. The pore structure is more and more disturbed as a function of the substitution of sand by the particles of polymer (due to the increase of large capillarities and air voids), excepted for the P25 sample that shows a porosity distribution close to the one of the reference sample, confirming the previous results of Figure 5.

Table 2: Distribution of total intruded volume of the mortar samples.

Range of porosity	Mortar samples (ratio of sand replaced by the PWP particles)				
	R	P25	P50	P75	P100
< 0.01 μm	6.5	4.3	3.8	4.0	4.3
0.01-0.05 μm	22.7	25.1	21.5	20.9	17.3
0.05-10 μm	65.8	65.6	69.0	68.2	71.0
> 10 μm	4.9	5.0	5.7	6.9	7.5

3.3 Study of the pores distribution using X-rays tomography

Based on the computational analysis of X-rays images, tomography allows characterizing the internal structure of the lightweight mortar samples. Figure 6 provides all the axial sections recorded during the study. The comparison between all the images allows mapping the distribution of the different materials (pores, cement, PWP and sand) according to the ratio of sand replaced by the polymer particles.

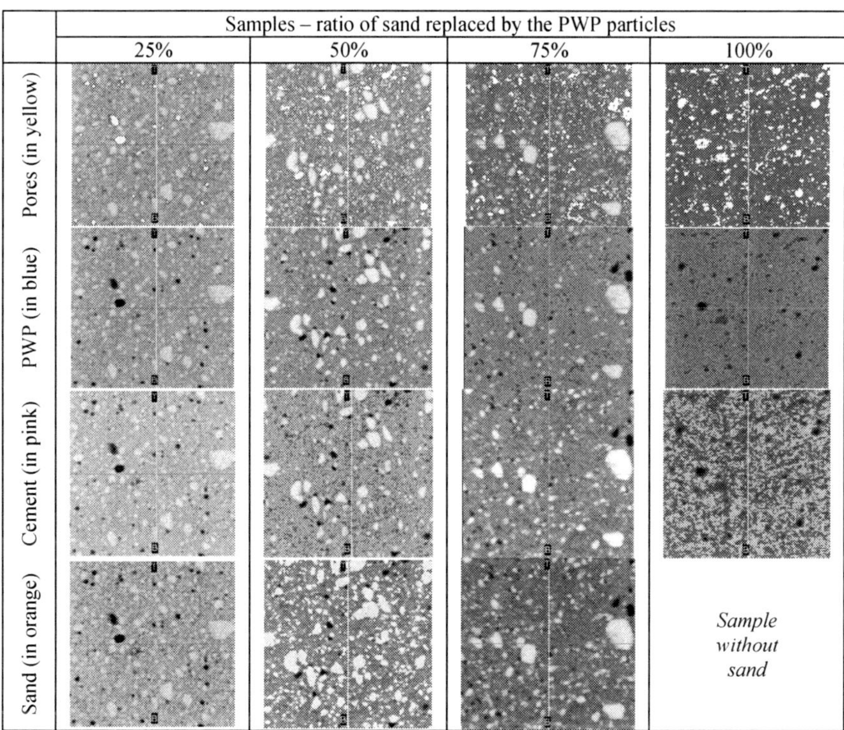

Figure 6: Tomography sections of the materials contained in the mortar samples.

As shown in Figure 7 and using the sequence of images dedicated to the pores distribution, it is possible to reconstruct the 3D images of the macro-porosity (pores sizes > 170 μm) according to the ratio of sand replaced by the polymer particles.

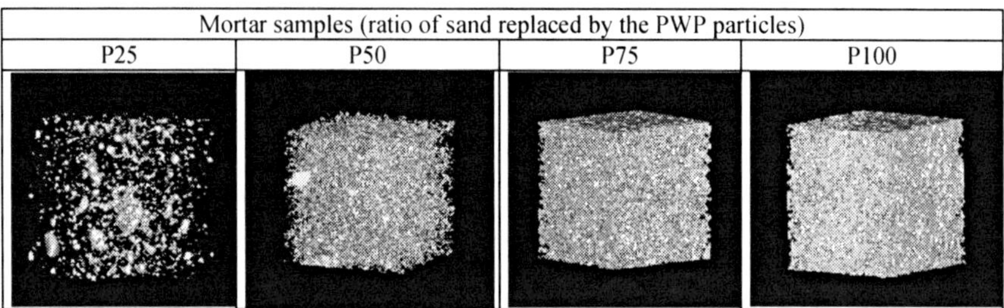

Figure 7: 3D tomography images of the pores distribution for each mortar sample containing PWP particles.

Figure 8(a) shows an example of axial section of the distribution of materials composing the mortar sample when the PWP particles are replacing 25% of sand. Each color represents one specific material: (i) blue for the polyamide particles; (ii) pink for the cement paste; (iii) orange for the natural sand aggregates and (iv) yellow for the pores. Figure 8(b) shows an example of 3D reconstruction done on the same samples.

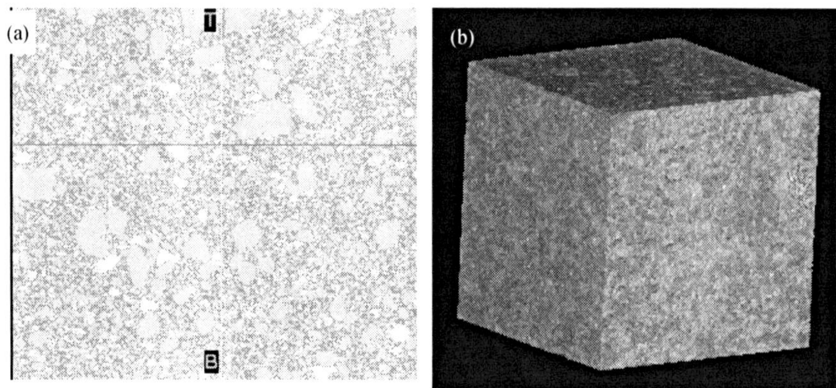

Figure 8: P25 mortar sample: (a) section of the materials distribution; (b) 3D reconstruction of the mortar sample.

4 Conclusions

Lightweight mortars can be manufactured using a high ratio of polymer waste particles (made of recycled polyamide) in place of sand. The investigation of the microstructure by several methods of characterization (SEM, EDS spectroscopy, MIP and X-ray tomography) allows obtaining different data.

(i) The polymer waste particles are homogeneously dispersed into the hydrated cement paste;
(ii) The composition of the polymer waste particles does not disturb the hydration of the cement paste (numerous C-S-H are detected);
(iii) The addition of polymer waste particles into the mortar mix increases the amounts of air voids and large capillarities pores;
(iv) The microstructure and the pores distribution are significantly disturbed when more than 50% of sand is replaced by the polymer waste particles;
(v) The replacement of 25% of sand by the polymer waste particles provides a microstructure and a pore distribution close to those of a reference sample.

References

[1] Siddique, R., Khatib, J. and Kaur, I., Use of recycled plastic in concrete: A review. *Cement and Concrete Composites*, **28**, pp. 1835-1852, 2008.

[2] Peng, J.H., Chen, M.F. and Zhang, J.X., Study on waste expanded polystyrene as lightweight aggregate for thermal insulating mortar. *Journal of Building Materials*, **5**, pp. 166, 2002.

[3] Wang, R. and Meyer, C., Performance of cement mortar made with recycled high impact polystyrene. *Cement and Concrete Composites,* **34**, pp. 975-981. 2012.

[4] Choi Y-W. and Moon D-J., Chung J-S., Cho S-K., Effects of waste PET bottles aggregate on the properties of concrete. Cement and Concrete Research, **35**, pp. 776–781, 2005.

[5] Bouvard, D., Chaix, J.M., Dendievel, R., Fazekas, A., Létang, J.M., Peix, G. and Quenard, D., Characterization and simulation of microstructure and properties of EPS lightweight concrete. *Cement and Concrete Research*, **37**, pp. 1666-1673, 2007.

[6] Gadea, J., Rodriguez A., Campos, P.L., Garabito, J. and Calderon, V., Lightweight mortar made with recycled polyurethane foam, *Cement and Concrete Composites*, **32**, pp. 672-677, 2010.

[7] Junco, C., Gadea, J., Rodriguez, A., Gutierrez-Gonzalez, S. and Calderon, V., Durability of lightweight masonry mortars made with white recycled polyurethane foam, *Cement and Concrete Composites*, **34**, pp. 1174-1179, 2012.

[8] EN 1936. Natural stone test methods. Determination of real density and apparent density and of total and open porosity; 2006.

[9] EN 1015-3. Methods of test for mortar for masonry. Part 3: Determination of consistence of fresh mortar (by flow table). 1999.

[10] EN 1015-6. Methods of test for mortar for masonry. Part 6: Determination of occluded air. 1999.

[11] EN 1015-10. Methods of test for mortar for masonry – Part 10. Determination of dry bulk density of hardened mortar; 2000.

[12] Taylor, H.F.W., Cement Chemistry, Thomas Telford Publishing, London, 2nd Edition, 1997.

www.witpress.com, ISSN 1743-3533 (on-line)

[13] Baruchel, J., X-ray Tomography in Material Science, Hermes-Science, Paris, 2000.
[14] Kak, A.C. and Slaney, M., Principles of Computerized Tomographic Imaging, IEEE Press, New-York, 1988.
[15] Grangeat, P., La Tomographie, Hermes Science, Traité IC2, Paris, 2002.
[16] Cui, L. and Cahyadi, J.H., Permeability pore structure of OPC paste. *Cement and Concrete Research*, **31**, pp. 277-282, 2001.
[17] Mindess, S., Young, J.F. and Darwin, D., Concrete 2nd Ed., Englewood Cliffs (NJ): Prentice Hall, 2002.
[18] Pipilikaki, P. and Beazi-Katsioti, M., The assessment of porosity and pore size distribution of limestone Portland cement pastes. *Construction and Building Materials*, **23**, pp. 1966-1970, 2009.

An improved criterion to minimize FE mesh-dependency in concrete structures subjected to blast and impact loading

H. G. Kwak & H. G. Gang
Department of Civil and Environmental Engineering, KAIST, Korea

Abstract

In the context of an increasing need for reliability and safety in concrete structures under blast and impact loading conditions, the behavior of concrete under high strain rate condition has been an important issue. Since concrete subjected to impact loading associated with high strain rate shows quite different material behavior from that in the static state, several material models are proposed and used to describe the high strain rate behavior under blast and impact loading. In the process of modelling, in advance, mesh dependency in the used finite element (FE) is the key problem because simulation results under high strain-rate condition are quite sensitive to applied FE mesh size. It means that the accuracy of simulation results may be deeply dependent on FE mesh size in simulations. This paper introduces an improved criterion which can minimize the mesh-dependency of simulation results on the basis of the fracture energy concept, and HJC (Holmquist Johnson Cook), CSC (Continuous Surface Cap) and K&C (Karagozian & Case) models are examined to trace their relative sensitivity to the used FE mesh size. To coincide with the purpose of the penetration test with a concrete plate under a projectile (bullet), the residual velocities of projectile after penetration are compared. The correlation studies between analytical results and the parametric studies associated with them show that the variation of residual velocity with the used FE mesh size is quite reduced by applying a unique failure strain value determined according to the proposed criterion.
Keywords: high strain rate concrete, penetration simulation, failure strain, mesh-dependency, fracture energy.

WIT Transactions on Engineering Sciences, Vol 77, © 2013 WIT Press
www.witpress.com, ISSN 1743-3533 (on-line)
doi:10.2495/MC130241

1 Introduction

Under hard impact, blast and impulsive load condition, concrete shows quite different behavior compared to that in static state. Many material properties of concrete such as strength, tangent modulus and critical strain are changed with strain rate. There are two main reasons for this unique behavior. First reason is inertia resistance effect. Concrete is a brittle material which fails due to internal microcrack creation and propagation. Decrease in the amount of microcrack at certain stress level reduces the stiffness of material. Under high strain-rate condition, there is not enough time to generate and propagate microcracks. This lack of microcracking decreases the amount of deformation along the loading direction. This means that stiffness of material in loading direction increases. Therefore, higher stress level than static condition is needed to propagate enough microcracks for failure in enhanced stiffness direction. Second reason is lateral inertia confinement effect. Lateral expansion caused by Poisson's ratio effect is limited during impact loading due to lack of time to expand. This limitation shows similar effects like lateral confining stress which increase strength of the material. In these reasons, strength and critical strain increases as strain-rate increases [1]. Therefore, several material models are proposed and used to describe the high strain rate behavior under blast and impact loading. In the process of modelling, mesh dependency in the used finite element (FE) is the key problem because simulation results under high strain-rate conditions are quite sensitive to applied FE mesh size. This paper suggests a criterion that minimizes mesh-dependency in applied FE mesh size on the basis of fracture energy concept. The typical high strain rate models (HJC [2], CSC [3], K&C [4]) are examined with penetration simulation to check relative sensitivity to the used FE mesh size. Furthermore, improvement in accuracy of simulation results is investigated by applying a unique failure strain value determined according to the proposed criterion.

2 Size effect on behavior of concrete

Since concrete is a brittle material which fails due to creation and propagation of internal microcrack in fracture process zone [5], concrete shows different failure behavior according to pattern of internal microcracking. Size of concrete specimen is the factor which makes difference in pattern of internal microcracking. As can be seen in Figure 1 [6], internal crack pattern changes according to size of specimens. And this difference in crack pattern makes the difference in strength and critical strain shown in Figure 2 [6]. The reason of this difference in crack pattern is that shear force cannot reach to bottom of specimen, as height of specimen get longer.

Therefore, size of the concrete specimen should be considered in estimation of fracture energy. This size effect is also observed in element mesh size of FE analysis. When large finite elements are used, each element has a large effect on the structural stiffness. When a single element cracks, the stiffness of the entire structure is largely reduced to fails easily [7]. By focusing on this size effect on failure behavior of concrete in FE mesh and specimen, criterion is proposed based on the fracture energy concept and numerical experiment for the size effect is conducted at the next chapter.

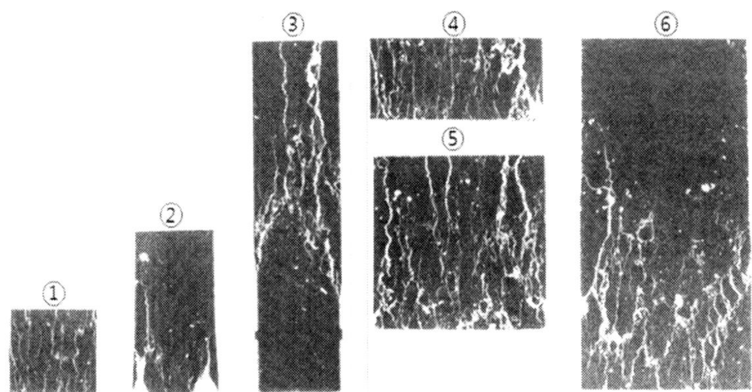

Figure 1: Crack patterns for different size of specimens.

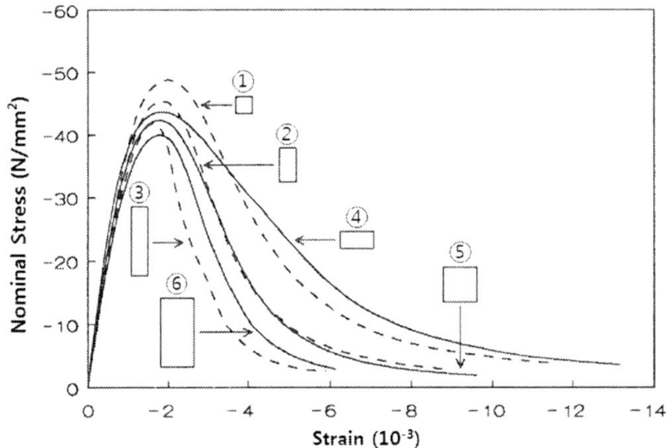

Figure 2: Stress-Strain curve for different size of specimens.

3 Introduction of a fracture energy based criterion

3.1 Estimation of fracture energy

As can be seen in Figure 3(a) [6], fracture energy is divided into continuum damage energy part and local fracture energy part. Continuum damage energy represents strain hardening portion which varies with size of specimen and local fracture energy represents strain softening portion which has constant value regardless of specimen size [7]. Since the main purpose of this paper is

modification of failure strain according to FE mesh size, local fracture energy which can convert size effect into difference of failure strain is used to propose a criterion. Local fracture energy part can be approximated by the area of a triangle in Figure 3(b) and be fomulated by adding *F(x)* the size effect function which has length scale. As a result, eqn (1) is the fomulated equation for local fracture energy where ε_0 is the failure strain and ε_c is the critical strain.

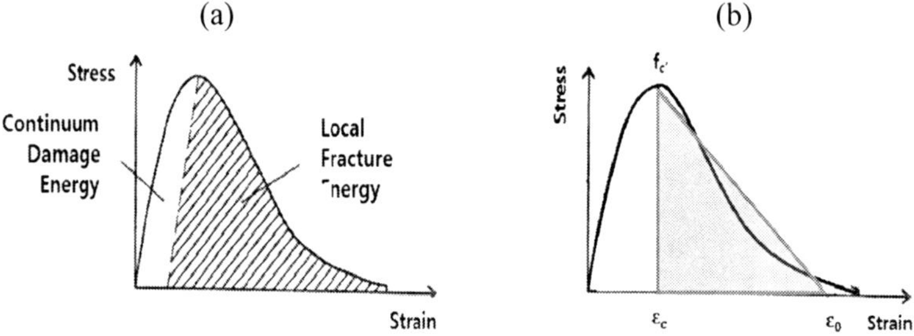

Figure 3: Fracture energy (a) Configuration, (b) Estimation.

$$G_f = \frac{1}{2}(\varepsilon_0 - \varepsilon_c) f_c' F(x) \tag{1}$$

3.2 Criterion proposal

The size effect function *F(x)* consists of *F(b)* and *F(h)*. *F(b)* is the function of specimen width and *F(h)* is the function of specimen length. First of all, *F(b)* is estimated. Assuming crack pattern as an exponential function, crack pattern is formulated with $f(x)=\alpha e^{\beta x}$ [7]. When b is width of specimen and center of the specimen is x=0, boundary conditions are $f(0)=1$ and $f(b/2)=(b_0/b)^{\gamma}$ [9]. Using two boundary conditions, $\alpha=1$ and $\beta=(2/b)\ ^{\gamma}\ ln(b_0/b)$ are gained. *F(b)* is calculated as eqn (2) by integrating *f(x)* with b and inserting maximum length $b_0=6(mm)$ that uniform crack pattern occurs. Fracture energy is given by eqn (3) by subsituting *F(b)* into eqn (1).

$$F(b) = \frac{2}{b_0}\int_0^{b/2} e^{-\left(\frac{2}{b}\right)\gamma\, ln\left(\frac{b}{b_0}\right)x} dx = \frac{\left(\left(\frac{b}{6}\right)^{\gamma}-1\right)}{\gamma ln\left(\frac{b}{6}\right)} \tag{2}$$

$$G_f = \frac{1}{2}(\varepsilon_0 - \varepsilon_c) f_c' \frac{\left(\left(\frac{b}{6}\right)^{\gamma}-1\right)}{\gamma\, ln\left(\frac{b}{6}\right)} F(h) \tag{3}$$

Regression analysis has been carried out to fomulate *F(h)* with experiment data of failure strain value according to specimen size. In this procedure, $\gamma = 1$ is assumed. As can be seen in Table 1 [6, 8], failure strain is rarely affected by the width of specimen b. Subsequently, *F(b)* is set to the constant value *F(50)*. *F(h)* is govern by eqn (4), where critical strain $\varepsilon_c = 0.0135$ that is normal value for 48(Mpa) concrete and fracture energy G_f=8.5(Nmm/mm^2) given by experiment [6]. Figure 4 shows result of regression analysis with the data from Table 1.

$$F(h) = \frac{2G_f}{(\varepsilon_0 - \varepsilon_c) f_c' F(50)} \tag{4}$$

Table 1: Measured failure strains according to size of specimens.

Height of specimen (mm)	Width of specimen (mm)	Failure strain (ε_0)
50	50	0.0103
50	100	0.01
100	100	0.0069
100	50	0.0071
200	50	0.0052
200	100	0.005

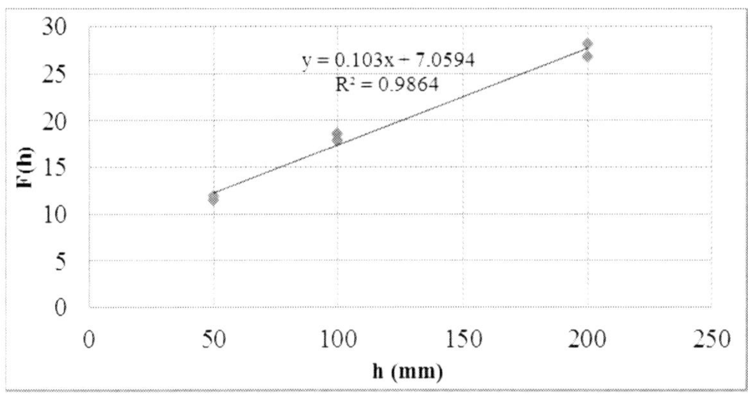

Figure 4: Regression analysis for *F(h)*.

The trend line follows *F(h)* function. Finally *F(h)* is governed by eqn (5) and criterion is proposed by eqn (6).

$$F(h) = 0.0103h + 7.06 \tag{5}$$

$$\varepsilon_0 = \frac{2G_f \gamma \ln\left(\frac{b}{b_0}\right)}{f_c'\left(\left(\frac{b}{b_0}\right)^{\gamma} - 1\right)(0.103h + 7.06)} + \varepsilon_c \tag{6}$$

Unique failure strain for each mesh size is proposed by this criterion eqn (6) and applied to FE analysis under high strain rate condition. b is defined as mesh width size and h is defined as mesh height to apply for FE analysis. The optimized value of $\gamma = 4.9$, which shows good agreement with being applied to mesh, is assumed.

4 Analytical approach

4.1 Penetration simulation

In this section, penetration simulation is performed for HJC, CSC, K&C models using LS-DYNA commercial FE analysis program. Projectile radius is 0.0127 (mm) and depth is 144 (mm). Plate size is 305 (mm) x 305 (mm) x 178 (mm). Projectile is placed on plate like figure 5 and half modelling is used for saving time. Initial velocity of projectile is 400m/s [10]. Projectile mesh size is fixed to 4.87(mm^2) and plate mesh size is variable. Residual velocity of projectile after penetration is checked to find out mesh-dependency. Furthermore, after applying failure strain of criterion on concrete plate, residual velocity of projectile is rechecked to confirm improvement in mesh-dependency.

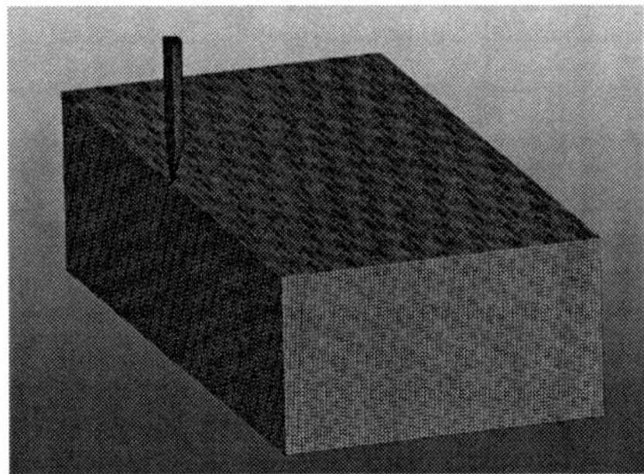

Figure 5: Configuration of penetration simulation.

4.2 Simulation results

Each model has 3 sets of cases which has different height of mesh. For case1, plate is divided into 5 pieces along the height direction, for case 2, 10 pieces, and for case 3, 25 pieces. All cases have 6 different width size of mesh. Residual velocity is measured according to mesh width size for each case. Figure 6, 7 and 8 shows the results of simulation for high strain-rate models, where *h* is height of mesh.

Black lines are case 1, grey lines are case 2, light grey lines are case3 and dashed lines represent the results of simulation that failure strain of criterion is applied to. Failure strain caculated by criterion is described in Table 2, where h is height of mesh and b is width of mesh. As can be seen in these figures, simulation results before applying criterion shows a significant mesh-dependency. HJC model has 201(m/s) difference in simulation results, CSC model has 198(m/s) and K&C model has 239(m/s) difference in simulation results according to mesh size. Reliablity of simulation results cannot be assured with these values. When failure strain determined by the criterion is applied to concrete plate in FE anlalysis, variation of simulation results are greatly decreases. As can be seen dashed lines in figure 6, 7, 8, difference in residual velocity is reduced. After criterion is applied to concrete plate, HJC model has

www.witpress.com, ISSN 1743-3533 (on-line)

121(m/s) difference in simulation results, CSC model has 131(m/s) and K&C model has 100(m/s) difference in simulation results according to mesh size. In conclusion, mesh dependency in penetration simulation accompanied with high strain rate is improved by using the criterion proposed in this paper.

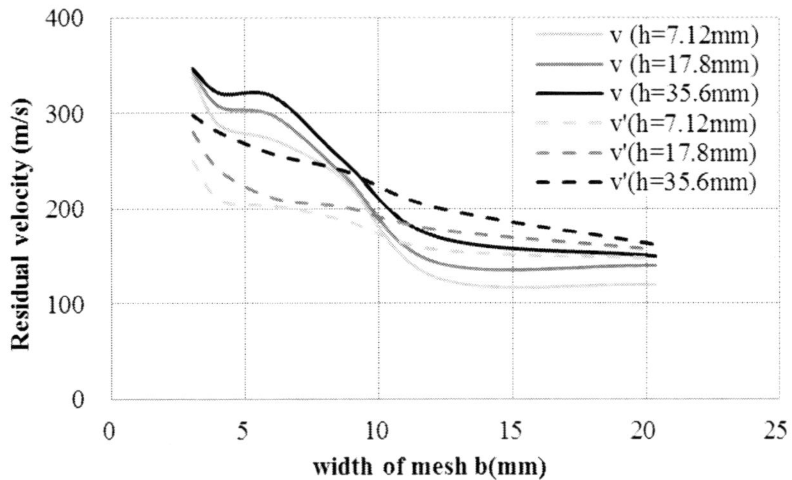

Figure 6: Residual velocity versus width of mesh (HJC model).

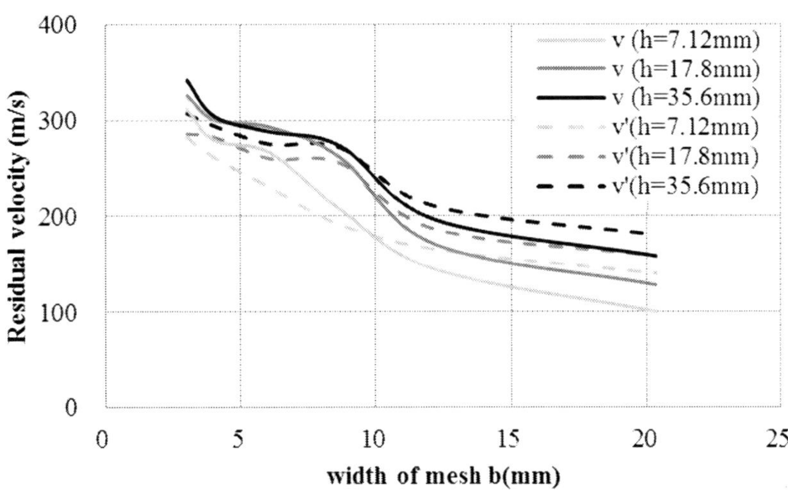

Figure 7: Residual velocity versus width of mesh (CSC model).

Table 2: Failure strain calculated by criterion according to size of mesh.

b (mm)	Case 1 (h=35.6mm)	Case 2 (h=17.8mm)	Case 3 (h=7.12mm)
3.05	0.0354	0.0414	0.0460
4.07	0.0354	0.0414	0.0460
6.1	0.034	0.0397	0.0445
8.71	0.0133	0.0154	0.0171
12.2	0.00512	0.00578	0.00634
20.3	0.00187	0.00196	0.00203

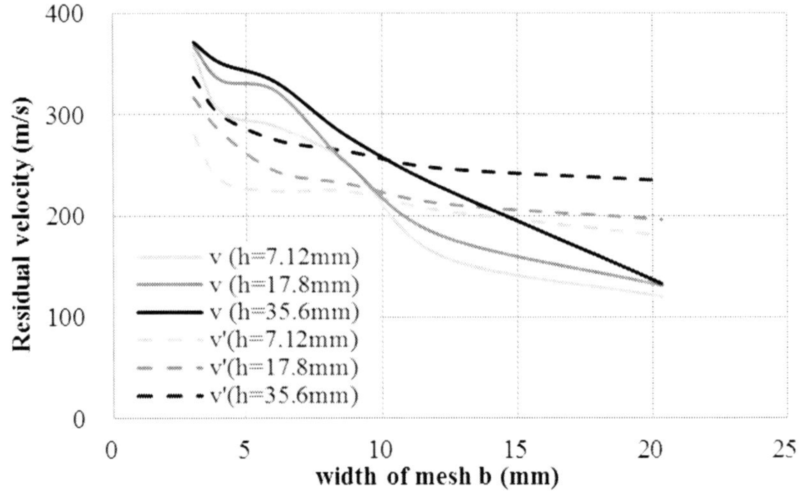

Figure 8: Residual velocity versus width of mesh (K&C model).

5 Conclusion

In this study, the mesh-dependency is investigated in the penetration simulations with high strain-rate models (HJC model, CSC model and K&C model). This study proposes a criterion based on the concept of fracture energy to minimize the mesh-dependency. The analytic approaches for penetration simulation of steel projectile into concrete plate is performed according to the FE mesh size of concrete plate. Simulation results show that the variation of residual velocity with the used FE mesh size is reduced by applying a unique failure strain value determined according to the proposed criterion. In conclusion, the mesh-dependency in FE analysis of concrete structure subjected to blast and impact loading can be minimized with the criterion proposed in this paper.

Acknowledgements

This work is financially supported by Korea Minister of Ministry of Land, Transport and Maritime Affairs (MLTM) as U-City Master and Doctor Course Grant Program, and the Defense Research Laboratory Program of the Defense Acquisition Program Administration and the Agency for Defense Development of Republic of Korea.

References

[1] Biscoff, P.H. and Perry, S.H., Compressive behavior of concrete at high strain rates. *Materials and Structures*, 24, pp. 425-450, 1991.

[2] Holmquist, T.J., Johnson, G.R., and Cook, W.H., A computational constitutive model for concrete subjected to large strains, high strain rates, and high pressure. *14th Internal Symposium on Ballistics*, Quebec, Canada, pp. 591-600, 1993.

[3] Malvar, L.J., Crawford, J.E., Wesevich, J.W. and Simons, D., A Plasticity Concrete Material Model for DYNA3D. *International Journal of Impact Engineering*, 19, pp. 847-873, 1997.

[4] Schwer, L.E. and Murray, Y.D., A three-invariant smooth cap model with mixed hardening. *International Journal of Numerical and Analytic Methods in Geomechanics*, 18, pp. 657-688, 1994.

[5] Taylor, L.M., Er-Ping Chen and Kuszmaul, J.S., Microcrack-Induced Damage Accumulation in Brittle Rock under Dynamic Loading. *Computer Methods in Applied Mechanics and Engineering*, 55, pp. 301-320, 1986.

[6] Vonk, R.A., A Micromechanical Investigation of Softening of Concrete Loaded in Compression. *Heron*, 38(3), 1993.

[7] Kwak, H.G and Filippou, F.C, *Finite element analysis of reinforced concrete structures under monotonic loads*, Department of civil engineering university of California Berkeley, California, 1990.

[8] Van Mier and J.G.M, Multiaxial Strain-Softening of Concrete, Part 1: Fracture. *Materials and Structures*, 19(111), pp. 179-190, 1986.

[9] Bazant, Z.P. and Oh, B.H., Crack Band Theory for Fracture of Concrete. *Materials and Structures,* RILEM, Paris, Vol. 16, pp. 155-176, 1983.

[10] Hanchak, S.J., Forrestal, M.J., Young, E.R. and Ehargott, J.Q., Perforation of Concrete Slabs with 48 Mpa (7 ksi) and 140 MPa (20 ksi) Unconfined Compressive Strengths. *International Journal of Impact Engineering*, 12(1), pp. 1-7, 1992.

Visualization of the healing process on reinforced concrete beams by application of Digital Image Correlation (DIC)

E. Tsangouri[1], K. Van Tittelboom[2], D.Van Hemelrijck[1] & N. De Belie[2]
[1]*Department of Mechanics of Materials of Constructions (MeMC), Vrije Universiteit Brussel (VUB), Belgium*
[2]*Magnel Laboratory for Concrete Research, Ghent University, Belgium*

Abstract

Fabrication of concrete with self-healing capabilities has recently become a hot research topic. In general, material science is focused on the development of smart engineering concrete and cementitious composites with an extended service life. Indeed, materials that remain durable and keep their mechanical performance, damage mechanisms occurring should heal by themselves. In the case of this study, formation of damage and recovery of the mechanical properties is investigated by application of an encapsulated healing agent. On an experimental level, it is imperative to implement an optical, non-contact and on-line technique to visualize and compare the crack propagation at the loading and reloading (when the initial cracks are filled by the healing agent) stage. For that reason, optical measurements by application of Digital Image Correlation (DIC) are performed during the tests. Processing images captured by a 4-digital cameras system during all the loading stages of four-point bending tests give a full- field view of the crack displacement and strain profiles. A step further, the visualization of the cracking phenomena by DIC offers a useful tool to apply fracture theories of concrete on healing systems.
Keywords: reinforced concrete, self-healing, encapsulation, cracking, DIC.

WIT Transactions on Engineering Sciences, Vol 77,
www.witpress.com, ISSN 1743-3533 (on-line)
doi:10.2495/MC130251

1 Introduction

The need for innovative building materials that resist to damage, time, weather and loading conditions and extend their service life is imperative. This demand has led several laboratories worldwide, following the philosophy of smart self-healing (SH) systems, to focus their research topic on advanced materials that can recover their mechanical properties after damage. More specific, in concrete technology, the recent 20 years, investigation is done in autogenous and autonomous healing concepts. In both cases, the quantitative characterization of the healing process is a hot area of research.

Experimentally, a method is required that gives a full-field view of the crack formation (main cause of damage), visualizes the mechanical behavior and highlights the recovery of material properties due to the healing procedure. In the case of this study, the healing phenomena appeared on an already successfully applied SH system is investigated in practice by application of Digital Image Correlation (DIC), an already well-established and promising Non Destructive Testing (NDT) optical technique.

2 Self-healing system

In recent years, the research team of the Magnel Laboratory at the University of Ghent has been working on the encapsulation of a polyurethane-based 2-component healing agent into cementitious materials [1, 2]. Till now, hollow glass or ceramics tubes filled with healing agent were embedded into cementitious mortar beams of small dimensions. The healing efficiency was measured with different methods (μ-CT, X-Ray, Acoustic Emission) and the optimized SH system set-up was defined. A step further in this study, this encapsulated system was applied on reinforced concrete beams and the success of it on real-scale tests was investigated. According to the main idea of the aforementioned smart material systems, the formation of cracks, during four-point bending in our case, is breaking the embedded capsules, releasing the healing agent. The open cracking area is filled with the agent and the defected area reverts after a certain curing time. Later on, reloading of the healed element gave a clear view of the regain in mechanical properties [1, 2].

3 Digital Image Correlation

DIC was developed in the 1980s and since then has been well improved and established. The basic principles and theoretical background of this on-line and full-field view optical method are well-known and described in detail in literature [3]. The DIC system consists of 2 cameras (-CCD) and a data acquisition system and requires a random speckle pattern attached to the material. A number of images are captured by cameras and via a digitizing board, the analog signal is converted into an integer gray level intensity value for each pixel of the image. The light intensity is measured on a range between 0 and 255 (black to white). The grey scale images of an object during loading are compared by an image correlation piece of software which calculates the displacement of regions in each image relatively to the reference image. Then, the strains are derived from the displacement gradients [3]. DIC capturing options appear promising in the case of this study, since crack opening can be measured as the deformation of certain points of the image at several loading stages and the strain profiles of loading and reloading stage indicate the damaged area and the potential recovery of it after healing.

4 Testing material

Rectangular-shaped concrete beams with dimensions of 650x100x100 mm^3 were manufactured. Steel bars applied as bending reinforcement (at a level of 100 mm above the bottom of the beam) and short steel fibers randomly distributed at the whole beam prevent extension of shear cracks that may appear during loading. The geometry of the beam and the reinforcement were chosen to create cracks, during loading, wide enough to break the capsules and at the same time to restrict the crack opening in less than 1.00 mm so that the areas will be easily filled with the limited amount of agent carried by capsules. Short and longer glass or ceramic hollow tubes (capsules), carrying the healing agent, were placed at the steel bar reinforcement level randomly distributed at the whole length of the beam. Different series of beams were manufactured depending on the capsules material (glass/ceramic), the length of the tubes, and the cover of capsules (mortar layer, cement paste bar or nothing). A series of beams, made without any encapsulation system were loaded and reloaded and are used for reference. The material characteristics are summarized in the following Table 1.

Table 1: Material characteristics.

Concrete	
CEM I, 52.5 N, w/c = 0.5	Compressive strength= 58.6 N/mm^2
Reinforcement	
2 steel bars	d =6 mm at height= 10 mm from the bottom
Steel fibers	Randomly distributed, d =0.71 mm
Healing system	
Glass/ceramic tubes	Short (60 mm)/Long (400 mm)
Healing agent	Polyurethane-based 2 component

5 Experimental set- up

5.1 Four-point bending test set-up

All the beams were tested under four-point bending (upper loading span of 200 mm). The displacement-controlled loading was stopped when 4.00 mm of

www.witpress.com, ISSN 1743-3533 (on-line)

deflection was reached. The beams were cured in room temperature for 24 hours and then reloaded following the same procedure. Apart from DIC on-line monitoring, Acoustic emission measurements were done. An 8-sensors system was attached at the surfaces of each beam and then a 3D location analysis was done to locate the capsules breakage and characterize on en energy level the crack formation of the virgin and the healed beams. Furthermore, after testing, permeability tests were performed to compare the water penetrating into the initial and healed cracks. Analysis of the results obtained from the aforementioned techniques, combined with the load-displacement curves derived out of the INSTRON testing machine give an integrated overview of the healing procedure. Due to the limited size of this publication, acoustic emission and permeability test results are not presented in this paper.

Finalizing the experimental investigation, the series of beams tested are categorized into three different groups according to the DIC analysis giving a representative overview of the healing efficiency.

5.2 Digital Image Correlation set-up

Two different DIC systems, containing 2-CCD cameras each, are used to visualize the crack formation. One of the DIC cameras systems was monitoring deformation phenomena at the bottom of the beams. The other, was facing the side of the beam giving an overview of the cracks formatting between the two upper loading supports (Fig. 1). The technical characteristics and the set-up of the cameras are shown in the following table (Table 2) and the figure respectively. Almost 1000 images were captured during each loading test. Deformation analysis was conducted using the software package VIC3D-2009, which compared the speckle pattern of the deformed beam obtained during loading or unloading with a reference pattern. During post-processing analysis, the speckle pattern area under investigation is defined as Area of Interest (AOI). Every pixel of this area stores a certain grey scale value, corresponding to the intensity of the light reflected by the surface. The varied grey value pattern is obtained on different square groups of pixels (subsets). The image correlation software locates every subset of the initial image in the deformed image. In this way, deformation measurements were transformed into digital correlation calculation and the displacement vector of each subset was obtained by subtracting the new coordinates from the original ones.

During the bending test, displacement fields (U, V, W (mm)) at three directions (X, Y, Z), the coordinates of each point of the area of interest (AOI) measured in mm and/or pixels, the sigma error of the correlation and the load of the INSTRON machine were derived [3]. Furthermore, requesting for strain profiles, the displacement fields are differentiated to calculate the normal and shear strains (exx, eyy, exy (%, unity, micro strain)) and the principal strains (e1, e2 based on different criteria).In the case of this study, only deformation (X) and strain (exx -unity) results out of the DIC-system at the side of the beam are presented.

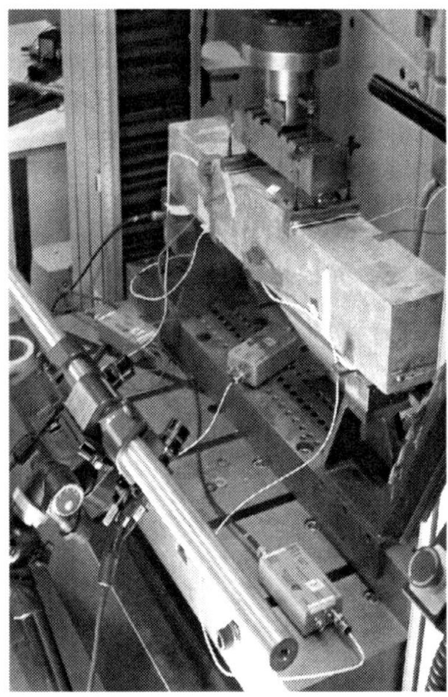

Figure 1: Digital Image Correlation 2-cameras system set-up.

Table 2: Digital Image Correlation technical characteristics.

CCD camera type:	AVT Stingray
Lenses:	23 mm
Resolution:	2456 x 2058
Focused area:	bottom system: 130–150 mm side system: 250 mm
Time capture:	every 2 seconds

6 Experimental results

Taking into account the load- displacement curves measured by INSTRON machine and the strain and crack opening profiles of DIC, beams that exhibited healing can be categorized into the following groups:

- Healed beam – case1: not reopening of the initial crack and a new crack is formatted during reloading
- Healed beam – case2: crack reopening during reloading but less wider than at the loading stage

These two healing phenomena are compared with the reference beam behavior. Since the choice of the appropriate SH encapsulated system is not the main purpose of this study, representative results from one beam each time are used to explain in practice the healing process for both aforementioned categories.

In the following graphs (Figs. 2–4), the load-displacement curves as captured by the INSTRON machine are presented for the reference beam and the two cases of healed beams.

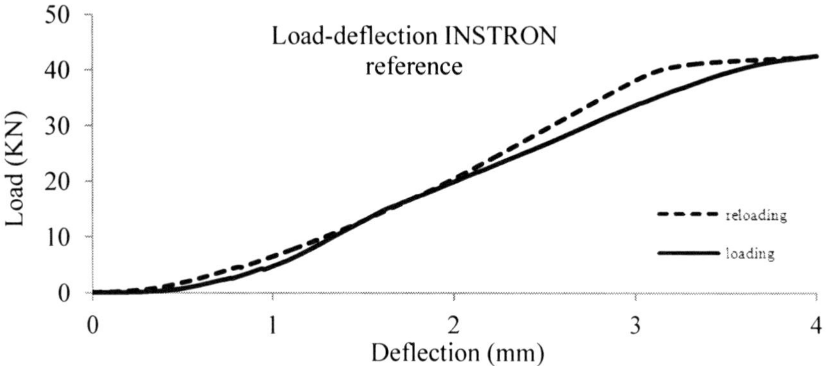

Figure 2: Load-deflection curve of the reference beam.

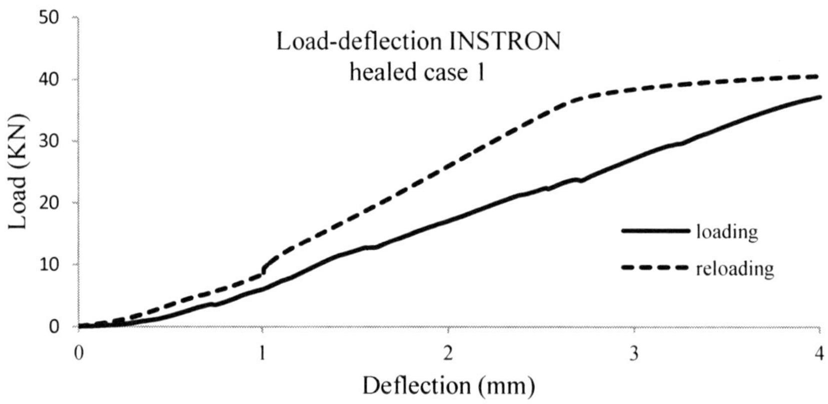

Figure 3: Load-deflection curve of the healed beam – case 1.

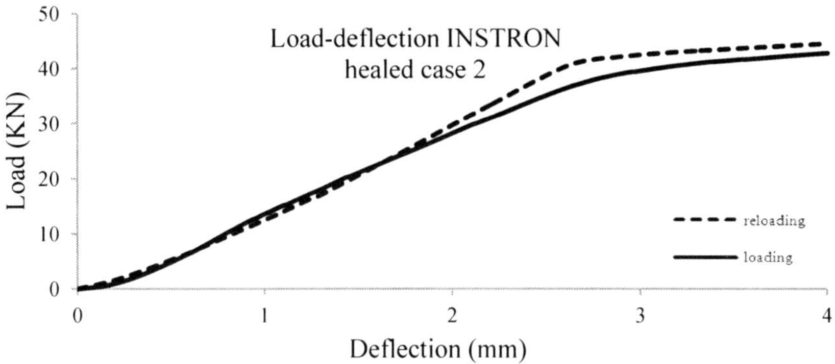

Figure 4: Load-deflection curve of the healed beam – case 2.

6.1 Reference beams

As is shown in the images below (Fig. 5), the bending behavior of the loading and reloading stage of the reference beam are quite similar. The reopening of the already existed cracks is verified by DIC analysis. Applying a straight line parallel and few millimeters above the bottom of the processed images, DIC software determine values of displacement and strain in (xx) direction (parallel to the steel reinforcement) for 200 points defined on this line. In the graph of crack opening related to the position of the 200 points (xx direction) that follows, one can see that the width of cracks is getting higher values at the case of reloading compared to the loading ones (Fig. 6).

Furthermore, there are no new cracks appearing at the reloading stage. Finally, as it is shown at the following strain profile of the beam (Fig. 7), the peak values of strain at the area of crack formation are higher in the case of reloading.

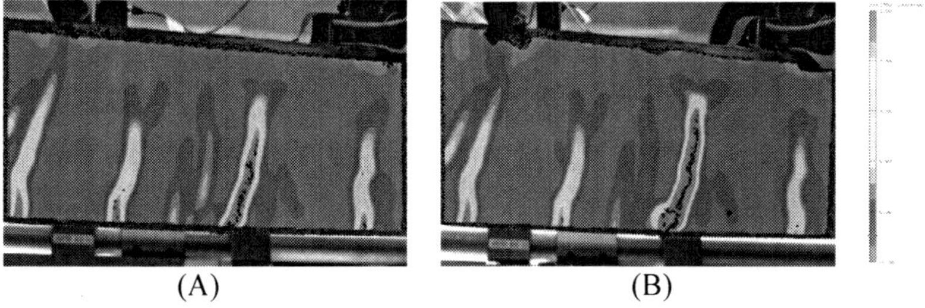

Figure 5: Crack pattern after loading (A) and reloading (B) of the reference beam.

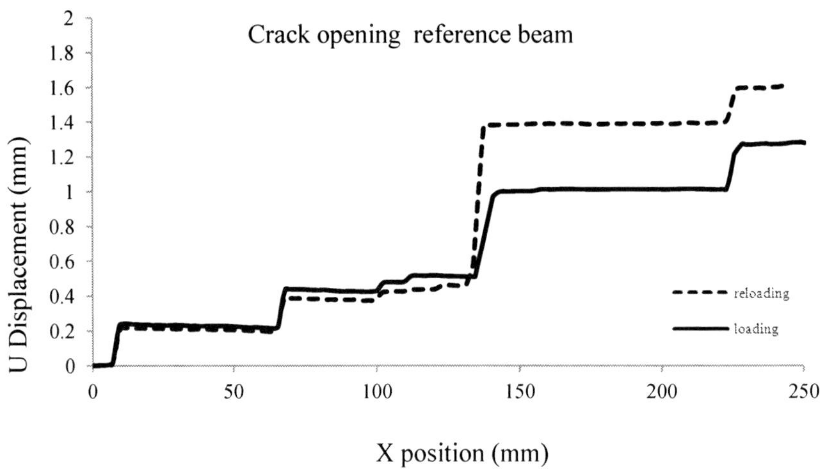

Figure 6: Crack opening at the bottom of the reference beam.

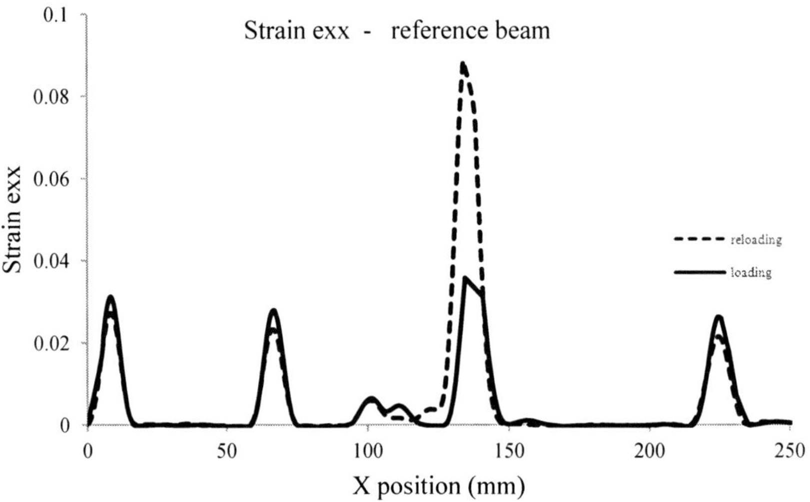

Figure 7: Strain distribution at the bottom of the reference beam.

6.2 Healed beam: case 1

In few of the beams a new crack seems to appear during reloading. This new crack formation is not visible by the INSTRON loading curve, but can be easily visualized at the following crack pattern (Fig. 8) of the loading and reloading stage of testing (healing system: short ceramic tube encapsulation protected by cement paste bar). Moreover, as it is noticed, there is a crack that is not formatted anymore at the stage of reloading. On the other hand, capsule breakage is filling this crack making the cracking area more resistant than before. Due to this reason, another weak point next to the initial crack is generating a new crack.

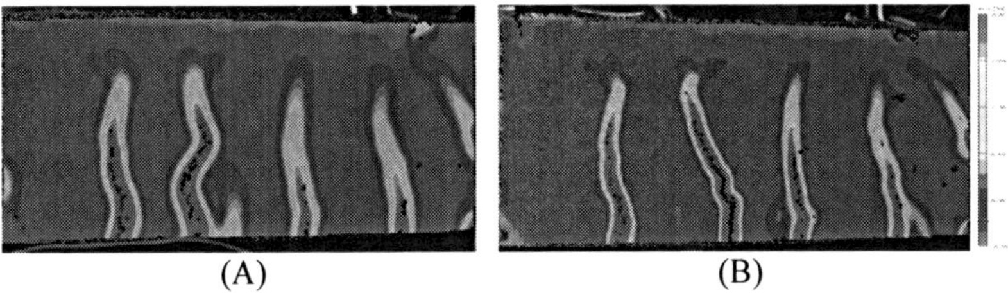

Figure 8: Crack pattern after loading (A) – reloading (B) of the healed beam – case 1.

Quantitatively, the new crack opening is shown at the graph of displacement for 200 points of the line fixed at the bottom of the speckle pattern on the DIC analysis. The position in which the crack is formed at the reloading stage is changed obviously. Similarly, at the strain- position graph there is a peak of strain appearing on a different area of the beam compared to the initial loading stage (Figs. 9 and 10). Finally, the total opening of the cracks remains the same (since all the other cracks seem not to be healed by capsules breaking).

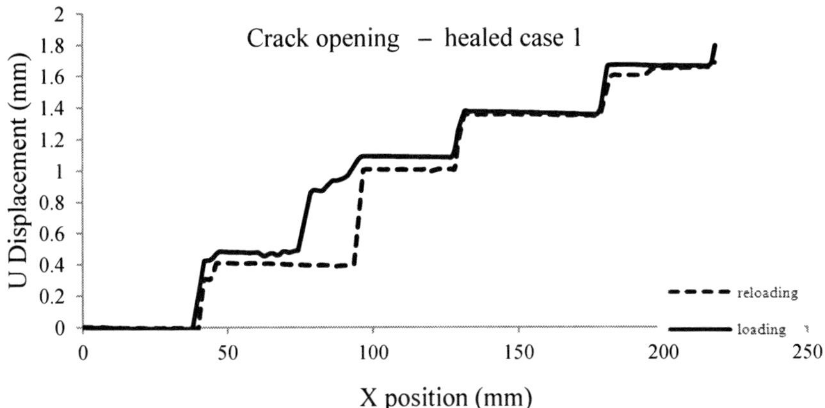

Figure 9: Crack opening at the bottom of the healed-case 2 beam.

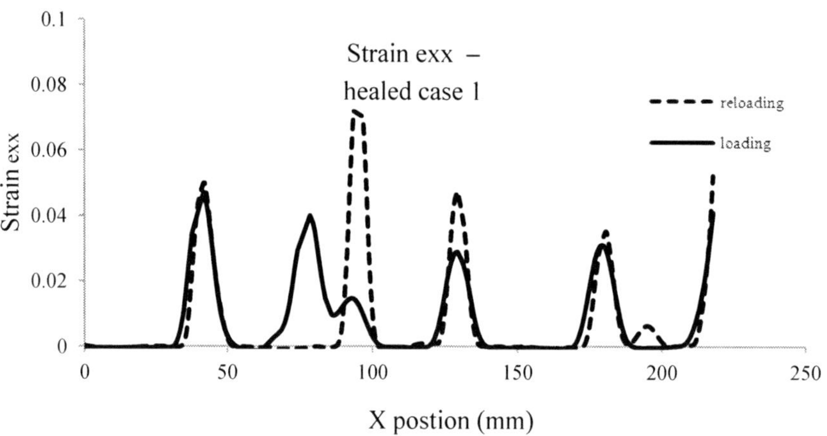

Figure 10: Strain distribution at the bottom of the healed beam – case 1.

6.3 Healed beam: case 2

At most of the beams from different series (as mentioned and defined above according to the geometry and the type of capsule/ healing system) during reloading the already existed cracks reopen, but the width at the end of reloading got values lower than the width at the end of loading (Fig. 11). Apparently, cracks are filled by healing agent after capsule breakage but the already damaged cracking area is not strong enough to afford reloading and generates cracking again. The change in stiffness of this representative beam is obvious at the curve of loading-displacement derived from the INSTRON machine (healing system: short glass tube encapsulation protected by cement paste bar). The formation of less wide cracks was noticed during the DIC analysis taking into account the following crack patterns and graphs (Figs. 11–13).

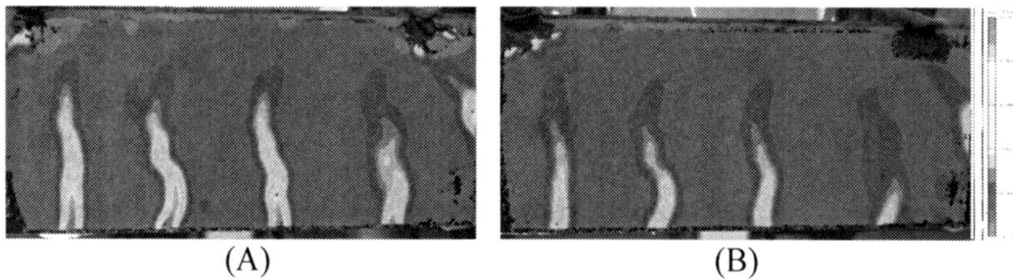

Figure 11: Crack pattern after loading and reloading of the healed beam – case 2.

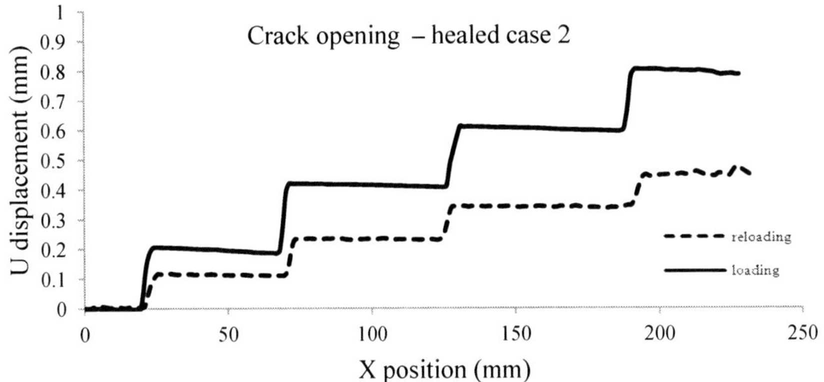

Figure 12: Crack opening at the bottom of the healed – case 2 beam.

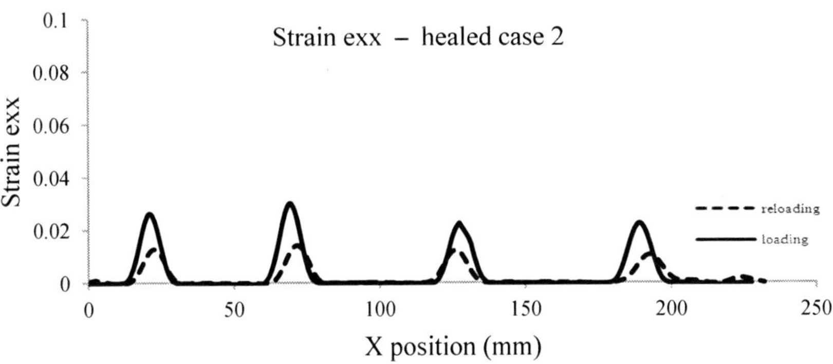

Figure 13: Strain distribution at the bottom of the healed – case 2 beam.

In the graph of crack opening it is shown that the sum of the crack width is taking values smaller at the reloading stage compared to the initial loading, but cracks seem to appear at the same defected area as before reloading (Fig. 12). Summarized and presented at the following table, the cracking width is getting the half of the loading value during reloading for all the cracks of this certain beam.

Table 3: Crack opening of the healed – case 2 beam.

	crack opening (μm)-loading	crack opening (μm)-reloading
1st crack	205	101
2nd crack	206,05	120,05
3rd crack	203,34	105
4th crack	200,63	109,48

www.witpress.com, ISSN 1743-3533 (on-line)

Further, strain peaks at the graph of strain-position (200 points of the bottom line) are getting lower values at the reloading stage (Fig. 13).Strain concentration phenomena around the healed cracks is further investigated in the near future.

7 Conclusions: future work – fracture mechanics

Full field DIC-images were used to investigate in full detail the cracking behavior of concrete with self-healing capabilities. It was found that, further conclusions cannot be drawn due to the limited number of beams tested in which healing phenomena appear.

In the near future, focusing at the phenomena around the crack tip will be combined with the fracture mechanics theories. The main aim of this study will be to prove the regain of mechanical properties due to healing by monitoring cracking phenomena.

Moreover, analysis of the promising results of the healing system applied can be used to obtain an experimental set-up that can successfully characterize mechanically the healing procedure.

Acknowledgement

The authors wish to acknowledge the IWT for financial support under the SIM-Engineered Self-Healing Materials project.

References

[1] Van Tittelboom, K., De Belie, N., Self-healing concrete by the internal release of adhesive from hollow glass fibres embedded in the matrix, *2nd International conference on Self-Healing Materials,* USA, 2009.

[2] Van Tittelboom, K., De Belie, N., Van Loo, D., Jacobs, P., Self-healing efficiency of cementitious materials containing tubular capsules filled with healing agent. *Cement and Concrete Composites*, 2011.

[3] Sutton, A, Orteu, J., Schreier, W., *Image Correlation for Shape, Motion and Deformation Measurements*, USA, 2009.

Section 6
Advances in composites

Utilising implanted carbon fibre as a resistive heating element in wind turbine blade anti-icing systems

A. Maheri
Faculty of Engineering and Environment, Northumbria University, UK

Abstract

Wind turbines installed in cold climates are at risk of blade icing and required to be equipped with systems preventing the build-up of ice and/or ice removal. This paper presents the results of a preliminary investigation on the feasibility and suitability of using carbon fibres as resistive heating elements implanted in the structure of the composite blades. The proof of the concept of using carbon fibre implanted in a glass fibre composite as heating elements for de-icing purpose is demonstrated. Moreover, using a genetic algorithm (GA) optimiser the optimum depth of implanted carbon fibres and the optimum magnitude of heat source which minimise the energy consumption of the system whilst subjected to manufacturing and controllability constraints are obtained. A finite difference model has been also developed to perform the transient heat transfer analysis which is to be used as the evaluator for the optimiser module. The effect of heated fibres on the fibre-matrix debonding strength is also examined. It is shown that, heating the carbon fibres up to 95°C does not affect the fibre-matrix debonding strength. However, the rapture of the carbon fibre tows uncovered in the polymer matrix is the dominant failure mode.
Keywords: wind turbine, blade icing, resistive heating de-icing, resistive heating ant-icing, carbon fibre, GA optimisation, fibre-matrix de-bonding strength.

1 Introduction

The increasing demand for wind energy has led to more wind turbines being deployed in remote areas, many of which are subjected to extreme cold temperatures. It is estimated that wind power generation in 2010 in cold climate locations is approximately 20% of the installed capacity [1]. At colder

WIT Transactions on Engineering Sciences, Vol 77, © 2013 WIT Press
www.witpress.com, ISSN 1743-3533 (on-line)
doi:10.2495/MC130261

temperatures air has a considerably higher density. Hence there is more power available for conversion. However, wind turbines installed in cold climates are at risk of blade icing for large parts of the year. Laakso *et al.* [1] calculated an estimated cost for wind turbine energy loss due to icing and showed that it can reduce the energy yield between 15–25% for a period of icing between 30 and 60 days of the year.

Additional problems caused by ice include the reduced fatigue life, changed natural frequencies and consequently the risk of flutter instability, and potential threat to human safety and damage to nearby structures due to projected ice. The factors which directly influence the fatigue life are asymmetric masses causing rotor unbalance and increased edgewise bending moment fluctuation amplitude due to ice accretion [2].

Many manufacturers are already applying measures to help prevent the build-up of ice (anti-icing) or ice removal (de-icing) by using specially engineered rotor blade coatings, microwave emitters and resistive heating. A resistive heating anti-icing system uses electrical power to heat elements at regular intervals such that the surface temperature is maintained at a level which prevents ice formation and adhesion. A de-icing system, on the other hand, will only be activated at a point when a certain amount of ice has accumulated on the surface of the rotor blade. De-icing includes an inherent downtime for the wind turbine but an ant-icing system does not. The cost effectiveness of either type depends on the energy consumption of the system and the power loss due to wind turbine downtime.

Copper wires and graphite sheets are the most popular resistive heating elements employed in wind turbine blades. Systems using implanted copper wires have the problem of scaling up to MW wind turbines. Systems using graphite sheet fitted to the leading edge are an aftermarket solution and add weight to the blade.

The current composition for most modern wind turbine blades is either epoxy or polyester matrix composites reinforced with glass and/or carbon fibres. Due to monetary reasons however the majority of modern turbine blades are made from glass fibres, although epoxy with carbon fibres has superior mechanical properties with higher strength and stiffness. Another less conventional application of carbon fibre is to use it for its electrical properties [3]. Carbon fibres are supplied in a variety of different forms, from continuous filament tows to short chopped fibres and mats. Tows of continuous filament contain thousands of individual filaments of carbon fibre. These tows can be woven into fabrics which can then be used to reinforce polymers in a composite. The highest strength for a carbon fibre composite is obtained by using unidirectional continuous reinforcement, with the strength lying in the direction of the continuous fibres.

This paper, firstly, demonstrates the proof of concept of using carbon fibre implanted in a glass fibre composite as heating elements for de-icing purpose. Secondly, using a GA the optimum depth and the optimum heating power of carbon fibres as heating elements for a typical wind turbine blade at a general span location are obtained. Finally, the effect of implanted carbon fibres as heating elements on the fibre-matrix debonding strength of the composite specimen is examined.

2 Resistive heating using carbon fibres, de-icing

The specimen shown in Figure 1 is made of 4 layers of ±45° glass fibre epoxy embedding a fifth layer of five tow of carbon fibre located at about 20mm spacing at angle 0. A K-type thermocouple was placed with the measurement junction resting on the middle carbon fibre tow. Both ends of each tow are covered by a conductive silver paint.

The thermogram of Figure 1(b) shows a temperature rise across the whole composite when connected to a source of DC power, but as anticipated the highest temperatures are directly above the heating elements.

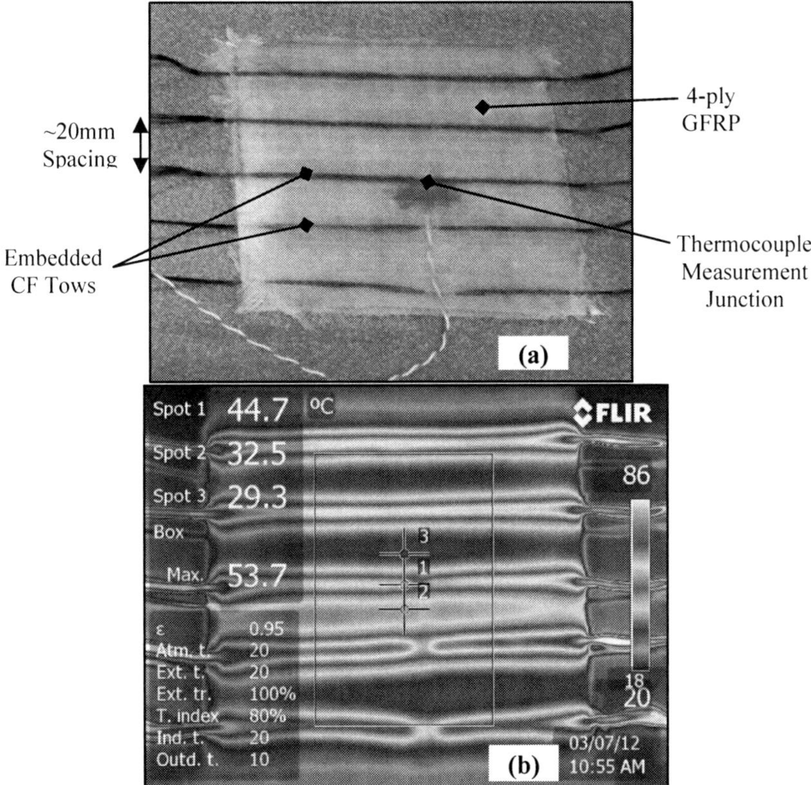

Figure 1: (a) Specimen with carbon fibre tows as heating elements, (b) thermogram.

In order to evaluate the effectiveness of the embedded carbon fibre heating elements at de-icing of the surface of the composite material, the specimen was merged in water and was frozen at -15°C (Figure 2(a)) with an ice thickness of

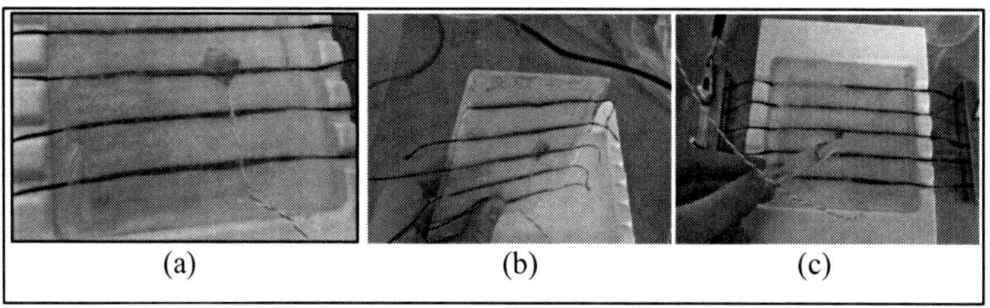

Figure 2: De-icing using implanted carbon fibres.

approximately 3 mm each side (Figure 2(b)). After the start of de-icing process the ice started to come off after approximately 9 minutes (Figure 3(c)).

3 Performance optimisation, anti-icing

Figure 3 shows the cross-section of a typical wind turbine blade rotating with an angular speed of Ω and the coordinate systems $n-s-r$ and $x-y-r$, where r is the radial location along the blade span.

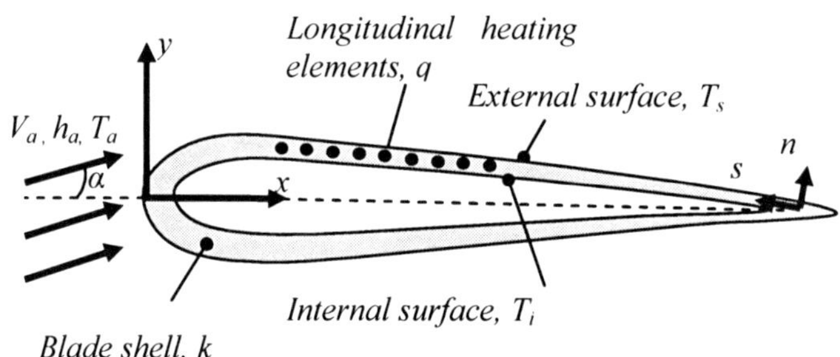

Figure 3: Implanted heating elements in wind turbine blade.

In this figure, $V_a = \sqrt{V_w^2 + r\Omega^2}$ is the relative velocity seen by the blade at a general radial location r, V_w is wind speed, h_a and T_a are the air convection heat transfer coefficient and temperature respectively, α is the angle of attack (AOA), T_i and T_s are the internal and external shell surface temperature respectively, and k stands for the blade shell conductive heat transfer coefficient.

In developing the heat transfer model the following assumptions are made:

- The heat transfer along s axis is neglected. This assumption is valid if the distance between the heat sources (carbon fibres) distributed along s axis is small and the strength of each heat source is controllable independently.

- The heat transfer along r axis is neglected. Although the relative air velocity seen by the blade and consequently the convective heat transfer coefficient depend on the span location, modular design of heating elements along the r axis makes it possible to control the strength of each module proportional to heat transfer rate, keeping the temperature distribution along the blade span uniform.
- Wind speed, rotor speed and consequently the AOA remain constant for the duration of the transient heat transfer analysis.
- The internal surface of the wind turbine blade is assumed to be fully insulated. Therefore the heat loss due to natural convection is assumed to be zero. The validity of this assumption depends on the position of the heating elements and the shell thickness. Using ANSYS based on a heat flux of $q = 1500W / m^2$, the predicted steady state surface temperatures found for the steady state external surface temperature achieved for the insulated and natural convection scenarios were 46.16°C and 49.32°C for the natural convection and insulated internal surfaces cases respectively. This shows about 6% over-prediction when the internal surface is assumed to be insulated. This error reduces as the shell thickness increases and the heat source is positioned closer to the blade surface. Table 2 shows the rest of the parameters used for this analysis.

Taking into account the effect of the AOA, Wang *et al.* [4] correlated the experimental data with respect to the Nusselt and Reynolds numbers as follows:

$$Nu_x = 0.0943(C_1 + D_1\alpha)\mathrm{Re}_c^{0.636}\,\mathrm{Pr}^{1/3} \quad for\ \mathrm{Re}_c > 5\times10^5 \qquad (1)$$

$$Nu_x = 2.482(C_1 + D_2\alpha)\mathrm{Re}_c^{0.389}\,\mathrm{Pr}^{1/3} \quad for\ \mathrm{Re}_c \leq 5\times10^5 \qquad (2)$$

where Re_c represents the chord length specific Reynolds number, $C_1 = 1$ for zero AOA and $C_1 = 0.75$ for non-zero AOA, $D_1 = \pi / 180$ and $D_2 = C_1 D_1$.

Using finite difference methods, the one-dimensional transient heat transfer model was coded in MATLAB. Commercial packages such as ANSYS are suitable for providing solutions for problems with a single set of inputs. However they are not efficient as an evaluation module within an optimisation process, in which design variables, and control parameters are changed through an iterative search algorithm. The purpose of the developed module in MATLAB is to provide efficient evaluation of design candidates during the optimisation procedure. Figure 4 shows the external surface temperature as a function of time obtained by the evaluator module and the results obtained by ANSYS. The data used are as shown in Table 1. A heat flux of $q = 3000W / m^2$ was used. In the evaluator module a time step of 0.1 second was used.

It can be seen from Figure 4 that the surface temperature reached steady state at 76.62°C when using the evaluator module coded in MATLAB against 76.61°C when ANSYS is employed. This figure shows an excellent agreement with the results obtained by the two software tools with a difference of less than 0.02%.

Table 1: Problem settings.

Parameter	Value
Radial location, r	9.5 m
Rotor speed, Ω	50 rpm
Wind speed, V_w	15 m/s
Angle of attack, α	10°
Matrix initial temperature, T_i	1°C
Initial internal cavity temperature, T_i	1°C
Ambient temperature, T_a	-20°C
Air heat capacity, $c_{p,a}$	1005 J/kgK
Air thermal conductivity, k_a	0.023W/mK
Shell thickness, h	10 mm
Chord length, C	1 m
Heat source depth, h_q	2 mm
Shell heat capacity matrix, $c_{p,m}$	960 J/kgK
Shell thermal conductivity, k_m	0.5 W/mK
Shell density	1800 kg/ m^3

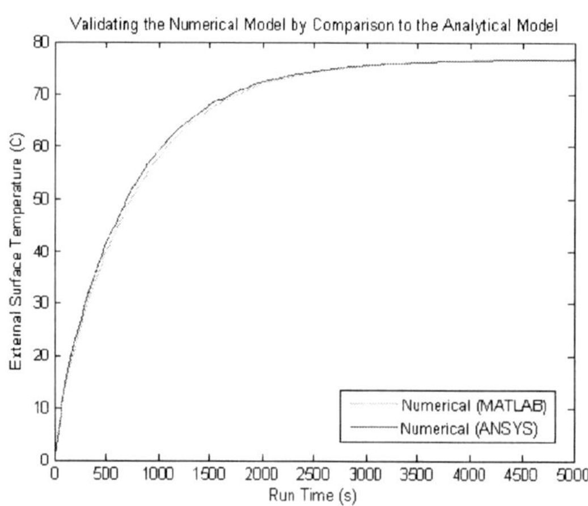

Figure 4: Validation of the developed evaluator module against ANSYS package.

3.1 Optimisation model

3.1.1 Design qualities and design variables

The first stage of the optimisation process is to identify the design qualities of the system. In this instance there are four main qualities to consider: Energy consumption, capital cost, maintenance cost and the system controllability.

The energy consumption is proportional to the power required to heat the surface. The capital costs are a function of material selection, material volume, and design for manufacture. Process control can be controlled be adding restraints at the design stage. Response time, number of activations, and nodal temperature control limits are all examples of process control.

Since the proposed anti-icing system is at conceptual design phase, energy consumption is chosen as the primary objective for the feasibility study while parameters corresponding to the system controllability are treated as constraints.

Here the design variables are the depth of the heat source below the surface h_q, and its strength, q.

3.1.2 Objective and constraints

A time period of 1 hour is chosen to allow compensation for variation in energy consumed during the start-up period. The objective function for energy consumed at a specific location is given by:

$$E = \int_0^{3600} q_i dt \quad (3)$$

In order to avoid the risk of de-lamination or reducing the structural strength of the composite material due to the thermal stresses and the thermal fatigue, the maximum nodal temperature at any location within the blade material is limited to 20°C.

To ensure the system operates as an anti-icing system capable of preventing ice build-up, the external surface temperature is constrained between 3°C and 4°C. The heat source is to be activated when the external surface temperature drops below 3°C and de-activates when the upper limit of 4°C is reached. The response time for the surface to reach the chosen temperature, the overshoot temperature, and the number of activations are three parameters affecting the performance and controllability of the system. The overshoot temperature is defined as the temperature that the system is permitted to reach beyond the upper and lower surface temperature limits.

While the heat flux mainly affects the response time and the overshoot temperature, the depth of the heat source has significant influence on the number of activations (Figure 5).

The heat flux tests illustrated in Figures 5(a) and 5(b) show that the system has a start-up response time of 41 seconds for a heat flux of q = 1000W/m^2, compared to 17 seconds with a q = 2000W/m^2 heat flux. The latent heat energy stored within the matrix material could be responsible for the increase in overshoot temperature when the greater heat flux is applied. Furthermore, the time taken from activation to de-activation of the power source once the system has stabilised is 15 seconds for the 1000W/m^2 system and 8 seconds for the 2000W/m^2 system. This indicates that although the power supplied is increased by a factor of 2, the 'on time' required for the system to maintain the set temperatures is decreased.

Figures 5(a) and 5(c) show that the number of activations is reduced from 20 to 8 by locating the heat source at a depth of 4mm. This can justified by Fourier's Law and observing that at a greater depth, a greater proportion of the

heat is directed away from the external surface and serves to provide the system with latent heat. When the heat source is located at a distance further from the external surface it is evident that more latent heat is stored within the matrix. When the latent heat is added to the supplied heat it results in a greater over shoot temperature which decreases the level of control over the system.

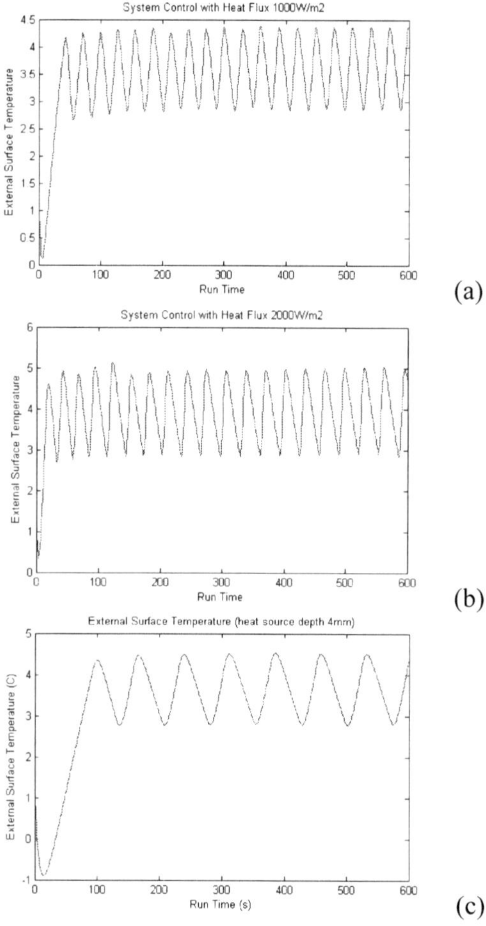

Figure 5: Surface temperature for (a) h_q = 2mm and q = 1000W/m^2; (b) h_q = 2mm and q = 2000W/m^2; and (c) h_q = 4mm and q = 1000W/m^2.

The response time, overshoot temperature and the number of activations are not independent parameters. Figure 6 shows the relationship between the number of activations and the response time. The relationship between response time and the overshoot temperature is also plotted in Figure 6. These results are corresponding to various heat flux values located at a depth of h_q = 2mm. The results in Figure 6 show that the relationship between the response and number of activations is inversely linear. According to this figure it is not possible to achieve a reduction in response time and a reduced number of activations simultaneously. A reduction in response time produces the undesired result of a system with a higher number of activations.

Figure 6 shows that a response time reduction from 300 seconds to 100 seconds does not yield a large increase in overshoot temperature. However, as the response time is reduced below 60 seconds, the over-shoot temperature increases rapidly.

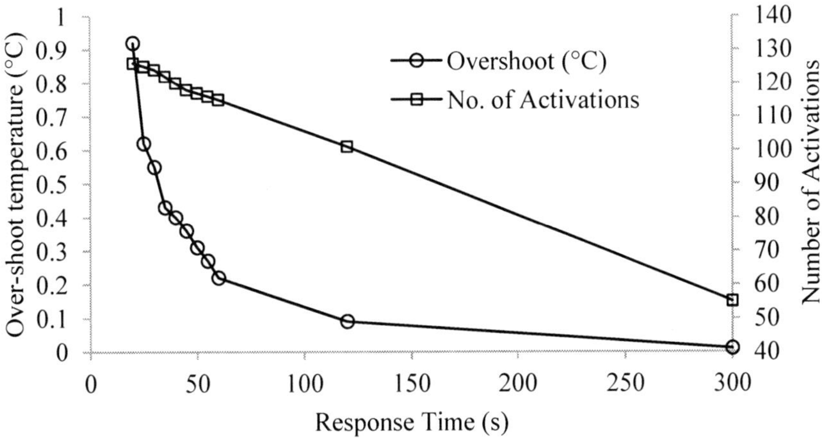

Figure 6: Number of activation and overshoot temperature versus response time.

Since not all control qualities can be achieved simultaneously, in view of Figure 6 a maximum response time of 60 seconds and a maximum overshoot temperature of 0.5°C is used for this study.

3.1.3 Genetic algorithm optimisation

The initial population is generated randomly between limits that are set based on observations of the results from the initial investigation of the finite difference model. A real number encoding is used with standard mutation and arithmetic crossover operators. The infeasible solutions obtained by crossover and mutation operators are rejected. The fitness is defined as:

$$fitness = 1 \Big/ \int_0^{3600} q_i dt \tag{4}$$

Using probability of crossover of 0.7, probability of mutation of 0.2 and a population size of 50, the optimum design variables obtained are: heat flux = 837W/m^2 at a location of 2 mm, which was the closest node to the surface. The results shown in Figure 7 give a maximum fitness of $fitness = 7.42 \times 10^{-4}$ equivalent to an energy consumption of $E = 1.35 kJ / m^2$. This is the energy required to maintain the surface temperature within the chosen limits.

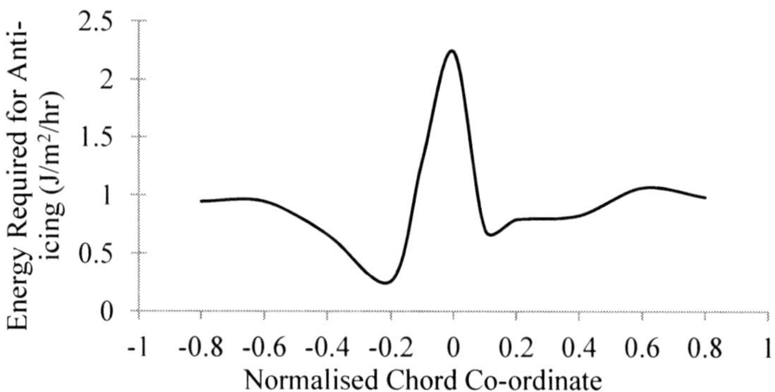

Figure 7: Energy required at normalised chord coordinates.

4 Structural integrity

Heated fibres within a polymer matrix-depending on the severity of heating, the length of the heating period and its variation trend- potentially may lead to fibre-matrix debonding and/or changes in fibre and matrix strengths and therefore an accelerated normal failure mode (delamination, fibre breakage, etc). While the effect of cyclic heating on the overall structural integrity of the blade needs comprehensive experimentations, in this paper the effect of heated fibres on the fibre-matrix debonding strength is examined. Figure 8 shows the test rig employed to measure the ultimate load leading to the breakage of the free fibre tows (uncovered by matrix). It was found that without heating, the uncovered fibres start breaking at a total force of about 280 N. Experiment was repeated by loading a similar specimen up to 260 N (close to the ultimate load) and then

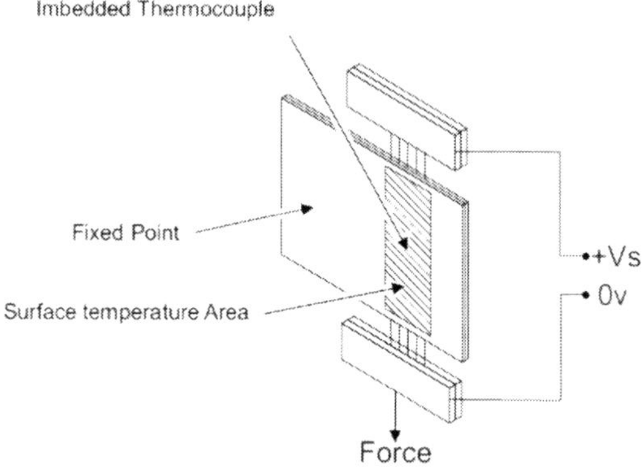

Figure 8: Schematic diagram of the test rig employed to measure the ultimate load leading to the breakage of the free fibre tows (uncovered by matrix).

the temperature of the fibre started to rise incrementally (about 5°C) to a maximum of 95°C (Figure 9). After each temperature rise, the temperature of the fibre was kept constant for about 30 minutes. No fibre-matrix debonding was observed during heating process. Increasing the force to 280 N again led to the breakage of the uncovered fibre. That is, while heating the carbon fibres up to 95°C does not affect the fibre-matrix debonding strength, the rapture of the uncovered fibre is the dominant failure mode.

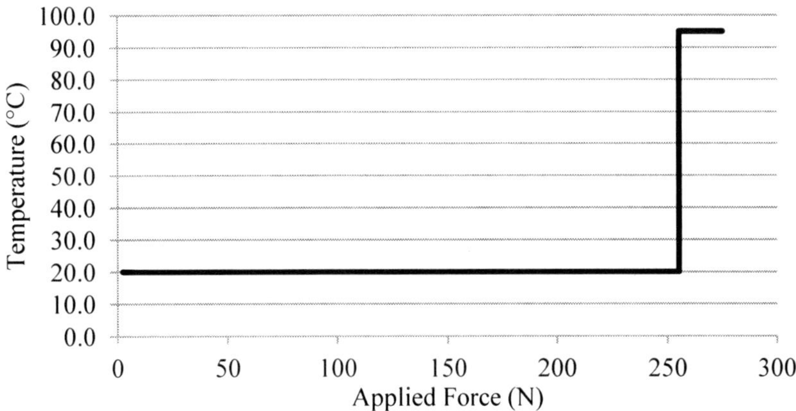

Figure 9: Incremental rise of fibre temperature and applied force.

5 Summary and conclusions

The proof of the concept of using carbon fibres implanted in a glass fibre composite as heating elements for de-icing and anti-icing systems was demonstrated. Employing a genetic algorithm optimisation tool with a one-dimensional finite difference heat transfer code, the optimum depth and the optimum heating power of the implanted carbon fibres for a typical wind turbine blade at a general span location were obtained. A parametric study was carried out to find the relationships between the system control factors. The genetic algorithm optimiser module was set with constraints chosen from the observations in the parametric study. Functioning as an anti-icing system, it was observed that the response time should be maintained above 60 seconds to prevent high overshoot temperatures. It was also showed that an increase in the heat source depth impacts the control of the system adversely because of an increase in the overshoot temperature. Similarly, an increase in the heat flux yields a significant increase in the overshoot temperature and the number of system activations. However, the response time decreases when the heat flux is increased. Finally, the effect of implanted carbon fibres as heating elements on the fibre-matrix debonding strength of the composite specimen was examined. It was shown, while heating the carbon fibres up to 95°C does not affect the fibre-matrix debonding strength, the rapture of the uncovered fibre is the dominant failure mode.

Acknowledgements

The author would like to thank the Northumbria University graduates Dale Whitehead, Benjamin Smith and James Brooke for their contribution to this research.

References

[1] Laakso, T., Talhaug, L., Vindteknik, K., Ronsten, G., Horbaty, R., Baring-Gould, I., Lacroix, A., Peltola, E., Wind Energy Projects In Cold Climates, Technical Research Centre, Biologinkuja, 2005.
[2] Dalili N., Edrisy A. and Carriveau R., A review of surface engineering issues critical to wind turbine performance, Renewable and Sustainable Energy Reviews, 13, pp. 428-438, 2009.
[3] Athanasopoulos N. and Kostopoulos V., Prediction and experimental validation of the electrical conductivity of dry carbon fiber unidirectional layers, Composites: Part B, 42, 1578-1587, 2011.
[4] Wang, X., Naterer, GF., Bibeau, E., Convective Heat Transfer From A NACA Airfoil At Varying Angles Of Attack, Journal of Thermophysics and Heat Transfer, 22(3), pp. 457-463, 2008.

Electrical properties of fiber and carbon nanotube reinforced polymer composites

H. Estrada[2], J. Trovillion[1], L. S. Lee[2], M. Tusz[1], A. Kumar[1] & L. D. Stephenson[1]
[1]*Construction Engineering Research Laboratory, US Army Engineer Research and Development Center, USA*
[2]*School of Engineering and Computer Science, University of the Pacific, USA*

Abstract

Heightened concerns for electromagnetic interference (EMI) in defense applications have led to developments of multi-functional materials that can provide structural performance while shielding electronic components from electromagnetic waves. Since EMI shielding effectiveness is highly dependent on electrical properties, the goal of this paper is to characterize the conductivity, permittivity, and resistivity of fiber and Carbon Nanotube (CNT) reinforced polymer composites (nanocomposites). The key component of the nanocomposite is a non-woven textile formed using synthetic carbon nanofibers. The material is proprietary and was obtained from Nanocomp Technologies, Inc (www.nanocomptech.com).Electrical testing of specimens was conducted using a low frequency impedance analyser in order to measure the conductivity and resistivity of the nanocomposite for frequencies ranging from 0.1 kHz to 1000 kHz. The nanocomposite material's EMI shielding effectiveness is estimated to be approximately 46 dB for a resistivity of 0.88 Ohms/square. Given the material's average area density of 12.8 g/m^2, these results can be considered promising since the Nanocomp material can be incorporated into conventional composites intended for load bearing applications in order to add EMI shielding to the material with little additional weight.
Keywords: carbon nanotube, composites, electrical properties, electromagnetic interference shielding.

WIT Transactions on Engineering Sciences, Vol 77,
www.witpress.com, ISSN 1743-3533 (on-line)
doi:10.2495/MC130271

1 Introduction

The use of carbon nanotubes has great potential in a number of military and civilian applications; particularly when combining CNTs with other materials to develop multifunctional structural materials that integrate multiple attributes synergistically to perform electrical, electromagnetic, thermal, and impact resistant functions, while maintaining minimum strength and stiffness requirements for a given structural application. In this study we concentrate on applications requiring structural performance and shielding from electromagnetic interference (EMI). These two functions can easily be enhanced by embedding CNTs in fiber reinforced polymer materials to produce multifunctional components. Since their discovery by Iijima [6] in 1991, CNTs have been investigated as possible reinforcing materials in polymer based composites [2]. CNTs have been shown to possess many desirable material properties, including high electrical conductivity, high tensile strength, high flexibility, high modulus, and high thermal conductivity. These properties inherent in CNTs allow for significant improvement in mechanical, electrical, and thermal properties of polymers and composites when these materials are loaded with CNTs. However, a number of obstacles exist towards full implementation of CNTs as reinforcing materials. The challenges include a limited understanding of the processes for combining CNTs with composite materials, the need for characterization of the interaction between primary reinforcement (e.g., carbon or glass fibers) and CNTs in a polymer matrix, and significant variations in the properties (physical and chemical) and purity of commercially available CNTs due to their many sources and fabrication techniques; an even larger obstacle to widespread adoption of CNTs is high material cost, which is likely to decrease as volumetric demand increases. Cost constraints aside, once issues related to material quality, as well as interaction, dispersion, and other characteristics have been adequately addressed, CNT additives can be used to enhance electromagnetic shielding of composite materials while improving available strength and stiffness.

1.1 Electromagnetic interference shielding

Electromagnetic interference (EMI) is partly caused by rapidly changing voltages and currents in various electrical and electronic devices. Ambient EMI can be considered electromagnetic pollution, most of which consists of spurious, conducted or radiated signals of electrical origin, such as radiation emitted from telecommunications equipment. This EM pollution can adversely alter the operation of sensitive circuits in solid state electronic components found in many defense and civilian instruments. One means of properly protecting this equipment is to utilize materials capable of EMI shielding as containment or housing. In this paper we investigate CNT nanocomposites as potential EMI shielding materials.

There are three mechanisms involved in EMI shielding: reflection, absorption, and multiple reflections of EMI radiation. The reflection mechanism requires materials to possess mobile charge carriers (electrons or holes) to

www.witpress.com, ISSN 1743-3533 (on-line)

interact with incoming electromagnetic radiation waves; that is, electrically conductive materials, which are only required to have a low level of conductivity (or its inverse, resistivity) to provide effective shielding. A volume resistivity of the order of 1.0 ohm-cm is typically sufficient [4] for effective shielding. The absorption mechanism requires the material to have electric dipoles (materials with high dielectric constant values) and/or magnetic dipoles (materials with high magnetic permeability values) to interact with incoming electromagnetic radiation waves, and is highly dependent on the thickness of the shield. The multiple reflections mechanism requires material to have large surface or interface areas where multiple reflections of radiation waves can occur. In all three mechanisms, shielding is achieved by electromagnetic radiation losses, which are controlled by a material's electrical conductivity and/or magnetic permeability.

Polymers and fibers (glass or carbon) generally contain negligible concentrations of mobile charge carriers or electric dipoles. This lack of mobile charge carriers makes polymer based composites non-conductive (electrically insulating) and thus transparent to electromagnetic radiation. In order for fiber reinforced polymer (FRP) composites to achieve any level of EMI shielding capacity, they must be made electrically conductive by incorporating (loading) intrinsically conductive fillers, such as carbon black, metal particulates, or nickel-coated short carbon fibers [5]. An additional potential polymer filler is CNTs; more specifically, multiwall CNTs (MWCNTs). Previous studies have shown that MWCNT polymer nanocomposites' shielding mechanism is primarily by absorption [7]. Also, the multiple-reflection mechanism's shielding effectiveness is controlled by the shield thickness; if the shield is thicker than the so called "skin depth" the contribution from this mechanism can be ignored [1]. The sum of electromagnetic radiation losses due to reflection, absorption and multiple reflections constitutes EMI shielding effectiveness. These losses are difficult to characterize for CNT nanocomposites since they depend on many factors that are difficult to control; this has been reported in a number of studies where shielding effectiveness values display large variability in similar materials [3]. Some of the more critical factors include processing synthesis, purity of CNTs, and dispersion of CNTs. There are two primary approaches used to predict EMI shielding effectiveness of CNT nanocomposites; one is experimental and one semi-empirical. In both cases, material conductivity plays a key role in EMI shielding effectiveness.

Intuitively, conductivity of polymer based composites should increase as conductive filler loading (filler concentration) is increased. However, full conductivity does not happen instantaneously, a critical filler concentration is needed for the material to display a dramatic increase in conductivity – essentially the point at which the material is converted from an insulator to a conductor. This critical concentration is known as the electrical percolation threshold concentration, or the percolation threshold. The percolation threshold is the point at which the filler particles form a continuous 3-D conductive network (percolating networks) throughout the resin matrix. These percolating networks allow electrons to tunnel from one filler particle to another in order to overcome the inherent high resistance of the resin matrix. The formation of percolating networks depends primarily on filler's intrinsic conductivity, the particles' geometric aspect-ratio and the distribution of the particles.

Since CNTs have high conductivity and high aspect-ratio, the resulting nanocomposites can be made conductive at a very low percolation threshold provided that the CNTs are uniformly distributed. The conductivity of nanocomposites is also affected by the type of polymer as well as CNT type, surface functionalization, and synthesis method [3]. The polymer type is typically chosen to address a function other than EMI shielding (e.g., structural performance or durability), or it can be selected based on availability and overall cost. CNTs tend to agglomerate because of inherent electrical charge, which adversely affect the uniform dispersion of CNTs throughout the resin materials. One approach used to provide uniform CNT spatial distribution is to chemically functionalize (surface chemical treatment) CNTs. However, this technique disrupts the extended π-conjugation of CNTs and reduces electrical conductivity. The use of an innovative approach to assemble CNTs into a non-woven textile form to uniformly distribute them in polymer composites is investigated in this study.

CNT based nanocomposites have a number of advantages over conventional metal-based and other EMI shielding materials; including light weight, corrosion resistance, flexibility, and ease of processing. In fact, the use of CNTs can lead to a significant reduction of filler loading required to achieve a desired level of EMI shielding. For instance, percolation thresholds in the range of 5–15% volume concentration are typical for carbon black filler, and are even higher for dispersed metal particle fillers, 10–30% [7]. By comparison, 0.3% weight percolation thresholds for CNTs were found by Trovillion *et al.* [8], which clearly have a great potential to reduce weight and overall costs if incorporated in FRP composites.

There are two primary approaches used to predict EMI shielding effectiveness of CNT based nanocomposites, one is entirely experimental and one semi-empirical. Both are discussed in the next section. The experimental approach is currently underway using a panel that was fabricated using Nanocomp material sandwiched between several glass fiber mats and EMI shielding effectiveness is measured directly. The panel has not been tested to date, but is expected to be tested in the future. In this paper, EMI shielding effectiveness is determined by experimentally measuring conductivity and utilizing a basic theoretical relationship to calculate shielding effectiveness.

1.2 Characterization of EMI shielding effectiveness

Direct experimental measurement of EMI shielding effectiveness is based on standard IEEE 299 – *Draft standard method for measuring the effectiveness of electromagnetic shielding enclosures*, which is specifically intended for enclosures with at least one dimension greater than 2 meters. This standard replaces MIL-STD 285 (1988) and is valid for a frequency range of 9 KHz to 18 GHz and extendable to 50 Hz and 100 GHz, respectively. There is also an ASTM standard that is similar to IEEE 299, ASTM D4935-10 – *Standard Test*

Method for Measuring the Electromagnetic Shielding Effectiveness of Planar Materials. EMI work summarized here is based on a procedure from IEEE 299, with a faraday enclosure and aperture that accommodates samples having dimensions of at least 25 inches (635 mm) by 50 inches (1270 mm). The primary objective of this test is to determine a panel's contribution to the shielding effectiveness of the faraday cage, assuming ideal behavior, by performing reference and effective measurements, the difference of which is the shielding effectiveness.

Traditionally, EMI shielding effectiveness (or signal attenuation) has been measured using a logarithmic scale, expressed in decibels (dB). This is an effective way to characterize quantities that vary over several orders of magnitude. Using this scale, shielding effectiveness (SE) can be expressed as,

$$SE = 20\log_{10}\left(\frac{E_I}{E_T}\right) \tag{1}$$

where, E_I is the incident electric field (reference reading based on open enclosure) and E_T is the transmitted electric field (shielded enclosure). Values of SE in the range of 20 dB to 80 dB correspond to attenuation of the incident electric field by 90% to 99.99%; i.e., SE = 20 dB corresponds to an attenuation of 90%, SE = 40 dB corresponds to an attenuation of 99%, SE = 60 dB corresponds to an attenuation of 99.9%, and SE = 80 dB corresponds to an attenuation of 99.99% of the incident electromagnetic plane wave field. Electromagnetic waves with frequency in the range of 30 MHz to 100 MHz requires 35 dB to 45 dB for attenuation; and for critical applications (such as military applications), the attenuation requirements may be as high as 80 dB.

When the electric field measurements have been made using a logarithmic scale, the shielding effectiveness is computed from the reference and shielding measurements (that is, Equation (1) may be written as),

$$SE = |E_I|(dB) - |E_T|(dB) \tag{2}$$

where, $|E_I|(dB)$ is the reference measurement, with no panel, in dB and $|E_T|(dB)$ is the shielding measurement, with panel, in dB. In both cases, the measurements are performed by placing one antenna on the outside of the enclosure (the receiving antenna) and another antenna within the enclosure (the transmitting antenna). A signal is emitted from the transmitting antenna and is measured at the receiving antenna. The difference in performance from a reference (i.e., open aperture) to an experimental state (i.e., closed aperture with the panel being evaluated) is used to calculate the EMI shielding effectiveness. A schematic of the open aperture and closed aperture setup is shown in Figure 1. First, a reference (calibration) measurement is recorded with an open aperture (Figure 1(a)), which establishes a baseline measurement. Then, a measurement is taken again with the shielding panel closing the aperture (Figure 1(b)).

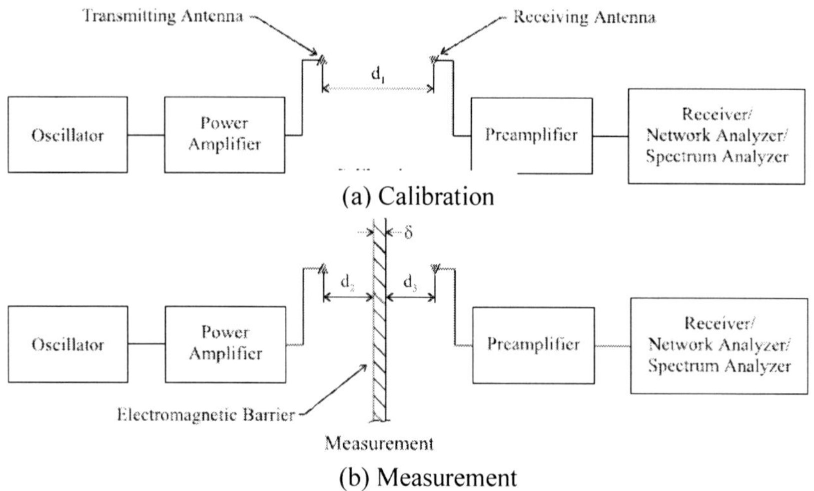

Figure 1: Schematic of experimental setup: (a) calibration and (b) measurement.

The panels of interest should be designed for high altitude electromagnetic pulse (HEMP) protection, which are typically required to operate in the range of 0.1–10,000 MHz. For this frequency sweep, an antenna configuration that includes low, mid, and high frequencies is required. A list of antennae and corresponding frequencies is shown in Figure 2. The test frequencies, type of measurement, antennas used for measurement, and antenna spacings (d_1, d_2, and d_3) are shown in Figure 1; and are listed in Table 1.

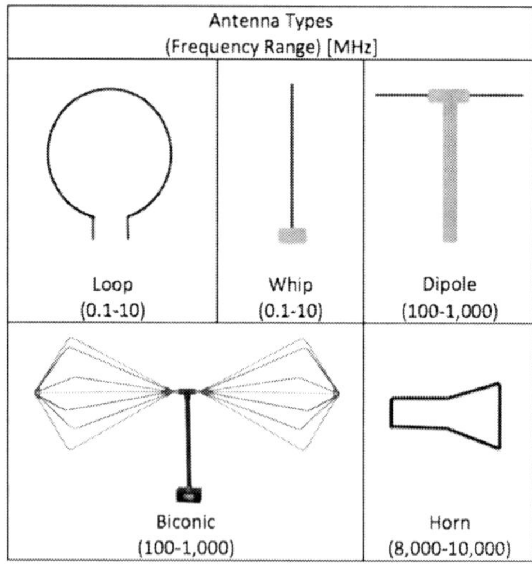

Figure 2: Types of antennae required for panel testing.

Table 1: Testing rubric for open and closed aperture tests.

Frequency (MHz)	Measurement Type	Antenna Type (transmit/receive)	d_1 (m)	d_2 and d_3 (m)
0.1 1 10	H-field, coplanar	Loop/Loop	0.6	0.3
0.1 1 10	E-Field, vertical	Whip/Whip	0.6	0.3
100 1000	Plane wave, horizontal	Biconic/Dipole	3	1
8000 9000 10000	Plane wave, horizontal	Horn/Horn	2.25	1.125

In order to properly seal the gap between aperture and panel, a metallic gasket is used. Also, copper tape is applied around the testing panel to ensure contact between the outer edge of the panel and the innermost conductive gasket.

Besides direct experimental measurement, EMI shielding effectiveness can be determined semi-empirically using a measure of a material's conductivity as discussed previously and emphasized in this paper. With the material's resistance, EMI shielding effectiveness can be obtained using the following basic equation [9],

$$SE = -20\log\left|\frac{2R_A\eta_0}{2R_A\eta_0 + \eta_0^2}\right| \tag{3}$$

where, R_A is the sheet resistance in Ohms/square(sheet resistance units) and η_0 is the wave impedance of free space, which is approximately equal to 377 ohms. This relationship is used in this study to approximate the EMI shielding effectiveness of the nanocomposite. In fact, using the electrical properties reported by Nanocomp Technologies, Inc (www.nanocomptech.com) for their synthetic carbon nanofibers in a non-woven textile form, average area density of 12.8 g/m^2 and a resistivity of 0.88 Ohms/square, a single sheet could theoretically produce 46 dB EMI shielding effectiveness. As previously discussed, this corresponds to well over 99% attenuation of the electromagnetic waves (SE = 40dB corresponds to an attenuation of 99%).

1.3 Materials

In addition to the nanocomp sheet, the samples included epoxy resin and fiber glass fabric. To achieve high quality nanocomposite specimens, a low viscosity polymer was used. For this investigation, SC 15 epoxy was selected due to its high strength, low viscosity, and extensive use in structural composite applications. Also, this epoxy system is compatible with the CNT sheet since previous studies have shown that CNTs develop good adhesion and wettability with this epoxy. This resin is a low viscosity epoxy resin, specifically developed for the vacuum-assisted resin transfer molding (VARTM) manufacturing processes – a manufacturing process that has been proven cost-effective for applications requiring a low-volume and high-degree of quality control. It is a two phase toughened, amine cured epoxy resin system that exhibits good cross-linking at room temperature. The manufacturer's suggested stoic mixture ratio of resin to hardener is 100:30. This resin is intended for structural and ballistic applications or any application that requires good damage resistance. It allows for a room temperature cure or a post-cure for four hours at 200°F to achieve maximum properties; including a modulus of elasticity of 390-ksi, elongation of 6.0%, tensile strength of 9.0-ksi, and dry glass transition temperature of 180°F. E-glass fiber reinforcement was obtained from CPIC Fiberglass (www.cpicfiber.com) in woven rovings having a unit weight of 800 g/m^2 and width of 1.27 m. The E-glass woven roving was selected for its excellent wet-out and impregnation characteristics, as well as high mechanical strength and adjustable construction. In general, E-glass fibers have mechanical properties that range from 69–72 GPa for longitudinal modulus, 3.45–3.9 GPa for tensile strength, and have fiber densities from 2570–2600 kg/m^3.

2 Sample preparation and testing

The CNT sheet was used to fabricate nanocomposite panels that were cut into specimens used for electrical testing, which was performed to determine the impedance and resistivity (conductivity) of the material. To produce high-quality specimens the VARTM technique was employed. In order to manufacture the composite panels, the plain-weave E-glass fabric was cut; six layers of the fabric were stacked in alternating warp-fill directions, then the nanocomp film was placed on top and an additional six layers of E-glass fabric were stocked on top. The setup was placed in a vacuum bagging system and infiltrated with SC-15 epoxy. The nanocomposite was then allowed to cure at room temperature for 24 hours, after which the panel was removed from the vacuum bag and post-cured at 100°C for one hour.

In this study, two fiber-reinforced panels were fabricated – one panel for electrical testing and one larger (dimensions 1422 mm by 813 mm or 56 in by 32 in) panel for EMI shielding effectiveness testing (the test has not been completed to date). A schematic illustration of the VARTM setup is shown in Figure 3.There were three challenges in manufacturing these panels: the static charges during layup caused the nanocomp sheet to crease, it was difficult to develop electrical contact (conductivity) in the panel because the resin is electrically isolating, and the VARTM process had to be modified because the nanocomp is impervious. The nanocomp creasing did not result in loss of electrical conductivity. The second issue was solved by painting the terminals with silver paint after polishing the sample ends using 400-grit sandpaper. For the large panel, this option was impractical and instead, a conductive medium was provided using copper tape along the edges of the nanocomp sheet before it

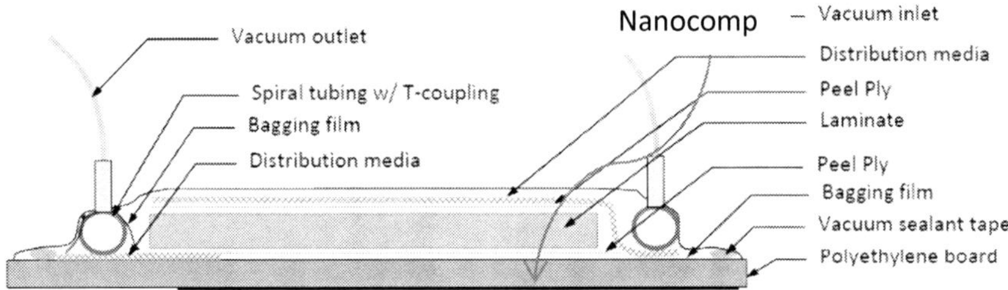

Figure 3: Schematic side view of VARTM fabrication setup.

was embedded in the E-glass layers and the composite impregnated with resin. This copper tape around the edges allows the nanocomp film to come in contact with the external enclosure of the EMI shielding testing equipment and ensures conductivity around the borders of the panel. This will also reduce high-frequency signal leakage through the EMI test aperture. The third issue required the panel to be infused two separate times one for the top six layers and the second time for the bottom six layers.

The CNT nanocomposite electrical properties were determined experimentally using impedance testing of three samples manufactured by VARTM as previously discussed. After curing, the first panel was removed and cut into three specimens of nominal dimensions 20 mm long, 10 mm wide, and 1 mm thick; the actual dimensions were taken at three points and averaged. The ends of each specimen were coated with silver paint to ensure connectivity with the impedance analyzer during measurements (see Figure 4).

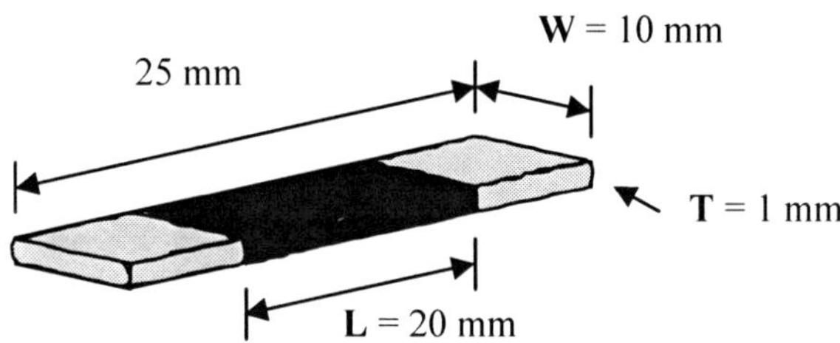

Figure 4: Sample dimensions and preparation procedure for impedance testing.

Electrical testing of specimens was conducted using an HP 4192A low frequency impedance analyzer, which is designed to measure properties such as electrical impedance, phase angle, resistance, conductance, inductance, and capacitance. Impedance (resistance in AC circuits) and resistivity were the only properties investigated since they are considered the primary factors that influence EMI shielding properties. The properties were measured for frequencies ranging from 0.1 kHz to 1000 kHz. At low frequencies, the material primarily reflects signals back to the outside, which is done separately for electric and magnetic fields. For EMI shielding of magnetic fields, the material needs to have high impedance, while for EMI shielding of electric fields the impedance needs to be low. At higher frequencies, the skin effect is more pronounced in EMI shielding, which means the thickness of the material absorbs the majority of the signal. This consequently requires high conductivity.

3 Results

Resistivity (conductivity) and impedance are the most critical electrical properties considered when characterizing EMI shielding effectiveness and the primary parameters investigated in this study. This section summarizes the results of a nanocomp CNT-film nanocomposite. The nanocomp material was used because it is easier to embed CNTs in sheet form than to mix them into the resin. The results for resistivity and impedance for nanocomp fabric laminate samples are shown in Figures 5 and 6, respectively. The figures include the average of a series of three tests and corresponding standard deviation, which is shown as an error bar. The resistivity varied with the orientation of the samples in the impedance analyzer test rig (up to 62% as shown by the error bar). However, when averaged, the resistivity values converged to results close to those of more repeatable samples.

Comparing these results to other CNT-loaded composite results with 0.8 wt % non-functionalized MWNT SC-15/SC-780/SC-1 experiments, the nanocomp film exhibits remarkable conductivity, close to that of copper and other highly conductive materials. The results indicate that at all frequencies the average impedance and resistivity are comparable; the low resistivity means that the material is highly conductive. This implies that the nanocomposite is an effective medium for shielding electrical and magnetic fields.

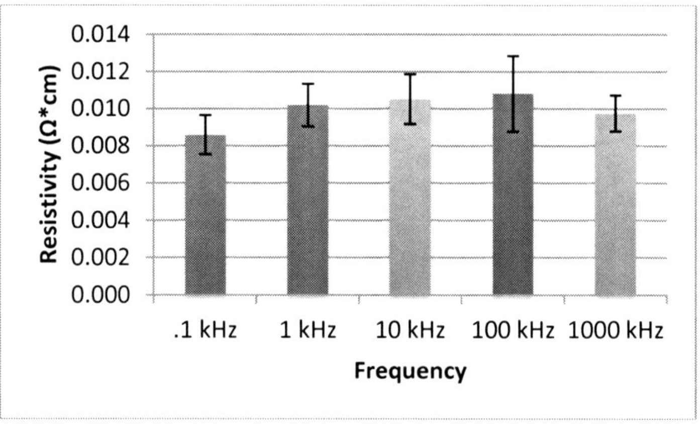

Figure 5: Resistivity of nanocomp composite panel.

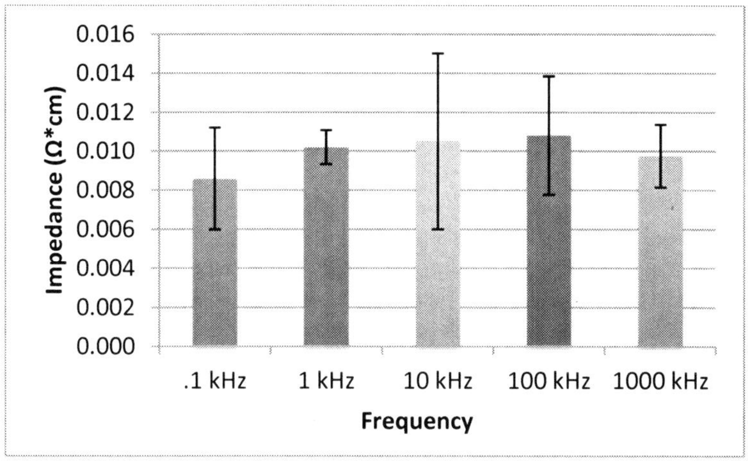

Figure 6: Impedance of nanocomp composite panel.

4 Conclusions

This paper presents an overview of the methodology for estimating the EMI shielding effectiveness of nanocomposites. The work emphasizes the experimental measurement of conductivity in order to calculate shielding effectiveness. The paper centers on the development of a material system with electrical properties that can effectively shield EMI waves, which requires a material to be electrically conductive. However, most fiber reinforced polymer composites are not electrically conductive and require that a separate conductive material be integrated within the material. Electrically conductive particulate materials such as CNTs can make epoxy conductive even at small concentrations since electron tunneling occurs even when particles are not in contact. The loading (proportion) of the conductive particulate material required to create electron tunneling and thus make the mixture conductive is the percolation threshold. CNTs only require a small fraction to attain percolation– as little as 0.3 weight percent [8]. However, CNTs must be properly distributed in the epoxy to create tunneling between particles for electrons to travel. This paper presents detailed experimental procedures of the specimen fabrication using nanocomp sheets to integrate the CNTs into the composite. Several conclusions regarding the conductivity of nanocomp based composites can be drawn from the results of this study:

1. One sheet of nanocomp can produce conductivities comparable to that of copper, and can provide up to 99% EMI shielding.
2. Fabrication of nanocomposites using nanocomp is rather complicated and requires modification of traditional manufacturing techniques, such as VARTM because the nanocomp sheet is impervious to the resin material.

3. Because the resistivity varied with the orientation of the samples in the impedance analyzer test rig (up to 62%), a standardized testing procedure needs to be developed.
4. Lastly, to ensure redundancy with respect to creases and conductivity of the nanocomp sheets in contact with the external enclosure, two layers of nanocomp sheets are recommended.

While these results show that nanocomposites have great promise for use in multifunctional applications, additional work needs to be completed in order to validate the behavior of practical infrastructure components and systems manufactured using these innovative materials. The primary issue is validation of EMI shielding effectiveness using experimental testing.

References

[1] Al-Saleh, M. H. and Sundararaj, U., Electromagnetic interference shielding mechanisms of CNT/polymer composites. *Carbon*, **47**, pp. 1738-1746, 2009.

[2] Chang, T. E., Kisliuk, A., Rhodes, S. M., Brittain, W. J., and Sokolov, A. P., Conductivity and mechanical properties of well-dispersed single-wall carbon nanotube/polystyrene composite, *Polymer,* **47**, pp. 7740-7746, 2006.

[3] Choudhary, V. and Gupta, A., Polymer/carbon nanotube nanocomposites (Chapter 4). *Carbon Nanotube – Polymer Nanocomposites*, ed. Yellampalli, S., InTech, intechweb.org, Rijeka, Croatia, 2011.

[4] Chung, D.D.L., Review electromagnetic interference shielding effectiveness of carbon materials. *Carbon*, **39**, pp. 279-285, 2001.

[5] De, S. K. and White, J. R., (eds). *Short fiber-polymer composites*. Woodhead Publishing Limited: Cambridge, 1996.

[6] Iijima, S., Helical microtubules of graphitic carbon. *Nature*, **354**, pp. 56-58, 1991.

[7] Mamunya, Y., Carbon nanotubes as conductive filler in segregated polymer composites - electrical properties (Chapter 9). *Carbon Nanotube – Polymer Nanocomposites*, ed. Yellampalli, S., InTech, intechweb.org, Rijeka, Croatia, 2011.

[8] Trovillion J., Estrada, H., Lee, L. S., Tusz, M., Kumar, A., and Stephenson, L. D. Electrical properties of carbon nanotube loaded resins, to be presented at the 2013 SAMPE Annual Conference in Long Beach, CA, May 2013.

[9] Vasquez, H., Espinaza, L., Lozano, K., Foltz, H. and Yang, S., Simple device for electromagnetic interference shielding effectiveness measurement, *Proceedings of the IEEE Transactions of Electromagnetic Compatibility; Practical Papers, Articles and Application Notes.* **220**, pp. 62-68, 2009.

Material homogenization of metal-ceramic composites by NTFA

E. Schnack, Y. Zhu & A. M. Rahman
Institute of Solid Mechanics, Karlsruhe Institute of Technology (KIT), Germany

Abstract

Metal matrix composites (MMC) are important lightweight materials because of their excellent mechanical properties. In this work, the 3D model microstructure of MMC with periodic spatial discretization is produced and used as the virtual material for a homogenization process. With an efficient description of the macroscopic thermo-mechanical behavior of metal-ceramic composites, the non-uniform transformation field analysis (NTFA) is adopted and extended by reformulation of the underlying equations. This method is an order reduction technique specifically designed for homogenization problems with micro-mechanical motivation. The implementation of NTFA is based on the finite element method. The homogenized material model is implemented into ABAQUS for structural analysis. By comparing with full-field simulation, numerical results at integration point level and in terms of structural problems highlight the efficiency of NTFA for three-dimensional homogenization problems.
Keywords: metal matrix composites, non-uniform transformation field analysis, homogenization.

1 Introduction

The objective of this study is to achieve the nonlinear numerical multi-field homogenization of particles reinforced light weight metal-ceramic composites. Due to the complexity of the microstructure, two-scales problem is considered: The micro-scale with the dimensions in the range of microstructure $(x, \sigma, u, \varepsilon)$ and the macro-scale in the range of the component size $(\bar{x}, \bar{\sigma}, \bar{u}, \bar{\varepsilon})$. Assuming small deformation, geometrically linear behavior and material continuity on small

WIT Transactions on Engineering Sciences, Vol 77, © 2013 WIT Press
www.witpress.com, ISSN 1743-3533 (on-line)
doi:10.2495/MC130281

scale (no pore and crack), the macroscopic and the microscopic strain-stress field are related by the averaging operator:

$$\begin{array}{ccc} Macro & \bar{\sigma}=\langle\sigma\rangle,\ \ \bar{\varepsilon}=\langle\varepsilon\rangle & Micro \\ \bar{x},\bar{\sigma},\bar{u},\bar{\varepsilon} & <------> & x,\sigma,u,\varepsilon \end{array} \tag{1}$$

Assuming that two scales are coupled by periodic boundary conditions, then the homogenization problem persists in the following simultaneous equations:

$$(\bar{P}):\begin{cases} div(\bar{\sigma})=0 \quad in \quad \bar{\Omega} \\ \bar{\sigma}\bar{n}=\bar{(t)}=t^{*} \quad on \quad \bar{\Gamma}_t \\ \bar{u}=u^{*} \quad \bar{\Gamma}_u \end{cases} \quad (\bar{P}):\begin{cases} div(\sigma)=0 \quad in \quad \Omega \\ \langle\varepsilon\rangle=\langle\bar{\varepsilon}\rangle \end{cases} \tag{2}$$

From the scale point, the aim of this work can be re-defined as developing a suitable method to determine the macroscopic thermo-mechanical behavior with respect to the physical processes at micro level. In this work, the artificial materials are generated which will be used to mimic the real metal ceramic composites. This artificial composite are parametrized by random variables such that the statistical material properties is close to that of real metal ceramic composites.

The metal ceramic composites show a pronounced heterogeneity at microscale that decides the macroscopic constitutive behavior. In order to exam the relationship between these two scales, thermo-mechanical homogenization should be investigated. The homogenization problem consists of the transmission of the adopted constitutive equations from micro scale to the macro scale. These constitutive equations are used to determine the elastic, inelastic and thermal behavior. By selecting appropriate methods, the information loss induced by the transition can be minimized. A choice of these methods is Nonuniform Transformation Field Analysis (NTFA) [1]. It is a method for the prediction of the behavior of heterogeneous materials in the presence of physical nonlinearity but in a geometrically linear context. It extends the Transformation Field Analysis (TFA) [2] in which phase-wise constant plastic strain fields were used in order to approximate the effective behavior of microheterogeneous materials. The NTFA was first mentioned by Michel and Suquet [3] and described in detail by Michel [1] for materials with isotropic constitutive properties on the microscale. In order to achieve realistic predictions of the structural behavior, the consideration of thermo-mechanics interaction and damage is often required. The NTFA is a numerical homogenization procedure with micromechanical motivation. More precisely, the homogenized (macroscopic) behavior of the micro-heterogeneous material is entirely defined based on approximation of microscopic information, where the solution is obtained using FEM.

2 NTFA method and implementation

2.1 Thermo-mechanical interaction

The basic idea of the NTFA is to approximate the space-time dependent inelastic strain using a N-dimensional basis of spatially heterogeneous fields and time-dependent coefficients according to:

$$\varepsilon^p\left(x,t\right) \approx \sum_{\alpha=1}^{N} \zeta_\alpha\left(t\right) \mu_\alpha(\mathrm{x}) \tag{3}$$

Here ε^p is inelastic strain, ζ_α time dependent coefficient and μ_α spatial heterogeneous field. By this equation the number of unknowns is reduced to N. Thereby a significant reduction of the computational cost can be achieved with respect to full-field simulations containing hundreds of thousands of degrees of freedom, while preserving a good accuracy of the macroscopic stress field. The main challenges of NTFA are (i) identifying the inelastic model μ_α and (ii) identifying the vector coefficients ζ_α. The inelastic model μ_α is described by Karhunen-Loeve decomposition [4]:

$$\mu_\alpha = \frac{\sum_{i=1}^{N} v_i^\alpha \varepsilon_i^p}{\left\langle \left\| \sum_{i=1}^{N} v_i^\alpha \varepsilon_i^p \right\| \right\rangle} \tag{4}$$

In the following, the formulation of ζ_α will be introduced. Assume the linearity of stress tensor σ with respect to $\varepsilon - \varepsilon^p$. Microscale quasi-static conditions are also assumed in order to describe the linear momentum. The elastic fourth order stiffness tensor in the microscopic scale is described as follows.

$$div\left[\mathbb{C}\left[\varepsilon(t,x) + (\theta-\theta_0)\beta(t,x) - \sum_{\alpha=1}^{N}\zeta_\alpha(t)\mu^\alpha(x)\right]\right] = 0 \tag{5}$$

$$\langle\varepsilon(t,x)\rangle = \bar{\varepsilon}(t) \tag{6}$$

Eqn. (5) and eqn. (6) have to be satisfied for arbitrary mode activity coefficients $\hat{\zeta}\left(t\right)$, arbitrary macroscopic strains $\bar{\varepsilon}\left(t\right)$ and for any temperature $\theta = \bar{\theta}$ in the considered setting. There are two steps in solving eqn. (5) and eqn. (6).

1st step: Zero mode activity is assumed, i.e. $\zeta_\alpha = 0, (\alpha = 1, ..., N)$. Thereby the linear thermo-elastic load cases are recovered. In general, if a d-dimensional set of independent boundary conditions $\mathcal{B}^\alpha$ is considered in a linear partial differential equation and the solutions to the individual problems are denoted u^α, then for any

linear combination of the boundary conditions:

$$\mathscr{B}_* = \sum_{\alpha=1}^{d} k_\alpha \mathscr{B}^\alpha, \quad k_\alpha \in \mathbb{R} \tag{7}$$

The displacement by superposition is expressed as:

$$u_*\left(\hat{k}, x\right) = \sum_{\alpha=1}^{d} k_\alpha u^\alpha \tag{8}$$

The linear thermo-elastic properties are determined based on elastic eigenstress problem. Here the case of zero prescribed strain and the thermal eigen stress field

$$\tau_\theta\left(x\right) = \beta_\theta\left(x\right)\Delta x_0 \tag{9}$$

Thereby induced linear-elastic eigenstress problem is

$$div\left(\mathbb{C}\left[\varepsilon_\theta + \tau_\theta\right]\right) = 0, \qquad \langle\varepsilon_\theta\rangle = 0 \tag{10}$$

The localization rule is as follows for displacement, strain and stress tensor with respect to a prescribed temperature increment $\Delta\bar{\theta} = \bar{\theta} - \theta_0$ from the reference temperature $\bar{\theta}$ and a prescribed macroscopic strain $\bar{\varepsilon}$.

$$u\left(x, \bar{\varepsilon}, \Delta\bar{\theta}\right) = \mathbb{Y}\left(x\right)\left[\bar{\varepsilon}\right] + u_\theta\left(x\right)\frac{\Delta\bar{\theta}}{\Delta\theta_0} \tag{11}$$

$$\varepsilon\left(x, \bar{\varepsilon}, \Delta\bar{\theta}\right) = \mathbb{A}\left(x\right)\left[\bar{\varepsilon}\right] + \varepsilon_\theta\left(x\right)\frac{\Delta\bar{\theta}}{\Delta\theta_0} \tag{12}$$

$$\sigma\left(x, \bar{\varepsilon}, \Delta\bar{\theta}\right) = \mathbb{C}\left(x\right)\mathbb{A}\left(x\right)\left[\bar{\varepsilon}\right] + \sigma_\theta\left(x\right)\frac{\Delta\bar{\theta}}{\Delta\theta_0} \tag{13}$$

with elastic localization operators $\mathbb{Y}$ for the displacement field and $\mathbb{A}$ for the strain field. For bulk materials, the exact form of the homogenized balance of energy in a thermo-elastic setting is:

$$-\bar{\theta}\left\langle\frac{\partial\sigma}{\partial\theta}\cdot\dot{\varepsilon}\right\rangle - div_{\bar{x}}\left(\bar{q}\right) + \bar{h} = \bar{\rho}\bar{c}_\theta\dot{\bar{\theta}} \tag{14}$$

This above procedure can be used to calculate the solution in terms of the elastic and thermal stresses $\sigma_e^\alpha\left(x\right)$, $\sigma_\theta\left(x\right)$, the strains $\varepsilon_e^\alpha\left(x\right)$, $\varepsilon_\theta\left(x\right)$ and the displacement fields $u_e^\alpha\left(x\right)$, $u_\theta\left(x\right)$ for $\alpha = 1, ..., 6$.

2nd step: The mode activity coefficient $\hat{\zeta}_\alpha \neq 0$ is constructed using the solution obtained for N auxiliary elastic eigenstress problems.

$$div\left[\mathbb{C}\left[\varepsilon_*^\alpha(x) - \mu^\alpha(x)\right]\right] = 0, \quad \langle\varepsilon_*^\alpha\rangle = 0 \quad (\alpha = 1, ...N) \tag{15}$$

The obtained stress fields are expressed as $\sigma_*^\alpha(x) = \mathbb{C}\left[\varepsilon_*^\alpha(x) - \mu^\alpha(x)\right]$. The displacement is denoted as $u_*^\alpha(x)$. The solution of the eqn. (5) and eqn. (6) is

www.witpress.com, ISSN 1743-3533 (on-line)

given by the following set of expressions.

$$u\left(x,\bar{\theta},\bar{\varepsilon},\hat{\zeta}\right) = \mathbb{Y}(x)[\bar{\varepsilon}] + u_{\theta}(x)\frac{\Delta\bar{\theta}}{\Delta\theta_0} + \sum_{\alpha=1}\zeta_{\alpha}u_*^{\alpha} \tag{16}$$

$$\varepsilon\left(x,\bar{\theta},\bar{\varepsilon},\hat{\zeta}\right) = \mathbb{A}(x)[\bar{\varepsilon}] + \varepsilon_{\theta}(x)\frac{\Delta\bar{\theta}}{\Delta\theta_0} + \sum_{\alpha=1}\zeta_{\alpha}\varepsilon_*^{\alpha} \tag{17}$$

$$\sigma\left(x,\bar{\theta},\bar{\varepsilon},\hat{\zeta}\right) = \mathbb{C}(x)\mathbb{A}(x)[\bar{\varepsilon}] + \sigma_{\theta}(x)\frac{\Delta\bar{\theta}}{\Delta\theta_0} + \sum_{\alpha=1}\zeta_{\alpha}\sigma_*^{\alpha} \tag{18}$$

The evolution of the internal variables is described by effective yield function for which the nonlinear isotropic hardening of the metal phase is described in term a von Mises-Type yield criterion with yield stress.

$$\bar{\varphi}\left(\bar{\tau}\right) = \left|\bar{\tau}\right|_2 - \sqrt{2/3}\,c\left(\sigma_0 + h\bar{q}\right) \tag{19}$$

where $\bar{q}$ an effective hardening variable resembling, c is the volume fraction of the inelastic material. Then the rates for $\hat{\zeta}$ and $\bar{q}$ are determined by

$$\bar{\varphi} \leq 0, \quad \dot{\gamma} \geq 0, \quad \bar{\varphi}\dot{\gamma} = 0, \quad \hat{\zeta} = \dot{\gamma}\frac{\partial\varphi}{\partial t}, \quad \dot{\bar{q}} = \sqrt{2/3}\,\dot{\gamma} \tag{20}$$

Note that based on eqn. (18) and with the effective linear elastic operator $\bar{\mathbb{C}} = \mathbb{C}\mathbb{A}$ and the macroscopic stress is a linear transformation of the macroscopic strain $\bar{\varepsilon}$ and the vector of mode activity coefficient $\hat{\zeta}$.

$$\bar{\sigma}\left(\bar{\varepsilon}(t), \hat{\zeta}(t)\right) = \mathbb{C}\left[\bar{\varepsilon}\right] + \sum_{j=1}^{N}\zeta_j(t)\left\langle\sigma_*^j\right\rangle \tag{21}$$

2.2 NTFA implementation

The step by step NTFA implementation with thermo-mechanical interaction is shown in the fig. 1.

1. The first step is the microscopic modeling with spatial discretization of the model microstructure. The microscopic definition of the elasto plastic behavior such as elastic module and the hardening parameters are described.
2. Numerical experiments by FEM are performed using kinetic or static load. This step is performed in order to determine the elastic properties, strain path and non-linear behavior.
3. The third step is to mode identification and to find possible characteristic plastic deformation fields.
4. In the fourth step, thermo-plastic properties and inelastic materials properties are calculated.
5. In the final step, the calculated homogenized material is subjected to structural problem using FEM.

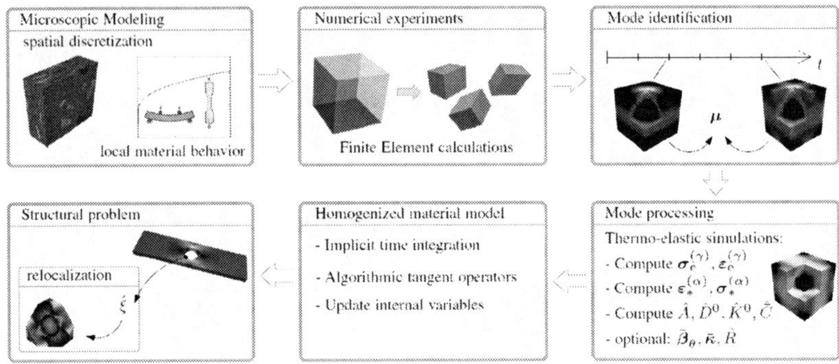

Figure 1: NTFA steps to solve structural problem with thermo-mechanical interaction.

3 Results and discussion

3.1 Model microstructure of MMC

Metal matrix composites are multi phase materials with interconnected phases and second phase particles. In order to find a suitable model microstructure, the following characteristic topological features are considered [5].

- The particles are assumed to be convex and polygonal in shape.
- The particles volume fraction is assumed to be vary a small value.
- Particles agglomeration is assumed with short distance heterogeneous distribution.
- Particle size variation is considered within a very small region.

The model microstructure is generated using Voronoi Tessellation. Particular microstructure is achieved by the erosion of the Voronoi cells by moving the cell walls homogeneously inward. The complete schematics of the model microstructure generation is presented in the fig. 2. The generated mesh is extended to be period based on the particles geometry.

3.2 Material characterization

The material examined in the following examples is Al/SiC. A unit cell problem containing 10 ceramic particles (SiC) with a total volume fraction of 18.2% embedded in a ductile aluminium matrix (Al) was considered. The shape of the particles is polyhedral. The discretization used in the following consists of a total of 219390 degrees of freedom and 209892 integration points. The material parameters of the constituents were chosen according to [3], i.e. the properties E=400GPa, ν=0.2 were assigned to the particles. For the aluminum matrix

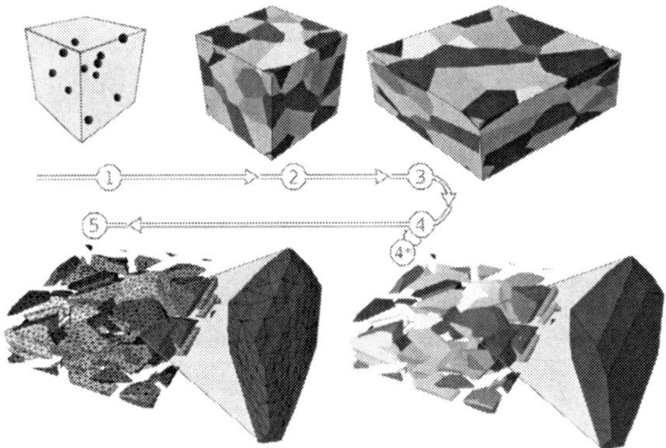

Figure 2: Schematic representation of the model microstructure generation.

material, the non-linear isotropic hardening law is chosen

$$\sigma_F(q) = \sigma_0 + hq^m \tag{22}$$

Here q is a strain like hardening variable resembling the accumulated plastic strain. The inelastic parameters are set to σ_0=75MPa,h=416.5MPa and m=0.3895. Young's modulus E=75GPa and Poisson's ratio ν=0.3 were used to characterize the elastic behavior of the matrix material [6].

Five inelastic strain path computations were carried out with 10 recorded time frames in each computation resulting in N_f=50 field frames. The total strain prescribed to the unit cells was approximately 5%. First, the new mode identification strategy was applied and resulted in a total of five inelastic modes for r_{min}=6.25e-6. The mode identification took 37 min on a single processor computer including the solution of the eigen problems. Notably, only 105 L^2 products had to be evaluated in the process. The large degree of non-uniformity of the plastic strain field, the displacement field and the stress field are exemplified in fig. 3, where selected components of the inelastic strain of two inelastic modes are shown.

3.3 NTFA and full field on unit cell level

The NTFA implementation in ABAQUS is applied to a MMC microstructure material. A strain-path consisting of two different loading directions is prescribed to (i) a full-field unit cell computation and (ii) the homogenized material model. The resulting macroscopic stress tensors are then compared. The strain rate of the process is:

$$\dot{\bar{\varepsilon}}_1 = \begin{pmatrix} 1 & 0 & 0 \\ & -1 & 0 \\ & & 0 \end{pmatrix} \frac{e_i \otimes e_j}{100s}, \qquad \dot{\bar{\varepsilon}}_2 = \begin{pmatrix} & & 2 \\ & 0.5 & 0 \\ 2 & 0 & -0.5 \end{pmatrix} \frac{e_i \otimes e_j}{100s} \tag{23}$$

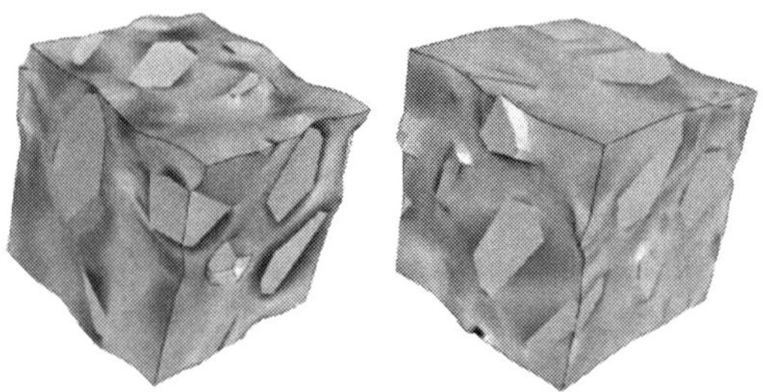

Figure 3: ε_{11}^{p} in Mode 1(left) and ε_{22}^{p} in Mode 2(right).

Each of the two rates is held constant for 1 s. The time history of macroscopic stress tensor of the full-field computation $\bar{\sigma}^{ref}$ and $\bar{\sigma}^{NTFA}$ are compared in fig. 4.The diagonal components (fig. 4; left) and the shear components (fig. 4; right) have been separated for clarity. A considerable agreement was found for the components of the macroscopic stress despite the partial load reversal (particularly in the $e_2 \otimes e_2$-direction). The components of $\bar{\sigma}^{NTFA}$ show only a small absolute discrepancy versus the reference computation. Additionally, it has to be pointed out that the relative errors with respect to $\|\bar{\sigma}\|_2$ are in the order of 2% and smaller. Therefore, the method can be considered to yield acceptable results for multi-axial loading and for partial load reversal. The agreement between NTFA and full field simulation is striking as the NTFA method has only several hundreds of freedoms while the full field simulation has thousands on them.

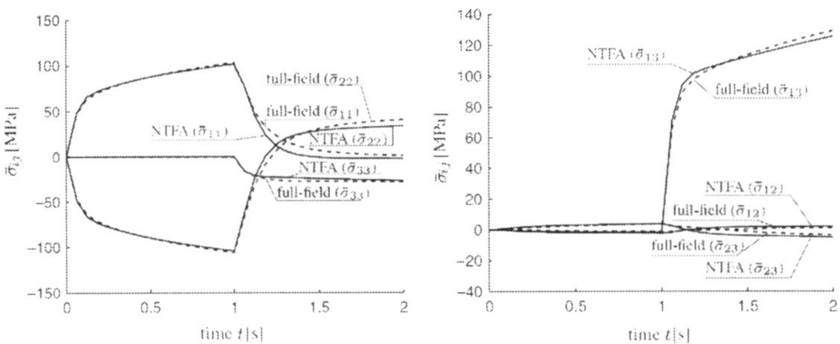

Figure 4: Time history of stress components for $\bar{\sigma}^{ref}$ and $\bar{\sigma}^{NTFA}$.

3.4 Structural application

3.4.1 Tension test

The homogenized material has been used to model a tension test on the three dimensional specimen shown in fig. 5. Since the homogenized material is no longer

isotropic, it is not possible to reduce the model of the axis-symmetric specimen to two dimensions. The discretization of the virtual specimen (fig. 5, middle) consists of approximately 70 000 nodes, i.e. a total of 210 000 degrees of freedom were considered in the structural problem. Kinematic boundary conditions (fig. 5, right) were prescribed to replicate a standard tension test in e_2-direction. The total displacement u_y at the end of the load step was set to 1mm equal to 5% of the measured section. The force displacement curve of the virtual experiment is shown in fig. 6. A large degree of non-linearity can be observed.

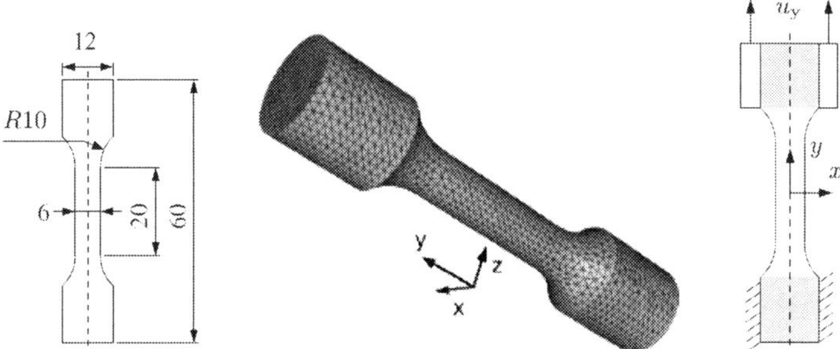

Figure 5: Tensile specimen(unit: mm): geometry (left), spatial discretization (middle) and boundary conditions (right).

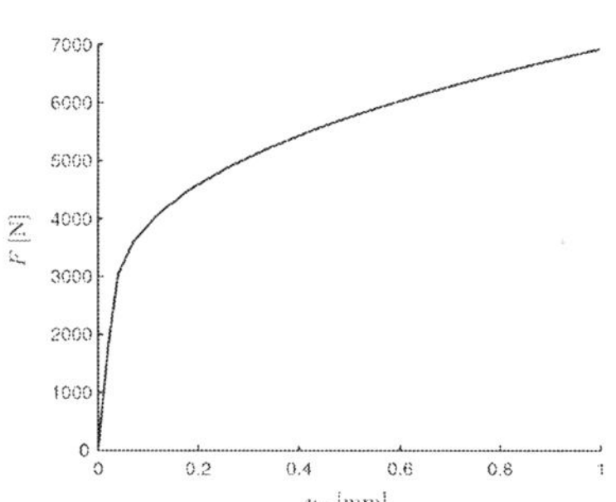

Figure 6: Force-displacement relationship.

The computation was used to validate the implementation of the homogenized material model into the finite element code ABAQUS/STANDARD. Particularly, the tangent stiffness operator was tested. During the analysis the stresses, the strains and internal variables computed for the homogenized material model were recorded for the integration point. The recorded strain-path was then used to model a non-proportional loading path on the unit cell level. A full-field simulation on the

www.witpress.com, ISSN 1743-3533 (on-line)

microscale was carried out in order to compare the macroscopic stress of the full-field simulation to the one predicted from the NTFA. The results are shown in fig. 7.

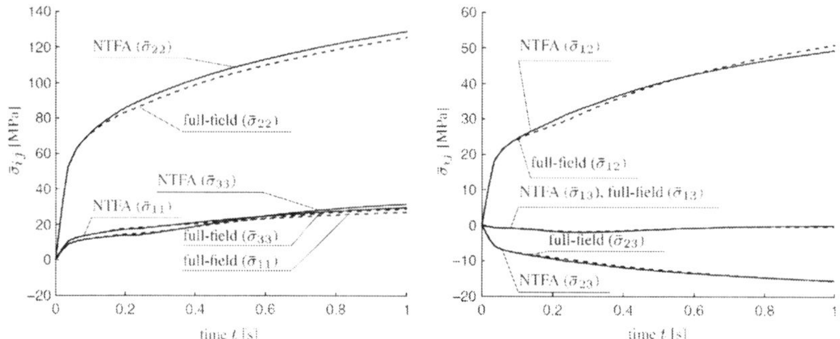

Figure 7: Time history of $\bar{\sigma}_{11}$, $\bar{\sigma}_{22}$, $\bar{\sigma}_{33}$(left) and $\bar{\sigma}_{12}$, $\bar{\sigma}_{13}$, $\bar{\sigma}_{23}$(right) for tension experiment.

3.4.2 Indentation test

In order to verify the method for problems with complex loading a semi-spherical indentation problem was modeled. The indenter was assumed to be linear elastic with the isotropic material properties of steel (Young's modulus E =210GPa, Poisson ratio ν=0.3).The tip of the indenter has a diameter of 5mm and is centered over a quadratic bi-material of edge-length 10mm and total height 7 mm. The bi-material consists of two layers: (i) the top layer modeled with the homogenized material model representing the MMC and (ii) an elasto-plastic base layer with the properties of the matrix material. The indenter is subjected to a total displacement of 0.1mm equaling 0.5% of the thickness of the MMC layer (fig. 8). Hard contact conditions were assumed in normal direction and a friction coefficient of 0.1 was estimated.

Similarly to the previous example, the time history of the stress and strain tensor was recorded. The integration point data were used to run a full-field simulation on the entire unit cell. The components of the macroscopic stress tensor of the full-field simulation and the homogenized material model are plotted in fig. 9 over time. Owing to the position of the integration point out of the center of the contact region, the deformation in the first few steps of the analysis was dominated by the neighboring regions that have already been in contact. Then the area containing nodes directly associated with the integration point get into contact and a load reversal is observed (fig. 9, e.g.$\bar{\sigma}_{22}$, $\bar{\sigma}_{12}$, $\bar{\sigma}_{13}$). Although only five modes were used in the analysis, the stress predicted by the homogenized material model shows considerable agreement to the full-field simulation. During the simulation of the contact problem, the algorithmic stiffness operator allowed for good convergence

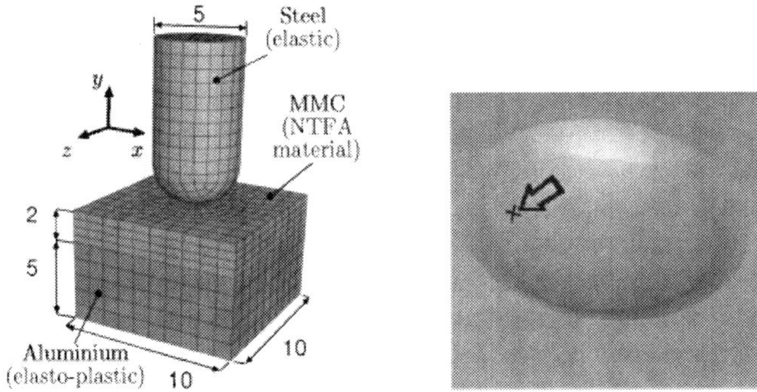

Figure 8: Indentation test geometry and integration point.

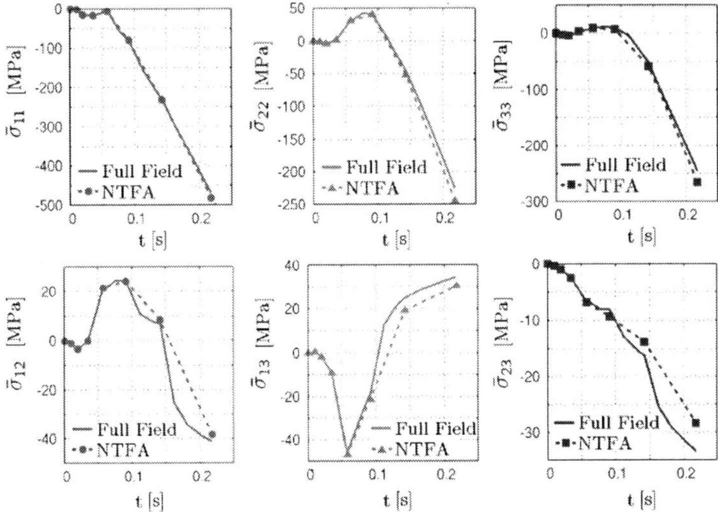

Figure 9: Components of the macroscopic stress: full-field simulation versus NTFA model.

rates. Notably, the homogenized model produced almost the same results as the full-field simulation, whereas the latter required significantly more load increments in order to converge. This is due to the local deformation getting excessive. The previous observations underline the robustness of the method in practical applications.

4 Conclusion

In this work, the extension of nonuniform transformation fields analysis (NTFA) to thermo-mechanical interaction is presented. A finite element based implementation of the NTFA was developed. Two major differences with respect to the prior implementations [1, 3, 4] are introduced: (i) three-dimensional problems are considered and (ii) the FEM was used to approximate the solution of the full-field simulations on the microscale. The computational scheme used for the implicit time integration procedure of the constitutive equations of the homogenized material is provided. It allows for a fast numerical solution and it can be used to explicitly provide a formulation for the algorithmic stiffness operator of the non-linear effective material. Different numerical examples are presented for a two-phase composite material consisting of a ductile elasto-plastic matrix and linear elastic particles with complex three dimensional microstructure. Numerical experiments are used to compare the results of full-field simulations to the homogenized model under non-proportional loading conditions. Further, the model is used in three-dimensional structural problems, one including contact formulations. In both cases, integration point data were extracted and used to prescribe the observed strain history to unit cell simulations. The components of the macroscopic stress tensors showed a good agreement to full-field simulations.

Acknowledgements

The financial support by the German Research Foundation (DFG Schn 245/30) is gratefully acknowledged.

References

[1] J.C. Michel, P.S., Computational analysis of nonlinear composite structures using the nonuniform transformation field analysis. *Computer Methods in Applied Mechanics and Engineering*, **193(48-51)**, pp. 5477–5502, 2004.

[2] Dvorak, G.J. & Benveniste, Y., On transformation strains and uniform fields in multiphase elastic media. *Proceedings of the Mathematical and Physical Sciences*, The Royal Society: London, pp. 291–310, 1992.

[3] Michel, J.C. & Suquet, P., Nonuniform transformation field analysis. *International Journal of Solids and Structures*, **40(25)**, pp. 6937–6955, 2003.

[4] S. Roussette, P.S., J.C. Michel, Nonuniform transformation field analysis of elastic-iscoplastic composites. *Composites Science and Technology*, **69(1)**, pp. 22–27, 2009.

[5] Fritzen, F. & Bohlke, T., Periodic three-dimensional mesh generation for particle reinforced composites with application to metal matrix composites. *International Journal of Solids and Structures*, **48(5)**, pp. 706–718, 2011.

[6] Felix Fritzen, T.B., Three-dimensional finite element implementation of the nonuniform transformation field analysis. *International Journal for Numerical Methods in Engineering*, **84(7)**, pp. 803–829, 2010.

The effects of natural weathering exposure on the properties of pultruded natural fibre reinforced unsaturated polyester composites

M. H. Affzan, H. M. Akil, Z. A. M. Ishak & A. A. Bakar
School of Materials and Mineral Resources Engineering, Universiti Sains Malaysia, Malaysia

Abstract

Due to the demand for green technology, the development of fibre reinforced composites has built-up an interest in finding out new and reliable composites. The natural fibre reinforced composites produced by the pultrusion technique may eventually replace the conventional fibre composite production technique used today. In considering outdoor application, experiments have been performed to study the effects of weathering exposure on pultruded natural fibre reinforced composites. Kenaf fibre reinforced composites (KFRC) were produced using the pultrusion technique at 70% fibre loading. The composites were then subjected to natural aging, which involved exposing specimens to an outdoor natural environment for a period of time. After a few weeks, the specimen showed some degradation and discolouring. Subsequently, compression and flexural tests were carried out with the aim of determining the mechanical properties after exposure. The decrease in the value of these properties was verified and discussed. The repeated process of sun-heating and the invasion of moisture from natural atmosphere such as rain and dew everyday had led to debonding and weakening in the fibre-matrix interfaces of the composite.
Keywords: fibres, polymer-matrix composites (PMCs), weathering exposure, pultrusion.

1 Introduction

Historically, natural fibres were used as reinforcement in composites long ago by the Egyptians. The first composite materials known was made with clay and

WIT Transactions on Engineering Sciences, Vol 77, © 2013 WIT Press
www.witpress.com, ISSN 1743-3533 (on-line)
doi:10.2495/MC130291

natural bamboo straw to built walls in Egypt 3000 years ago. However, starting from 20th century, modern composites such as glass fibre reinforced composites or simply called fibreglass, have been used to make boats and aircrafts in the 1930s (Taj *et al.* [1]).

Since the 1970s, the development of new fibres such as boron, aramid and carbon makes the application of composites wares widely increased. These high-tech synthetic fibres had entered and dominated the composites market due to their superior mechanical and thermal properties (Hapuarachchi *et al.* [2]). But recently, the environmental legislation as well as consumer awareness of recyclability has opened a huge chance to make natural fibres once again appear as an ideal reinforcement as a replacement for those synthetic fibres. The emerging demand of green economy or technology throughout the past few years had shown how important it was to produce a new material with approximately the same properties as common materials used today.

Currently, many types of natural fibres are available and are continuously being studied to be used as reinforcement in polymer composite including kenaf, jute, rice husk, ramie, sisal, coir, hemp, pineapple leaf fibre and many more. Compared to traditional glass fibre, natural fibre has lots of acceptable advantages such as abundant availability of raw materials from renewable resources (Kim *et al.* [3]), low density, good specific strength and modulus, economical viability, reduced tool wear, enhanced energy recovery, reduced dermal and respiratory irritation and good biodegradability (Dhakal *et al.* [4]). However, natural fibre reinforced composites experience some disadvantages including incompatibility between the hydrophilic natural fibres and hydrophobic thermoplastic and thermoset matrices requiring appropriate use of physical and chemical treatments to enhanced the adhesion between fibre and the matrix (Gassan and Gutowski [5]).

Kenaf (*Hibiscus cannabinus L.*) has been used as a cordage crop to produce twine, rope and sackcloth (Edeerozey *et al.* [6]), but recently, the development of this fibre as a reinforcement in polymer composite materials is catching the attention among researchers. Various techniques and treatment have been used to improve the mechanical and physical properties of these natural fibres. This lignocellulosic natural fibre can be produced in direct roving forms, which make it easier to be used as raw materials in pultrusion technique. Pultrusion technique principally producing a constant and continuous cross-section shapes fibre-reinforced polymer composite (Starr [7]).

The main concern of this study was to observe the effects of absorption of moisture to the pultruded KFRC when exposed to natural weathering and its effects towards the mechanical and physical properties. Basically, all polymer composites tend to absorb moisture in humid atmosphere which may lead to degradation of fibre-matrix interface region creating poor stress transfer efficiencies resulting in a reduction of mechanical and dimensional properties (Yang *et al.* [8]). Moisture diffusion in polymeric composites is known to be governed by three different mechanisms. The first involves diffusion of water molecules inside the micro gaps between polymer chains. The second involves capillary transport into the gaps and flaws at the interfaces between fibre and the matrix. The third involves transport of micro cracks in the matrix arising from the swelling of fibres (particularly in the case of natural fibre composites) (Hazizan *et al.* [9]).

This paper focused on development and evaluation of a pultruded KFRC. Standard half an inch diameter rod samples of KFRC produced by pultrusion technique had been used in this study. The samples were left to natural weather for a certain period of time and tests were conducted to evaluate the mechanical and physical properties.

2 Experimental

2.1 Materials

Kenaf fibre was in direct roving form and supplied by JUTEKO Bangladesh Pvt. Ltd. While the unsaturated polyester resin (Crystic P9901) for pultrusion grade was purchased from Rivertex Company, Malaysia. The applied unit for yarn count varies between the fibre types, but the standard unit is "tex" which is defined as mass per unit length (g/1000m) of roving, tow, yarn or strand (ASTM D2260). Tex also specifies roving linear density (Madsen [10]).

2.2 Preparation of pultruded composites

KFRC was prepared using pultrusion technique. Unidirectional kenaf fibre strands were placed on a creel of bookcase-type shelves, which were equipped with roving guider to lead the strands to the resin bath. Roving guider was used to ensure the strands did not scrape across one another as this would generate considerable static and caused "fuzz-balls" to build up in the resin bath, raising its viscosity (Meyer [11]). The continuous natural fibres were first impregnated with pultrusion grade unsaturated polyester (USP) resin in the resin impregnation tank. The pulling device worked as a pulling mechanism to pull the impregnated natural fibre through the heated steel die. The pulling device drew the stock through the die and determined the production speed (William [12]). Curing process was carried out in a heated die, which was precisely machined to impart the final shape. Finally, a cut-off mechanism was carried out to cut the continuous pultruded composites into the desired length. The average diameter of all composites rod was 12.7 mm and the fibre content used for all the samples in this study was 70% by weight. The processing parameters are given in Table 1.

Table 1: Processing parameter of pultrusion technique on KFRC.

Parameter	Pulling speed (mm/min)	Curing temperature range (°C)
KFRC	195–210	90–110

2.3 Material characterization

2.3.1 Water absorption investigation

Specimens produced by the pultrusion technique were cut into specific lengths and dried in an oven at 100°C for 24 hours to remove any moisture trapped inside. After measuring the dry weight of the specimens using an electronic balance accurate to 10^{-4} g, the specimens were left out to the natural weathering exposure to study the behaviour of the composites towards moisture uptake. The weight change was monitored as a function of time until 4800 h (200 days). The moisture content, *M(t)* absorbed by each specimen was calculated according to Tsai *et al.* [13]:

$$M(t) = 100\left(\frac{w_t - w_0}{w_0}\right) \tag{1}$$

where w_o is the dry weight and w_t is the weight after being exposed to natural environment for several times.

The diffusion coefficient, *D* is evaluated from the first term of Eq. 5.25 from the book *The Mathematics of Diffusion* (Crank [14]), which is:

$$D = \pi\left(\frac{a}{4M_\infty}\right)^2\left(\frac{M_t}{\sqrt{t}}\right)^2 \tag{2}$$

where a is the radius of the specimen and M_∞ is the saturation level of water absorption which is assumed as the maximum moisture content (M_m) absorbed by the specimen. The *D* value is calculated by using the moisture content, M_t at the certain time range, t which only being applied only at the beginning stages of the diffusion. Time was chosen at a very early stage of moisture process, so that the weight change can still be measured to vary linearly with the square root of time.

2.3.2 Flexural testing

Flexural test was carried out using Instron 8802 according to the standard ASTM D4476-03. Specimens were cut into two parts so that the cross section of each part was smaller than a half-round section. The specimen length is 16 to 24 times its thickness or depth, with at least 20% of the support span to allow a minimum of 10% overhang at the supports. The crosshead speed for flexural test was set at 5 mm/min. Three specimens for each condition were used to obtain a satisfactory result.

2.3.3 Compression testing

Compression test was carried out using Instron 8802 in accordance to the standard ASTM D 695-02a. The diameter and length of the specimen were 12.7mm and 25.4mm respectively. The crosshead speed for flexural test was set at 1.5 mm/min. Three specimens for each condition were used to obtain a satisfactory result.

3 Results and discussions

3.1 Water absorption investigation

Moisture uptakes or water absorption behaviour is the main concern of structural composite in various outdoor applications. Every composite system has a unique behaviour upon moisture uptakes characteristic which depends on several factors such as the content of the fibre, fibre orientation, environmental temperature, exposed surface area, permeability of fibre, void content, and hydrophilicity of each individual component (Nosbi *et al.* [15]). Figure 1 shows the percentage of moisture content absorbed by three different tex of KFRC specimens within 200 days exposure to natural weathering. The moisture content, M (%) absorbed by each specimen was calculated as Equation (1) earlier.

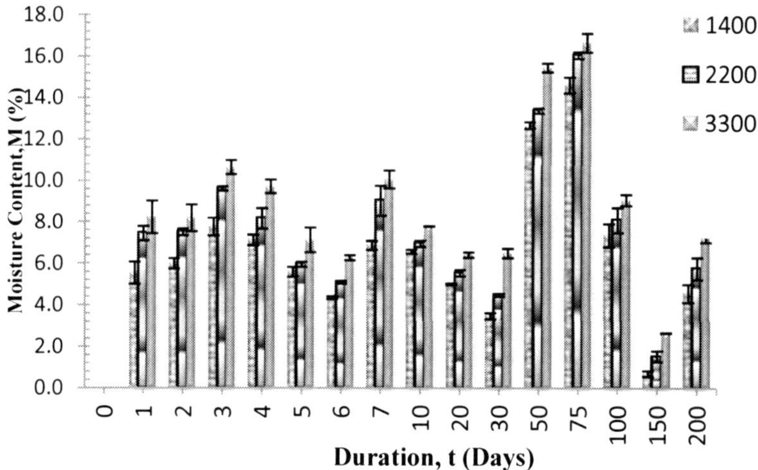

Figure 1: Moisture content absorbed after natural weathering exposure of KFRC.

Figure 1 shows the inconsistency of moisture content absorbed by KFRC from each day of data collection. The highest moisture content was taken on the 75th day of exposure while the lowest was on day 150. This inconsistency occurred from the changes of weather with different amount of UV from sun and moistures from atmosphere every day.

However, this inconsistency shows the trend between the three tex of KFRC. For every data taken, KFRC 1400 tex has the lowest moisture content followed by KFRC 2200 tex and the highest was KFRC 3300 tex. After 200 days, the data can be simplified (as shown in Figure 2).

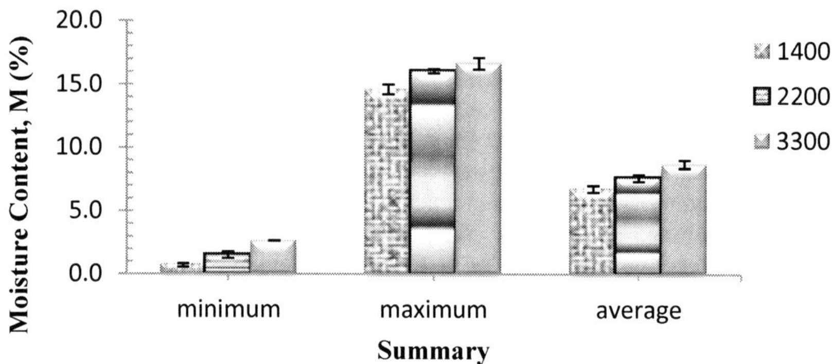

Figure 2: Summary of moisture content absorbed during natural weathering exposure of KFRC.

From Figure 2, the minimum, maximum and average percentage of moisture content within 200 days of exposure can clearly be seen. The minimum and maximum moisture content uptakes were 1–4% and 14–17% respectively of each tex. The average moisture content uptake for 200 days of exposure was ranged between 6 and 9%.

The inconsistent results also came from the moisture diffusivity of KFRC for each tex. Diffusion coefficient (D) and maximum moisture content (Mm) for all tex of KFRC specimens was calculated using Equation (2) above and is summarized in Table 2.

Table 2: Maximum moisture content and diffusivity of KFRC for different tex.

	1400 tex	2200 tex	3300 tex
Maximum moisture content, M_m (%)	14.6143	16.0630	16.6576
Diffusion coefficient, D (m^2/s)	2.7535×10^{-13}	3.2820×10^{-13}	4.0865×10^{-13}

From the summarized value of D and Mm as shown on Table 2, it is clear that KFRC with 3300 tex has greater maximum moisture content and diffusion coefficient compared to 1400 and 2200 tex. This difference can be attributed to the exposed area of the fibre itself to the surroundings, which can make contact to the water molecules. The greater amount of fibre area exposed to the surrounding, the higher the tendency of absorbing water. However, the molecular and microstructure aspects such as polarity, the extent of crystallinity of polymers and the presence of residual hardeners or other water attractive species may also effect the moisture diffusion (Nosbi *et al.* [16]).

3.2 Flexural testing

Variation of flexural strength and flexural modulus of KFRC after being exposed for 200 days to natural weathering are summarized in Figures 3 and 4 respectively. Each value represents an average data of three different tex of specimens and the flexural strength for polyester resin alone is 48 MPa (Munikenche Gowda *et al.* [17]).

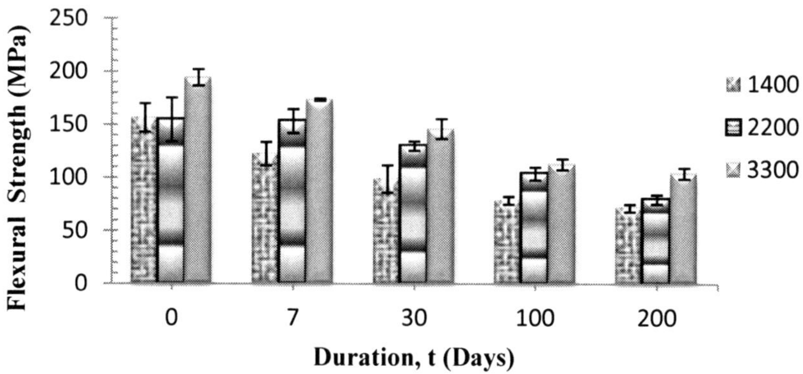

Figure 3: Flexural strength of KFRC after natural weathering exposure.

From Figures 3 and 4, the flexural strength and flexural modulus for all tex of KFRC was decrease with increasing time of exposure to natural weathering. But still, the KFRC with 3300 tex has the highest value among all for each day of data taken. It is assumed that the stress transfer between fibre and matrix interface is less effective due to the presence of moisture. Besides, moisture also causes the formation of hydrogen bonding between the cellulose fibre and water molecules (Nosbi *et al.* [16]).

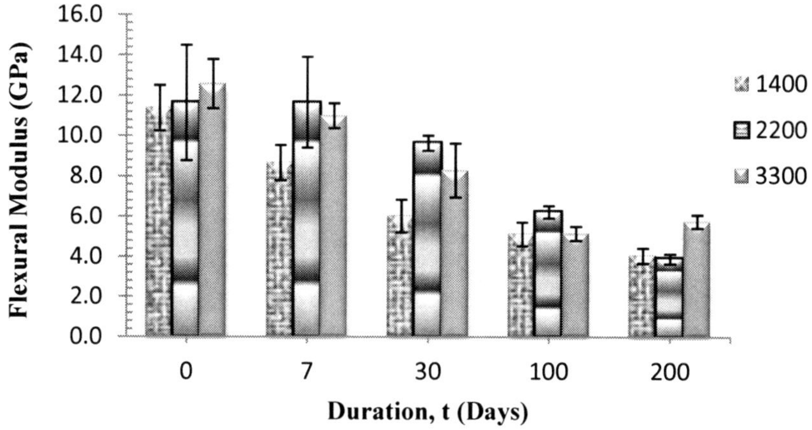

Figure 4: Flexural modulus of KFRC after natural weathering exposure.

3.3 Compression testing

Figures 5 and represent the compressive strength and compressive modulus of KFRC after exposure of 200 days to natural weathering. Each value represents an average data of three different tex of specimens.

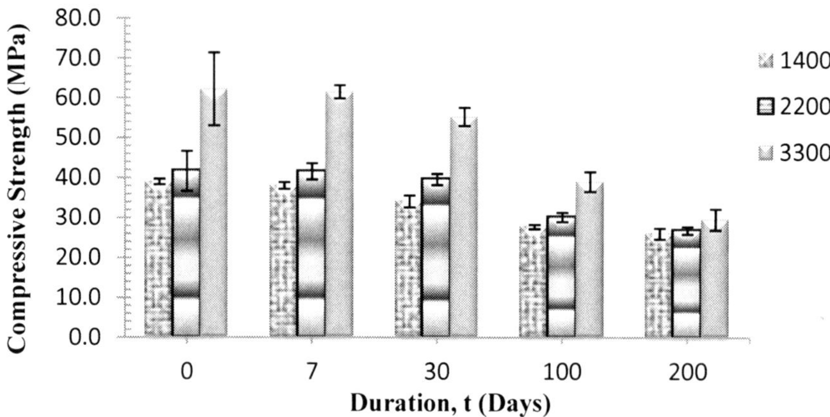

Figure 5: Compression strength of KFRC after natural weathering exposure.

As shown in Figure 5, the compression strength was decreased with increasing exposure time. This result was similar with the flexural properties as being discussed earlier. The compression strength for 3300 tex of KFRC showed reduction from 63 MPa to 31 MPa after 200 days of exposure. After 200 days of exposure, the compression strength for all different tex of KFRC seems to have almost the same value range from 29 to 31 MPa. However, in Figure 6, the inconsistent data was collected on the compression modulus in different tex of KFRC. The compressive modulus increased at the middle stage and finally drops at the end of exposure.

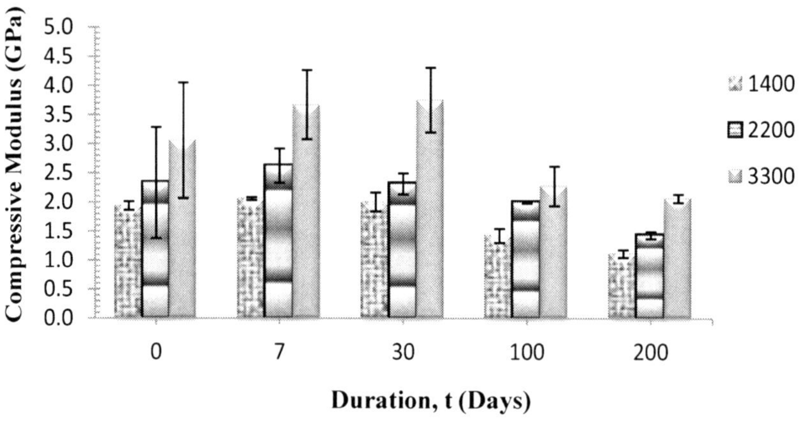

Figure 6: Compression modulus of KFRC after natural weathering exposure.

Both observations on flexural and compressive strength proved that moisture content can attack the fibre-matrix interface and significantly reducing the strength of the composites. Aside from the effect from moisture, micro cracks are also the sources of failure for the strength properties. Rapid changes of weather everyday caused the micro cracks to develop on the surface and the inside of the composites (Nosbi *et al.* [16]).

4 Conclusion

KFRC has been successfully produced using pultrusion method. From the study, a few conclusions can be drawn. Firstly, the natural weathering exposure study showed that KFRC could still be degraded throughout a period of time by several factors such as moisture and temperature changes. Prolonged exposure of KFRC to natural weathering may interfere the fibre matrix interface and causes degradation which significantly reduces the flexural and compression properties.

Acknowledgements

The authors are grateful to the School of Materials and Mineral Resources Engineering and Institute of Postgraduate Studies of Universiti Sains Malaysia (USM-814023 and 8640013) and Construction Industry Development Board of Malaysia (CIDB) for their assistance and contribution that has resulted in this article.

References

[1] Taj, S., M. Munawar, and S. Khan, *Natural fiber-reinforced polymer composites.* Proceedings Pakistan Academy of Sciences, **44**(2): pp. 129, 2007.

[2] Hapuarachchi, T.D., et al., *Fire retardancy of natural fibre reinforced sheet moulding compound.* Applied Composite Materials, **14**(4): pp. 251-264, 2007.

[3] Kim, S., et al., *Mechanical properties of polypropylene/natural fiber composites: Comparison of wood fiber and cotton fiber.* Polymer Testing, **27**(7): pp. 801-806, 2008.

[4] Dhakal, H., Z. Zhang, and M. Richardson, *Effect of water absorption on the mechanical properties of hemp fibre reinforced unsaturated polyester composites.* Composites Science and Technology, **67**(78): pp. 1674-1683, 2007.

[5] Gassan, J. and V. Gutowski, *Effects of corona discharge and UV treatment on the properties of jute fibre epoxy composites.* Composites Science and Technology, **60**(15): pp. 2857-2863, 2000.

[6] Edeerozey, A., et al., *Chemical modification of kenaf fibers.* Materials Letters, **61**(10): pp.2023-2025, 2007.

[7] Starr, T., *Pultrusion for engineers.* Woodhead Publishing, 2000.

[8] Yang, G., et al., *Relation of modification and tensile properties of sisal fiber.* Acta Sci Nat Uni Sunyatseni, **35**: pp. 53-70, 1996.
[9] Hazizan, M.A., et al., *Water absorption study on pultruded e-glass fibre reinforced unsaturated polyester composites.* Advanced Composites Letters, **19**(3): pp. 101 107, 2010.
[10] Madsen, B., *Properties of Plant Fiber Yarn Polymer Composites: An Experimental Study.* Technical University of Denmark, Department of Civil Engineering, Sektionen for Bygningsmaterialer og Geoteknik, 2004.
[11] Meyer, R., *Handbook of pultrusion technology.* Chapman and Hall New York, 1985.
[12] William Jr, D., *Materials science and engineering: an introduction.* John Wiley & Sons, 2003.
[13] Tsai, Y., et al., *Influence of hygrothermal environment on thermal and mechanical properties of carbon fiber/fiberglass hybrid composites.* Composites Science and Technology, **69**(3-4): pp. 432–437, 2009.
[14] Crank, J., *The mathematics of diffusion. 2nd.* Ed., Clarendon Press, Oxford, 1975.
[15] Nosbi, N., et al., *Degradation of compressive properties of pultruded kenaf fiber reinforced composites after immersion in various solutions.* Materials & Design, **31**(10): pp. 4960–4964, 2010.
[16] Nosbi, N., et al., *Effect of Water Absorption on the Mechanical Properties of Pultruded Kenaf Fibre Reinforced Polyester Composites* Advanced Composites Letters, **1**(20): pp. 5–10, 2011.
[17] Munikenche Gowda, T., A. Naidu, and R. Chhaya, *Some mechanical properties of untreated jute fabric-reinforced polyester composites.* Composites Part A: applied science and manufacturing, **30**(3): pp. 277–284, 1999.

Author Index

Aluminium Alloy Corrosion of Aircraft Structures

Modelling and Simulation

Edited by: ***J.A. DEROSE*** *and* ***T. SUTER****, EMPA (Swiss Federal Institute of Materials Science & Technology), Switzerland;* ***T. HACK****, EADS Innovation Works, Germany and* ***R. ADEY****, CM BEASY, UK*

Bringing together the latest research this book applies new modelling techniques to corrosion issues in aircraft structures. It describes complex numerical models and simulations from the microscale to the macroscale for corrosion of the aluminium (Al) alloys that are typically used for aircraft construction, such as AA2024. The approach is also applicable to a range of other types of structures, such as automobiles and other forms of ground vehicles.

The main motivation for developing the corrosion models and simulations was to make significant technical advancements in the fields of aircraft design (using current and new materials), surface protection systems (against corrosion and degradation) and maintenance. The corrosion models address pitting and intergranular corrosion (microscale) of Al alloys, crevice corrosion in occluded areas, such as joints (mesoscale), galvanic corrosion of aircraft structural elements (macroscale), as well as the effect of surface protection methods (anodisation, corrosion inhibitor release, clad layer, etc.).

The book describes the electrochemical basis for the models, their numerical implementation and experimental validation, and how the corrosion rate of the Al alloys at the various scales is influenced by its material properties and the surface protection methods. It will be of interest to scientists and engineers interested in corrosion modelling, aircraft corrosion, corrosion of other types of vehicle structures such as automobiles and ground vehicles, electrochemistry of corrosion, galvanic corrosion, crevice corrosion, and intergranular corrosion.

ISBN: 978-1-84564-752-0 eISBN: 978-1-84564-753-7
Published 2013 / 248pp / £148.00

Lightning Source UK Ltd.
Milton Keynes UK
UKOW04n0626150414

229996UK00001B/7/P

9 781845 647209